Theory of Light Scattering in Condensed Matter

Proceedings of the First Joint USA–USSR Symposium

Theory of Light Scattering in Condensed Matter

Edited by

Bernard Bendow

Solid State Sciences Division
Rome Air Development Center
Hanscom AFB, Massachusetts

Joseph L. Birman

Department of Physics
The City College of the City University of New York
New York, New York

and

Vladimir M. Agranovich

Institute of Spectroscopy
Academy of Sciences of the USSR
Akademgorodok, Moscow, USSR

PLENUM PRESS · NEW YORK AND LONDON

Library of Congress Cataloging in Publication Data

Joint USA–USSR Symposium on the Theory of Light Scattering in Condensed Matter, 1st, Moscow, 1975.
Theory of light scattering in condensed matter.

Includes indexes.
1. Solids–Optical properties–Congresses. 2. Light–Scattering–Congresses. I. Bendow, Bernard, 1942- II. Birman, Joseph Leon, 1927- III. Agranovich, Vladimir Moiseevich. IV. Title.
QC176.8.06J64 1975 530.4'1 76-47492
ISBN 0-306-30993-9

Proceedings of the First Joint USA–USSR Symposium
on the Theory of Light Scattering in Condensed Matter,
Moscow, USSR, May 26–30, 1975

A Division of Plenum Publishing Corporation
227 West 17th Street, New York, N. Y. 10011

Printed in the United States of America

FIRST JOINT USA-USSR SYMPOSIUM
ON THE
THEORY OF LIGHT SCATTERING IN CONDENSED MATTER

ORGANIZING COMMITTEES

USSR

V. M. Agranovich (Chairman)
V. L. Ginzburg
S. A. Akhmanov
A. S. Davydov
K. K. Rebane

USA

J. L. Birman (Chairman)
H. Z. Cummins, Advisory Committee
M. Lax, Advisory Committee

SUPPORT

USSR

Academy of Sciences, USSR,
The General Physics and Astronomy Section

Commission of Spectroscopy,
Academy of Sciences, USSR

USA

National Science Foundation, USA,
Division of International Programs

National Academy of Sciences, USA

USA PARTICIPANTS

Bernard Bendow	Air Force Cambridge Research Labs.
Joseph L. Birman	City College of City Univ. of New York
Eli Burstein	Univ. of Pennsylvania
Herman Cummins	City College of City Univ. of New York
Leo Falicov	Univ. of California, Berkeley
Richard Ferrell	Univ. of Maryland
Paul Fleury	Bell Telephone Labs., Murray Hill
Robert Hellwarth	Univ. of Southern California
Miles Klein	Univ. of Illinois, Urbana
Melvin Lax	City College of City Univ. of New York
Douglas Mills	Univ. of California, Irvine
Kai Ngai	US Naval Research Laboratory
Peter Pershan	Harvard Univ.
Philip Platzman	Bell Telephone Labs., Murray Hill
John Ruvalds	Univ. of Virginia
Y. R. Shen	Univ. of California, Berkeley
Michael Stephen	Rutgers Univ.
Chandra Varma	Bell Telephone Labs., Murray Hill
Peter Wolff	Mass. Institute of Technology
John Worlock	Bell Telephone Labs., Holmdel

USSR PARTICIPANTS

Afanas'ev, A. A.
Agranovich, V. M.
Akhmanov, A. S.
Apanasevich, P. A.
Bagaev, V. S.
Belyanin, V. B.
Bespalov, V. I.
Borovik-Romanov, A. S.
Bureeva, L. A.
Davydov, A. S.
Dzyaloshinskii, I. E.
Emel'yanov, V.
Fabelinskii, I. M.
Galanin, M. D.
Ginzburg, V. L.
Gorelik, V. S.
Hizhnyakov, V. V.
Ipatova, I. P.
Ivchenko, E. L.
Kagan, Yu. M.
Kaplyanskii, A. A.
Keldysh, L. V.
Khokhlov, R. V.
Krivoglaz, M. A.
Kulevskii, L. A.
Levanyuk, A. I.
Levinson, I. B.
Lifshitz, E. M.
Lifshitz, I. M.
Lozovik, Yu. E.
Mandelstam, S. L.
Mirlin, D. N.
Obreimov, I. V.
Ovander, L. N.
Peregudov, G. V.
Perlin, Yu. I.
Permagorov, S. A.
Pikus, G.
Pitaevskii, L. P.
Popkov, Yu. A.
Rashba, E. I.
Rebane, K. K.
Rebane, L. A.
Sterin, Kh. E.
Strizhevskii, V. L.
Sushchinskii, M. M.
Yakovlev, I. A.
Zavt, G. S.
Zel'dovich, Ya. B.

FIRST JOINT USA-USSR SYMPOSIUM
ON THE
THEORY OF LIGHT SCATTERING IN CONDENSED MATTER

OFFICIAL PROGRAM

Monday, May 26, 1975 (Scientists Palace, Moscow)

Session I. Phase Transitions and Liquid Crystals
Chairmen: V. M. Agranovich and H. Z. Cummins

1. V. L. Ginzburg and A. P. Levanyuk
Light Scattering in the Vicinity of Phase Transitions

2. P. A. Fleury
Light Scattering in the Region of Phase Transitions

3. A. A. Kaplayansky
Combinational Scattering of Light in the Vicinity of Phase Transitions in Hg_2Cl_2 and Hg_2Br_2 Crystals

4. P. Pershan
Light Scattering in Liquid Crystals

5. I. E. Dzyaloshinsky
Light Scattering Features in Liquid Crystals

Session II. Strong Anharmonicity Effects and Electron-Phonon Interaction in Crystals
Chairmen: J. L. Birman and I. M. Lifshitz

1. J. Ruvalds
   Theory of Interacting Excitations

2. A. P. Pitaevsky
   Weak Coupled States of Elementary Excitations in Crystals

3. V. S. Gorelik and M. M. Suschinsky
   Experimental Observations of Fermi Resonances and Coupled States of Phonons

4. M. A. Krivoglaz
   Light Scattering and Absorption by Anharmonic Local and Quasilocal Vibrations

5. Kai Ngai
   Electron-Phonon Interactions

Tuesday, May 27, 1975 (Spectroscopy Inst., Akadem Gorodok)

Session III. Scattering from Electrons and Electron-Hole Drops
Chairmen: P. Ferrell and L. V. Keldysh

1. V. S. Bagaev
   Light Scattering from Electron-Hole Drops in Semiconductors

2. J. Worlock
   Recent Studies of Electron-Hole Drops

3. P. Platzman
   Theory of X-Ray Raman Scattering

4. Yu. E. Lozovik
   X-Ray Scattering from Local Plasmons and Electron-Hole Drops

5. I. P. Ipatova and A. R. Subashiev
   Electron-Phonon Interaction Effects on the Shape of Combinational Scattering of Light in Metals and Highly-Doped Semiconductors

Wednesday, May 28, 1975 (Institute of Physical Problems, Moscow)

Session IV. High Intensity and Non-Linear Effects
Chairmen: R. V. Khokhlov and M. Lax

1. Ya. B. Zeldovich
Multiple Compton Scattering

2. R. Hellwarth
Non-Linear Susceptibilities and Light Scattering

3. L. V. Ovander
Light - Light Scattering in Crystals

4. S. A. Akhmanov
Active Laser Spectroscopy of Combinational Scattering of Light in Crystals, Comparison with Spontaneous Scattering Spectroscopy

5. I. B. Levinson
New Elementary Excitations and Combinational Scattering of Light in Nonequilibrium Phonon Systems

6. B. Bendow
Multiphonon Phenomena in Resonance Raman Scattering

7. V. L. Gurevich
Light Scattering Methods of the Investigation in the Transition from an Incoherent Phonon Beam into a Coherent Acoustic Signal

Thursday, May 29, 1975 (Scientists Palace, Moscow)

Session V. Bulk and Surface Polaritons
Chairmen: S. A. Akhmanov and E. Burstein

1. A. S. Davydov
Propagation of Light Through Media with Spatial Dispersion

2. V. M. Agranovich
Scattering and Absorption of Laser Photons by Neutrons in Crystals

3. L. M. Mirlin
Spatial Dispersion and Damping of Plasmo-Phonon Excitations in Semiconductors

4. D. Mills
Review of Scattering from Surfaces

5. M. Lax
Some Theoretical Aspects of Scattering from Surfaces

Session VI. <u>Resonance Scattering and Scattering from Local Excitations</u>
Chairmen: A. S. Davydov and L. Falicov

1. K. K. Rebane and V. V. Hizhnyakov
Review of Theoretical Aspects of Secondary Resonance Radiation: Scattering, Luminescence and Hot Luminescence

2. Y. R. Shen
Review of Recent Experiments of Resonance Scattering

3. J. L. Birman
Symmetry Effects in Resonance Scattering

4. E. I. Rashba
Light Scattering and Absorption from Local Dielectric Excitations

5. L. A. Rebane and G. S. Zavt
Experimental and Theoretical Research of Light Scattering from Local Centers

Friday, May 30, 1975 (Institute of Physical Problems, Moscow)

Session VII. Scattering from Electronic Excitations and Spin-Flip Processes
Chairmen: P. Platzman and K. K. Rebane

1. V. L. Strizhevsky
Theory of Light Scattering from Polaritons

2. M. Klein
Light Scattering from Electronic Excitations

3. P. Wolff
Spin-Flip Scattering Theory

4. V. V. Eremenko
Combinational and Mandelstam-Brillouin Scattering in $KMnF_3$ Crystals

Session VIII. Scattering by Superfluids
Chairmen: V. L. Ginzburg and C. Varma

1. M. Stephen
Review of Theory and Experiment of Light Scattering by Superfluids

2. R. Ferrell
Critical Scattering

3. A. P. Levanyuk
Effects of Impurities on Critical Scattering in Solids

Concluding Remarks: V. L. Ginzburg

Concluding Remarks: J. L. Birman

Preface

The First Binational USA-USSR Seminar-Symposium on the Theory of Light Scattering in Condensed Matter was held in Moscow 26-30 May 1975.

The initial conception for a light scattering seminar of about fifty scientists - half from each side, including theorists and experimenters "well versed in theory" - arose from discussions between Professor J. L. Birman and Professor K. K. Rebane at the 1971 Paris International Conference on Light Scattering in Solids. This conception won approval among the active scientists on both sides. After considerable planning and some delays, it received both material support and encouragement from the appropriate organizations on each side: in the USA: The National Science Foundation (Division of International Programs), and the National Academy of Sciences; in the USSR: the Academy of Sciences USSR.

A variety of reasons contributed to the positive response on both sides: for example, the considerable and high level of theoretical and experimental scientific activity on both sides in laser-related light scattering, optics, and generally - electrodynamics of condensed media - some along rather similiar lines; the impediments to free and easy communication and travel between USA and USSR scientists working on related problems; plus the desire to improve both contacts, and the free flow of information and individuals, to the mutual advantage of both sides. The recognition that frequent scientific exchanges are a sine qua non for progress, and the powerful desire on each side to meet personally with colleagues whom one knows of only by their published work, also played a role.

Among the basic areas of great interest currently are the spectroscopic investigation of elementary excitations in condensed matter and the interaction among them, and the basic nature of the radiation-matter interaction. The ultimate use of such results could be expected to contribute to the design of materials with desired optical properties, and the development of new physical instruments such as tunable lasers. For this reason the scope of the Seminar-Symposium encompassed "Light Scattering" in a general sense including "frontier" topics related to the optics and the electrodynamics of different forms of condensed media. Moreover, an additional objective of the Symposium was to improve understanding of the scientific approaches and methods utilized in the two countries. Restriction to narrow conventional lines would have been counterproductive in realizing this goal.

The participants in the First Seminar-Symposium expressed the opinion that the Seminar was highly successful, and represented a first step on a road which could lead to further exchanges of information and scientists of mutual benefit. Additional Seminar-Symposia in this series, alternating between USA-USSR were part of the initial conception, and the success of the First Seminar gives encouragement to the realization of this idea.

The present volume contains the Proceedings of the First Seminar-Symposium. In addition to the papers delivered at the Seminar, a selected few of the comments and discussion are included. Since the plans for publication developed during the Seminar, some of the papers published here are in a slightly different form than originally presented and only a few of the comments and discussion were submitted for publication. For completeness, the entire program of the First Seminar as well as the participants on both sides are listed. The Editors believe that these Proceedings will be very useful to the many active scientists throughout the world working on problems related to Light Scattering.

In the preparation of this volume one of the American Editors (B. B.) has assumed the responsibility of having all manuscripts typed in a uniform style suitable for photoreproduction. Certain minor editorial changes were made in various manuscripts and because of the desire to have expeditious publication these could not always be checked with the authors. The American Editors wish to warmly thank Professor V. M. Agranovich for providing the complete set of USSR manuscripts, for graciously assenting to the present publication, and for his general assistance.

Thanks are due to Mrs. Claire McCartney for her rapid and proficient typing of the scientific manuscripts, without which these proceedings could not have been a reality.

B. BENDOW

J. L. BIRMAN

August, 1976

Contents

SECTION II
STRONG ANHARMONICITY EFFECTS AND ELECTRON-PHONON INTERACTION IN CRYSTALS

SECTION III
SCATTERING FROM ELECTRONS AND ELECTRON-HOLE DROPS

SECTION V
BULK AND SURFACE POLARITONS

SECTION VI
RESONANCE SCATTERING AND SCATTERING FROM LOCAL EXCITATIONS

Section I

Phase Transitions and Liquid Crystals

LIGHT SCATTERING IN THE VICINITY OF PHASE TRANSITION POINTS IN SOLIDS*

V. L. Ginzburg and A. P. Levanyuk

Institute of Crystallography
USSR Academy of Sciences
Moscow USSR

The problem of light scattering in the vicinity of phase transition points and, particularly, near a tricritical point in solids, is discussed. The scattering intensity is shown to be highly dependent on the shear modulus, the latter fact being previously omitted by the authors.

The question of light scattering in the vicinity of phase transition points in solids and, specifically, of the possibility to observe critical opalescence has been under discussion for almost twenty years [1-12]. Unfortunately, the investigation of the problem is hampered because of both experimental difficulties and somewhat peculiar circumstances. In fact, the theory of light scattering developed without due regard to shear deformations [1-3, 4-7] leads, in particular, to the conclusion about the presence of very strong scattering in the vicinity of the tricritical point (i.e., near the Curie critical point, at which the first-order phase transition line turns into a 2nd-order one on the p, T-diagram). As the same phenomenon seemed to be observed [2] for the $\alpha \rightleftarrows \beta$ transition in quartz and since further experiments were not carried out for a long time, the development of a more detailed theory for scattering in solids was not stimulated. Meanwhile, the experiments described in [8] and some other papers provide certain reasons to believe that in [2] it was not critical opalescence that

*See also Physics Letters 47A, 345 (1974).

was observed, but virtually scattering on static heterogeneities (the question of whether it occurs at twin boundaries [8], α and β phase boundaries [12] or some other heterogeneities is not yet elucidated).

Consequently, in general the very possibility of critical opalescence in solids became doubtful (see, in particular, [11], [12]). On the other hand, there can be no doubt that fluctuations of the order parameter increase in the vicinity of the tricritical point. Specifically in quartz, near the $\alpha \rightleftarrows \beta$ transition, the scattering of x-rays [13] and neutrons [14] is observed, occuring, evidently, from thermal fluctuations (probably this refers also to light scattering in the vicinity of the transition point in $SrTiO_3$ [10]).

Thus, a question arises as to the reason for the discrepancy between the theory and observations. We noticed rather long ago that it might probably be attributed to the fact that in papers [1, 4-7] the influence of shear strains [15] was neglected; however, detailed calculations for different specific cases and, above all, the comparison of theoretical and experimental data, have not been completed yet. Nevertheless, in connection with the critical papers mentioned above, it seems appropriate here to throw some light on the problem in principle. For this purpose, similarly to our previous papers, we shall consider a phase transition with one order parameter η, taking no account of the crystal anisotropy, but with due regard to the strain (tensor U_{ik}). The theory of the self-consistent field (the Landau theory), which neglects fluctuations in the zeroth approximation,* will be used here.

Then the thermodynamic potential is given by

---

*The character of this approximation is evident, for instance, from [16-18]. It is of importance here that for the case of the tricritical point the Landau theory has especially wide range of applicability for calculating fluctuations. Moreover, while considering light scattering, we are interested only in a portion of the fluctuations (the characteristic value

$$q = 4\pi/\lambda_0 \sin \theta/2 << 2\pi/a \sim 10^8 \text{ cm}^{-1}),$$

allowing the applicability of one or another of the approximations [4] to broaden.

$$\Phi = \Phi_o + \frac{\alpha}{2}\eta^2 + \frac{\beta}{4}\eta^4 + \frac{\gamma}{6}\eta^6 + r\eta^2 U_{\ell\ell} + \frac{k}{2}U_{\ell\ell}^2 + \mu(U_{ik} - \frac{1}{3}\delta_{ik}U_{\ell\ell})^2 + \frac{1}{2}\delta(\text{grad }\eta)^2 , \tag{1}$$

where k is the bulk modulus and μ is the shear modulus. In equilibrium (here $\beta_1 = \beta - 2r^2/k$; it is assumed that the stresses $\sigma_{ik} = \partial\Phi/\partial U_{ik} = 0$):

$$\eta_0^2 = \frac{-\beta_1 + (\beta_1^2 - 4\alpha\gamma)^{\frac{1}{2}}}{2\gamma} \tag{2}$$

In the vicinity of the critical point T_{tc} the coefficients α and β_1 are proportional in our approximation to $(T_{tc} - T)$, while γ = const. Hence, close enough to T_{tc} we have $\eta_0^2 \sim (T_{tc} - T)^{\frac{1}{2}}$.

The mean square of the Fourier components η_q is as follows (see similar calculations in [4]: $\tilde{k} = k + 4\mu/3$)

$$\overline{|\eta_q|^2} = \frac{kT}{V}\{2\eta_0^2(\beta_1^2 - 4\alpha\gamma)^{\frac{1}{2}} + \frac{16\mu r^2\eta_0^2}{3k\tilde{k}} + \delta q^2\}^{-1} \tag{3}$$

If $\mu = 0$ and $T \to T_{tc}$, then $\overline{|\eta_q|^2} \sim (T_{tc} - T)^{-1}$, if the term δq^2 is neglected; but if $\mu \neq 0$, then $\overline{|\eta_q|^2} \sim (T_{tc} - T)^{-\frac{1}{2}}$. At the same time, the intensity of the "first order" scattering (see [4]) is proportional to $\eta_0^2\, \overline{|\eta_q|^2}$, if there is a quadratic relationship between $\Delta\epsilon$ and $\Delta\eta$. Under these conditions, when $\mu \to 0$, the intensity is $I \sim (T_{tc} - T)^{-\frac{1}{2}}$, but when $\mu \neq 0$ and T is approaching T_{tc}, the intensity exhibits only some final increase. This effect has to be observed in quartz. If, due to symmetry reasons, there is a change in permeability $\Delta\epsilon \sim \Delta\eta$ (see [3]), then $I \sim \overline{|\eta_q|^2} \sim (T_{tc} - T)^{-\frac{1}{2}}$ even at $\mu \neq 0$. Moreover, the "second-order" scattering which exists already in disordered (high temperature) phases (see [4]) may be considerable. Special attention should be paid to scattering in metastable phases near spinodes, i.e., lines corresponding to the loss of stability (see [5]), as well as to the scattering in the presence of stresses ("clamped" crystal, etc.) and an electric field. Thus, there exists a great many possibilities, though, generally speaking, it is naturally impossible to suppose $\mu = 0$ in a solid (in contrast, for example, to liquid crystals); this supposition would lead to the increase of critical opalescence, as compared with that being evaluated at $\mu \neq 0$. The other thing is that this opalescence in some cases, especially for the first-order transitions (though the latter are close to T_{tc}) may be masked by the scattering from quasistatic optical heterogeneities, which are also especially pronounced only in the vicinity of the phase transition point.

Finally, it should be noted, that light scattering near the phase-transition point must, and nowadays can, be considered more accurately ("scaling", etc.) than by the Landau theory. The appropriate analysis is of interest, but it does not alter the aforesaid as a whole.

The authors believe that the existence of critical scattering under discussion is a certainty, and also that there is no place for doubt as to the great interest of its investigation.

REFERENCES

1. V. L. Ginzburg, Dokl. Akad. Nauk SSSR 105, 240 (1955).
2. I. A. Yakovlev, T. S. Velichkina and L. F. Mikheeva, Dokl. Akad. Nauk SSSR 107, 675 (1956); Kristallographiya 1, 123 (1956), Usp. Fiz. Nauk 69, 411 (1957).
3. N. A. Krivoglas and S. A. Ribak. Zh. exp. teor. fiz. 33, 139 (1957).
4. V. L. Ginzburg and A. P. Levanyuk, J. Phys. Chem. Solids 6, 51 (1958); Memorial Volume to G. S. Landsberg, Acad. Sci. Publ. House p104; Moscow (1959).
5. V. L. Ginzburg and A. P. Levanyuk, Zh. exp. teor. fiz. 39, 191 (1960). Sov. Phys. JETP, 12, 128 (1961).
6. V. L. Ginzburg, Usp. Fiz. Nauk 77, 621 (1962); Sov. Phys. Uspekhi 5, 649 (1963).
7. A. P. Levanyuk and A. A. Sobyanin, Zh. exp. teor. fiz. 53, 1024 (1967).
8. S. M. Shapiro and H. Z. Cummins, Phys. Rev. Letters 21, 1578 (1968); 19, 361 (1967).
9. I. J. Fritz and H. Z. Cummins, Phys. Rev. Letters 28, 96 (1972).
10. E. P. Steigmeier, H. Auderset and C. Harbeke, Solid State Commun. 12, 1077 (1973).
11. F. J. Bartis, Phys. Letters A 43A, 61 (1973).

12. J. Bartis, J. Phys. C.: Solid State Phys. 6, 1295 (1973).
13. K. Ishida and G. Honje, J. Phys. Soc. Japan 26, 1558 (1969).
14. J. D. Axe and G. Shirane, Phys. Rev. B1, 342 (1970).
15. A. P. Levanyuk, Zh. exp. teor. fiz. 66, 2255 (1974).
16. V. L. Ginzburg, Fiz. tverdogo tela 2, 2031 (1960). Sov. Phys.-Solid State 2, 1824 (1960).
17. E. K. Riedel and F. J. Wegner, Phys. Rev. Letters 29, 339 (1972). A. M. Goldman, Phys. Rev. Letters 30, 1038 (1973).
18. R. Bausch, Z. Physik 254, 81 (1972); 258, 423 (1973).

COMMENT

(by H. Z. Cummins)

Professor Ginzburg has discussed some puzzling aspects of the structural phase transition in quartz which have not yet been completely resolved. This problem is of particular interest currently because of the discovery of "central peaks" in the neutron and light scattering spectra of a number of crystals during the last few years.

During the 1950's, Professor Ginzburg and his coworkers discussed light scattering from crystals undergoing structural phase transitions extensively, basing their analysis primarily on Landau's thermodynamic theory. They showed that there could be intense scattering if the transition is second order, but very close to the Curie critical point (or tricritical point) where the coefficients of the quadratic and quartic terms in the power series expansion of the free energy in terms of the order parameter η simultaneously go through zero. (See ref. 1 and literature cited therein.)

In 1956, Yakovlev, Velichkina and Mikheeva observed that when quartz passes through the alpha-beta structural transition at $573^{o}C$, there is a dramatic increase in the intensity of scattered light to $\sim 1.4 \times 10^4$ times the level at room temperature.[2] They also observed that the region of strong scattering, the "fog zone", was spatially localized around the phase boundary. The remarkably close agreement between the experimental intensity result and numerical predictions of the theory led to the identification of the observed scattering as critical opalescence.[1]

In ref. 1, Professor Ginzburg also discussed theoretical predictions for the temperature dependence of vibrational modes near the transition, based on his previous predictions of a "soft mode". Specifically, he argued that the frequency of the lattice vibrational mode corresponding to oscillation of the order parameter η should go to zero at a second-order transition, and that this would probably be the 207 cm^{-1} mode in quartz. This remarkable paper concluded with a call for spectroscopic studies of light scattering by crystals near points of second order phase transitions, and pointed out the applicability of lasers to such experiments!

A program of laser scattering spectroscopy of quartz was carried out between 1966 and 1968 in our laboratory, then at Johns Hopkins University, largely inspired by Professor Ginzburg's ideas. We first studied the temperature dependent Raman spectrum, finding that the 207 cm^{-1} mode did exhibit the largest temperature dependence initially, but that a feature at 147 cm^{-1} actually showed the ultimate soft mode behavior[3] reaching zero frequency at the transition. (J. F. Scott later showed that the 147 cm^{-1} feature is a second order peak which interacts anharmonically with the 207 cm^{-1} optic mode.[4])

Subsequent experiments by S. M. Shapiro showed, however, that the large increase in scattered light near the phase transition was apparently not due to the fluctuations of the soft mode because it was concentrated in a spectral region much narrower than the width of the soft mode.[5] In fact we never succeeded in measuring any width of the critical opalescence, and visual observation of the fog zone showed the granularity characteristic of static scattering of laser light. (Yakovlev et al. had not observed this granularity because their light source was spatially incoherent.) We therefore concluded that the intense scattering in the fog zone was due to essentially static inhomogeneities, perhaps consisting of small domains of Dauphiné twins, an idea previously suggested by R. A. Young, which is close to Professor Ginzburg's proposal that the fog zone consists of droplets of α quartz within the β phase.

In today's lecture Professor Ginzburg reviewed the 1974 theoretical paper in which he and A. P. Levanyuk showed that inclusion of strain energy in the total free energy expression for quartz would drastically reduce the predicted intensity of critical opalescence associated with the order parameter fluctuations, so that agreement between experiments and the earlier theory which neglected strain was not meaningful.[6]

The theoretical and experimental situation regarding central peaks was summarized at a conference on Anharmonic Lattices, Structural Transitions and Melting[7] in 1973 and has not changed significantly since. The major points of interest here are:

(1) In crystals showing the "central peak" effect, intense scattering is observed in a frequency range close to zero as the transition is approached. In most cases there is no unambiguous measurement of the width which may, in fact, be zero.

(2) The intensity of the central peak is effectively borrowed from the soft mode so that the integrated intensity (central peak plus soft mode) is proportional to the static susceptibility. Note that this corresponds to the results for quartz before the latest modification of Ginzburg and Levanyuk.[6]

(3) Of the various theoretical models proposed to explain the central peak, there is one of particular relevance, due to Jens Feder.[8] Feder states that "in the high temperature phase the order parameter vanishes on the average, but due to fluctuations one may find regions that have the low temperature structure..." which corresponds exactly to Professor Ginzburg's idea for quartz of droplets of α quartz within the β phase.

Thus quartz may turn out to be very closely related to other crystals exhibiting central peaks. The major theoretical problem to be investigated, it seems to me, is whether the dynamics of such droplets can be slow enough to produce the nearly static behavior observed in quartz.

REFERENCES

1. V. L. Ginzburg, USP. Fiz. Nauk. 77, 621 (1962) [Soviet Physics Uspekhi 5, 649 (1963)].
2. I. A. Yakovlev, T. S. Velichkina and L. F. Mikheeva, Kristalografiya 1, 123 (1956); I. A. Yakovlev and T. S. Velichkina, Uspekhi Fizicheskikh Nauk. 63, 411 (1957).
3. S. M. Shapiro, D. C. O'Shea and H. Z. Cummins, Phys. Rev. Letters 19, 361 (1967).

4. J. F. Scott, Phys. Rev. Letters 21, 907 (1968).
5. S. M. Shapiro and H. Z. Cummins, Phys. Rev. Letters 21, 1578 (1968).
6. V. L. Ginzburg and A. P. Levanyuk, Phys. Letters 47A, 345 (1974).
7. Anharmonic Lattices, Structural Transitions and Melting, edited by T. Riste (Noordhoff, Leiden, 1974).
8. J. Feder, in Ref. 7, p. 111.

CRITICAL OPALESCENCE AT STRUCTURAL PHASE TRANSITIONS

P. A. Fleury

Bell Laboratories

Murray Hill, New Jersey 07974 USA

The requirements for definitive observation of critical opalescence at structural phase transitions have long been known; but only recently have experiments been performed which demonstrate quantitative satisfaction of these requirements. The related singular, temperature dependence in the intensity, angular distribution, and spectral profile of scattered light are briefly reviewed. Complications associated with coupling between order parameter fluctuations and other degrees of freedom are considered. Results of high resolution ($\Delta\nu \leq 0.03$ cm^{-1}) light scattering experiments are discussed for 1) the ferroelectric phase transition in KDP and 2) the shear induced transformations at 118K and 151K in $PrAlO_3$. Critical opalescence is demonstrated for all three transitions. Mean field exponents appear to describe adequately the observed behavior. No measurable dynamics (linewidths) were found for the "central peaks" in any of the transitions.

The phenomenon of critical opalescence - the singular increase in the scattering of light by thermally driven fluctuations in the order parameter - is most often associated with the liquid-gas critical point. Here the so-called Rayleigh scattering from the isobaric density fluctuations has been used successfully, for example, to study the singular behavior of the specific heat and the thermal diffusivity near T_c.[1] In principle one should expect similar anomalous scattering from order parameter fluctuations at any second order phase transition, including structural transitions in solids provided, of course, that a finite coupling exists at T_c between the order parameter fluctuations and the electromagnetic field. Indeed, beginning with the observations of Iakovlev et al at the $\alpha \rightarrow \beta$ transition in quartz[2] there have been a number of reports of greatly enhanced light scattering efficiencies near structural transitions.[3-6] However, the question remains open in all these cases as to whether the observed effects are indeed due to critical opalescence, or arise merely from parasitic scattering by static inhomogeneities such as domain boundaries, etc.[7] This ambiguity remains because, in contrast to the experiments on liquids, none of these experiments have measured the second necessary feature for critical opalescence: the narrowing of the spectrum of the scattered light which must be quantitatively related to the intensity increase if it is indeed caused by critical fluctuations.[8] Professor Ginzburg's prophetic paper of 1962 clearly emphasized[9] the importance of determining the spectral composition as well as the intensity and polarization of the scattered light for a definitive observation of critical opalescence. In this same work he outlined the expected behavior of the spectral profile at $T \rightarrow T_c$ in what we now know as the quasi-harmonic "soft mode" approximation. Several years passed before the first light scattering experiments were done which demonstrated the soft mode behavior approaching a second order structural transition.[10] Progress has been rapid since then, however, and literally hundreds of light scattering studies have been performed on solids exhibiting structural transitions.[11] As these experiments have become more refined, the theoretical concern with spectral lineshapes and extraction of critical exponents has increased. More complicated response functions for the order parameter fluctuations, particularly accounting for their interactions with other degrees of freedom, were required which complicated the detailed behavior of the critical opalescence.

In this paper we shall briefly review the theoretical framework in which soft modes and critical opalescence are most conveniently discussed, and shall then examine the most recent and complete experimental investigations of critical opalescence in two different types of structural transition: 1) the ferroelectric transition in

KDP and 2) the shearing transitions at 151K and 118K in $PrAlO_3$.

In general the spectrum of scattered light is determined by the space and time Fourier transform of the polarizability autocorrelation function:

$$I_{ijk\ell}(\vec{r}-\vec{r}', t-t') \equiv \langle \alpha_{ij}(\vec{r}, t)\alpha_{k\ell}(\vec{r}', t')\rangle \tag{1}$$

The tensor function $\alpha_{ij}(\vec{r}, t)$ receives contributions from all types of excitations in the solid: acoustic and optic phonons, spin waves, excitons, plasmons, etc. as well as higher order combinations of these. We shall restrict our attention to those contributions to α_{ij} arising from fluctuations in the order parameter, and the other degrees of freedom to which it may be coupled. For simplicity we will suppress the tensor indices on α_{ij} and assume that the polarization conditions of the experiment are chosen to examine only fluctuations of the symmetry appropriate to the transition of interest. Under these assumptions we can write $\alpha_{ij}(\vec{r}, t) = C\delta_{ij}\Phi(\vec{r}, t)$, where $\Phi(\vec{r}, t)$ is the local instantaneous value of the order parameter and C is the polarizability modulus, which measures the light scattering efficiency. In the high symmetry phase $\langle\Phi\rangle$, the space-time average of the order parameter has a value of zero; in the low symmetry phase $\langle\Phi\rangle \neq 0$. The appropriate generalized susceptibility can be expressed quantum mechanically in terms of the retarded commutator[12]

$$\chi(\vec{r}-\vec{r}', t-t') \equiv \frac{i}{\hbar}\,\theta(t-t')\langle[\Phi(\vec{r}', t), \Phi(\vec{r}, t)]\rangle \tag{2}$$

The fluctuation-dissipation theorem relates the spectrum $(I(\vec{q}, \omega) = C^2 S(\vec{q}, \omega))$ to the Fourier transform of χ:

$$S(\vec{q}, \omega) = -\frac{\hbar}{\pi}\left(1 - \exp\left(\frac{-\hbar\omega}{kT}\right)\right)^{-1} \chi''(\vec{q}, \omega) \tag{3}$$

where χ'' denotes the imaginary part of χ. A set of sum rules useful for discussing critical exponents can be written for the frequency moments of the spectral response:

$$\int_{-\infty}^{\infty} \omega^{2N} S(\vec{q}, \omega)\,d\omega = \left\langle \frac{\partial^N \Phi(\vec{q}, t)}{\partial t^N}\,\frac{\partial^N \Phi(\vec{q}, t')}{\partial t'^N}\right\rangle_{t=t'} \tag{4}$$

In particular the N = 0 moment relates the integrated intensity to the average squared fluctuations and the static susceptibility:

$$I_{TOT} = \int_{-\infty}^{\infty} C^2 S(\vec{q},\omega)d\omega = C^2\langle \Phi(\vec{q})^2\rangle = C^2 \chi(\vec{q}, 0) \tag{5}$$

In this case as $T \to T_c$, I_{TOT} will diverge like the static susceptibility with the critical exponent γ: $I_{TOT} \sim \epsilon^{-\gamma}$; $\epsilon \equiv |T-T_c|/T_c$. This behavior assumes that the coefficient, C, relating the order parameter to the polarizability is independent of temperature.

In many cases, such as the displacive ferroelectric transition in cubic perovskites, or in general for unit cell multiplying transitions, the critical mode is not Raman active (i.e. carries no long wavelength polarizability) in the high symmetry phase. Above T_c, C = 0. And below T_c, C is replaced by some function of the static order parameter $\langle \Phi \rangle$. In the simplest case it is proportional to $\langle \Phi \rangle$; i.e. $C \to C'\langle \Phi \rangle$ where C' is a constant. Thus the critical opalescence is modified from Eq. (5) to[13]

$$I_{TOT} = C'^2 \langle \Phi \rangle^2 \langle \Phi^2 \rangle . \tag{6}$$

Since near T_c, the order parameter varies with the critical exponent β, Eq. (6) predicts $I_{TOT} \sim \epsilon^{(2\beta-\gamma)}$. Now in mean field theory ($\beta = 1/2$, $\gamma = 1$), <u>no divergence</u> in I_{TOT} is expected for this type of transition, since $2\beta - \gamma = 0$. For other model Hamiltonians (e.g. Heisenberg, Ising, X-Y) the values of $(2\beta - \gamma)$ are often negative ($\simeq -0.64$) and a divergence in I_{TOT} is expected. However, inasmuch as most second order structural transitions appear to exhibit mean field exponents, and hence would exhibit no divergence if they belong to the classes described by Eq. (6) the <u>best</u> candidates for observing critical opalescence are those where the soft mode is Raman active on both sides of T_c; i.e. C is a constant. This is the case for both the KDP and $PrAlO_3$ examples discussed below.

Before discussing the experiments, however, let us examine the spectral composition expected under various circumstances. The simple soft mode or quasiharmonic behavior is obtained when the form $\chi(\vec{q},\omega) = (\omega_{1q}^2 - \omega^2 + 2i\Gamma_{1q}\omega)^{-1}$ obtains, uncomplicated by mode coupling or frequency dependent damping effects. The static susceptibility diverges as $\chi(\vec{q}, \omega = 0) \sim \epsilon^{-\gamma}$, so that $\omega_{1q}^2 \sim \epsilon^{\gamma}$ must soften accordingly.[14, 15, 16] The critical opalescence and spectral profiles will then vary as suggested by Ginzburg.[9] The Stokes and anti-Stokes sidebands' displacements collapse continuously ($\omega_{1q}^2 \rightarrow 0$), while increasing in intensity as ω_{1q}^{-2}.

Until about 1971 available soft mode spectra could be adequately analyzed from this viewpoint with perhaps the addition of a, usually ad hoc, temperature dependence to Γ_{1q}. More refined experimental results, from both light and neutron scattering, revealed spectral behavior that could not be accounted for by the simple quasi-harmonic form of $\chi(\vec{q}, \omega)$ given above. For example the Raman-Brillouin spectra in ferroelectric $BaTiO_3$ illustrated the importance of coupling between the acoustic modes and the soft optic mode for the scattered lineshapes and strengths, as well as positions.[17] Essentially simultaneously high resolution inelastic neutron scattering[18] revealed a puzzling additional feature to the soft optic mode spectrum of the cell-doubling 110K transition in $SrTiO_3$. This additional feature was too narrow ($< 1\,cm^{-1}$) for its characteristic width or frequency to be resolved, but did increase markedly in intensity near T_c. It was soon realized that both complications could be described, at least phenomenologically, in terms of a general coupled mode formalism.

We can consider M pair-wise interacting excitations, noting that these may in general be multiphonon, as well as single phonon excitations. If U_i represents the amplitude of the ith excitation and F_i the field or external force which couples linearly to it, then the systems' equations of motion follow from $F_i = \sum_{j=i}^{M} a_{ij} U_j$. The a_{ij} are in general complex quantities which for $i \neq j$ express the mode coupling strengths. The a_{ii} is simply related to the uncoupled susceptibility for the ith excitation: $a_{ii} = (\chi_i^o)^{-1}$.

For light scattering, F_i is proportional to the polarizability of the ith excitation, so the scattered spectrum is determined by Eq. (3) with $\chi(\vec{q}, \omega)$ expressed as:

$$\chi(\vec{q}, \omega) = \sum_{i,j}^{M} F_i F_j \chi_{ij}(\vec{q}, \omega) \tag{7}$$

where the χ_{ij} are simply obtained from the equations of motion. In particular it is only necessary to invert the matrix of the a_{ij}: i.e. $[\chi]_{ij} = [a]_{ij}^{-1}$. In practice since the a_{ij} must often be taken as adjustable parameters, the extraction of unambiguous details from coupled mode spectra is prohibitively difficult for $M \gtrsim 3$. Fortunately the $M = 2$ case, even with $F_2 = 0$, contains the essential features for our purposes. In this case χ is just:

$$\chi(\vec{q}, \omega) = \frac{a_{22} F_1^2}{a_{11}a_{22} - a_{12}a_{21}} = \frac{\chi_1^o(\vec{q}, \omega) F_1^2}{1 - \chi_1^o(\vec{q}, \omega) \Sigma(\vec{q}, \omega)} \tag{8}$$

where $\Sigma \equiv a_{12}^2 \chi_2^o$ is called the self energy function and expresses the interaction between mode #1 and other degrees of freedom.

Depending upon the structure of Σ the observed spectrum can be more or less complicated. For example if Σ has a resonant form: $a_{12}^2(\omega_2^2 - \omega^2 + 2i\Gamma_2\omega)^{-1}$ the spectrum might look like that of superfluid helium, consisting of two doublets; one for first sound and the other for second sound excitations. If on the other hand Σ is relaxational rather than resonant:

$$\Sigma(\vec{q}, \omega) = i\omega\delta(1 - i\omega\tau)^{-1}; \tag{9}$$

a new central peak may appear in addition to the usual Mandelstam-Brillouin doublet. This complication was first noted for relaxing liquids by Rytov[19] and applied to structural transitions in solids by Levanyuk and Sobyanin.[20] It has been employed in the analysis of the neutron scattering results on $SrTiO_3$ and other structural transitions; implying that a relaxation process plays a significant role in the critical dynamics and opalescence of these transitions.[21] This interpretation, however, remains open to question because no dynamic (spectral) behavior has yet been resolved for these central peaks. Neutron experiments have insufficient frequency resolving power, and the light scattering technique encounters other difficulties (wave vector, or selection rule limitations) for many of the structural transitions in question.

For a quantitative exploration of critical opalescence and possible dynamic central peak behavior by light scattering, the structural transition in question must involve a long wavelength critical mode ($\vec{q}_c = 0$) which is Raman active both above and below T_c. The ferroelectric transition at 122K in KH_2PO_4 (KDP) exhibits these properties and is among the most thoroughly studied structural transitions. Although no neutron experiments have been done to reveal a central peak, one argues that if significant relaxing self energy behavior is a universal feature of structural transition dynamics, this material is a good candidate for study. Lakagos and Cummins[22] have carefully studied the temperature dependent lineshapes in coordinated Raman and Brillouin scattering experiments. They used a coupled mode analysis like that outlined above involving three modes: (1) an optic, (2) an acoustic and (3) a soft ferroelectric mode. With reasonable choices of parameters they could fit their spectra over the range 0 - 250 cm^{-1}, using simple quasi-harmonic forms for χ_1^0 and χ_2^0 and a self energy form for χ_3^0 as in Eq. (8). Their low frequency spectra for several temperatures near T_c are shown in Fig. 1. The narrow central peak whose intensity increases as $T \rightarrow T_c$ is not resolved here, but is assumed to arise from a relaxing self energy of the form of Eq. (9). Using a value for $\tau > 10^{-9}$ sec they could reproduce the agreement indicated by the solid curves in Fig. 1. However since no width was observed for the central peak, and since they could fit the entire spectrum, excluding that feature, by the above coupled mode analysis taking $\Sigma = 0$, one may say that the final word on the role of the central peak in critical opalescence is not yet clear. That is, it remains possible that the central peaks are due to scattering from defects, domains or other static sources.

The only second-order structural transition thus far examined which exhibits a central peak in neutron experiments and whose soft mode is Raman active on both sides of T_c is the 151K cooperative Jahn Teller transition in $PrAlO_3$. The transition involves a continuous change of direction from [101] toward [001] of the axis of octahedral rotation in the doubled perovskite unit cell.[23] This second order orthorhombic → tetragonal transition is driven by a soft exciton of B_1 symmetry[24] at $q = 0$. The exciton couples linearly to the [101] propagating, [10$\bar{1}$] polarized TA phonon

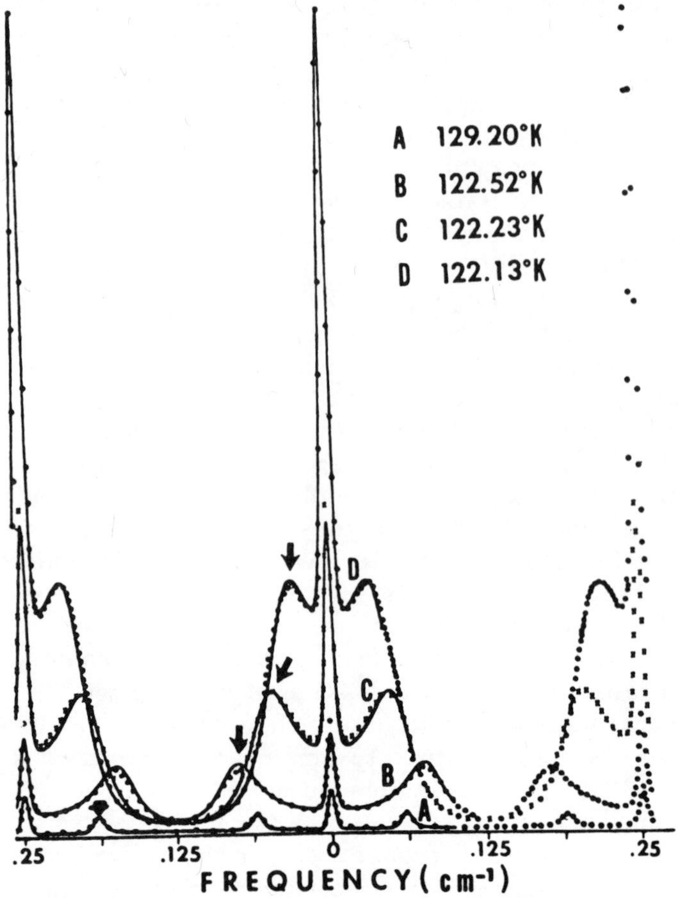

Fig. 1. Brillouin spectra of KDP at various temperatures near T_c = 122K. Arrows show positions of the Brillouin peaks of the interferometric order which is centered at zero frequency. Points represent data; solid curves are fits to coupled mode response functions, including a relaxing self-energy contribution to the soft ferroelectric mode. From Ref. 22.

which is the ultimate soft mode of the system. We have studied the Brillouin spectra[25] of this transition to check the validity of the Elliott et al model[26] for the temperature dependence of v_S as well as to search for a linewidth for the central peak observed in the neutron experiments.[24]

Parasitic scattering from domain walls, etc. is particularly severe in $PrAlO_3$, so that an additional means of reducing elastically scattered light is necessary. Fortunately molecular iodine vapor has a very sharp absorption line (halfwidth ~ 0.001 cm^{-1}) which lies within the gain curve for the 5145A line of the argon ion laser.[27] An intercavity etalon can be used to select the single longitudinal laser mode whose frequency coincides with the I_2 absorption frequency. Stabilization and control arrangements as shown in Fig. 2 thus permit rejection of elastically scattered

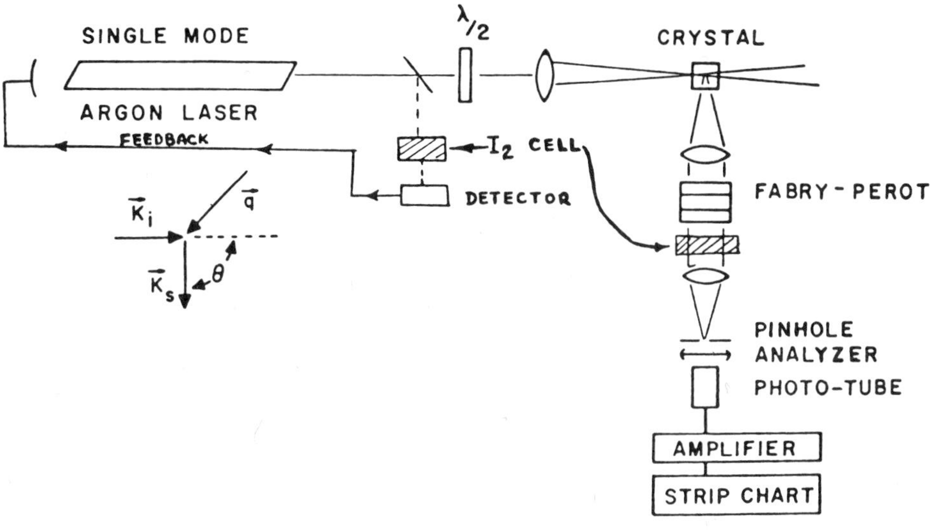

Fig. 2. Experimental arrangement used for $PrAlO_3$ studies. The single mode laser is tuned to the peak of a strong and narrow absorption band in molecular iodine vapor. A second I_2 cell in the path of the scattered light absorbs elastically scattered radiation, but passes inelastic spectra.

light with an extinction of ~ 10^{-5} or better. Using this technique we have obtained Brillouin spectra from the soft TA mode associated with the 151K cooperative Jahn Teller transition. Some selected spectra are shown in Fig. 3. Notice that the soft mode frequency decreases and its intensity increases as T_c is approached in just the manner expected for critical opalescence as was previously mentioned above. The relatively weak residual

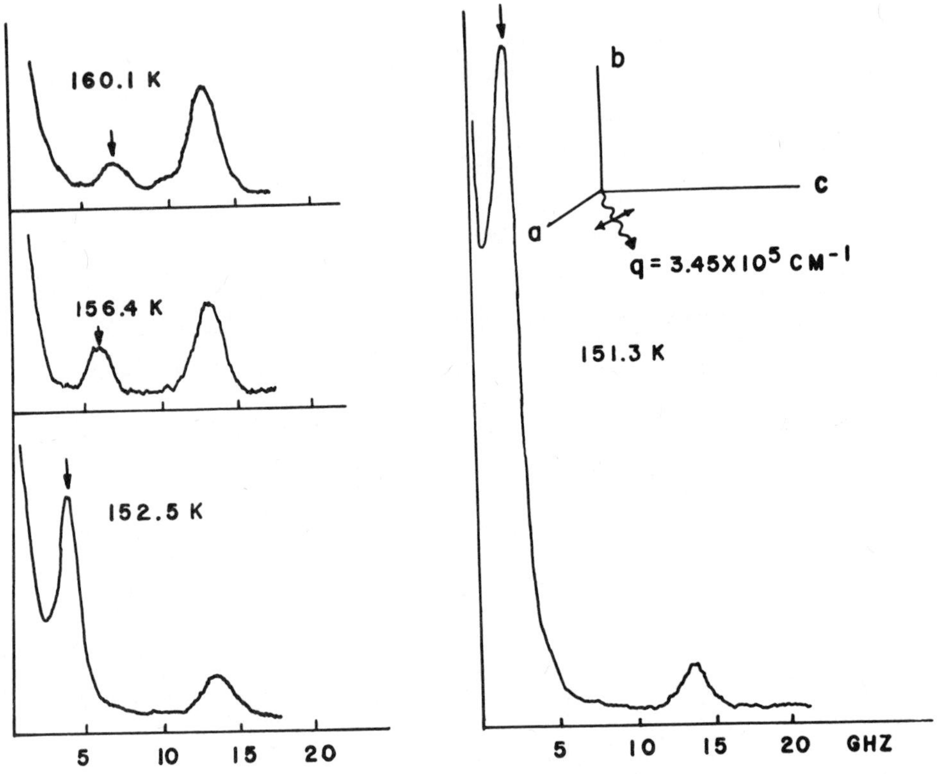

Fig. 3. Transverse mode Brillouin spectra of $PrAlO_3$ approaching the 151K cooperative Jahn Teller transition. Contrast the temperature independence of the companion TA mode at ~13 GHz to the simultaneous decrease in frequency and increase in intensity for the soft mode (arrows). Inset shows experimental geometry relative to the crystal's orthorhombic axes: a, b, and c. q is the soft mode wave vector [101]; polarized along [$\bar{1}$01].

scattering near zero frequency shift is presumably due to incomplete absorption by the iodine cell. No anomalous increase was observed in this residual scattering as T_c was approached. Nor was there any sign of a finite width central peak which gains strength at the expense of the Brillouin components near T_c. The instrumental resolution of our pressure-scanned Fabry-Perot interferometer (.05 cm^{-1}) permits us to set an upper limit for the dynamic width of any such central component at $\lesssim$ 0.03 cm^{-1}, or about thirty times less than that set by neutron scattering. The observed increase in scattered intensity, once the elastic contributions were eliminated by the I_2 cell, could be completely accounted for by the Brillouin components shown in Fig. 3. Thus the Brillouin spectra on the 151K transition in $PrAlO_3$ are consistent with the simple quasi-harmonic soft mode behavior for opalescence and require no relaxing self energy contributions.

Our experiments on $PrAlO_3$ led to the discovery of a new type of phase transition for the perovskite lattice,[25] a symmetry changing second-order structural transition for which strain appears to be the sole order parameter (in contrast for example to the 151K transition, in which strain, optic phonons, and electronic excitations are all strongly coupled). The soft mode frequency decreases with the mean field exponent on either side of T_c = 118K as shown in Fig. 4. Because below $\sim$ 130K, domain scattering becomes less severe in $PrAlO_3$, we were able to probe the near vicinity of the 118K transition even more thoroughly than the 151K transition. In particular not only could we use a 2.5X higher resolution on the Fabry-Perot, but we were also able to completely eliminate the elastic scattering with the iodine filter, as shown in Fig. 5. The latter result permits us to set a limit of $\lesssim$ 0.01 cm^{-1} on any dynamic central peak linewidth at the 118K transition in $PrAlO_3$.

Very close to T_c, it was noticed that the soft mode frequency does not precisely reach zero. There are several possible causes for this "saturation", one of which is the effect of a relaxing central peak. One consequence of a relaxing self energy for the spectral evolution near T_c is that the soft mode frequency does not reach zero, but saturates at a value ω_∞: $(\omega_\infty^2 = \omega_1^2 + \delta\tau^{-1})$. The central peak then begins to emerge, gaining strength at the expense of the soft mode sideband and exhibiting a linewidth $\sim \omega_1^2/\delta$. The very restrictive limits our experiments have set for any central peak width at the 118K transition, lead us to

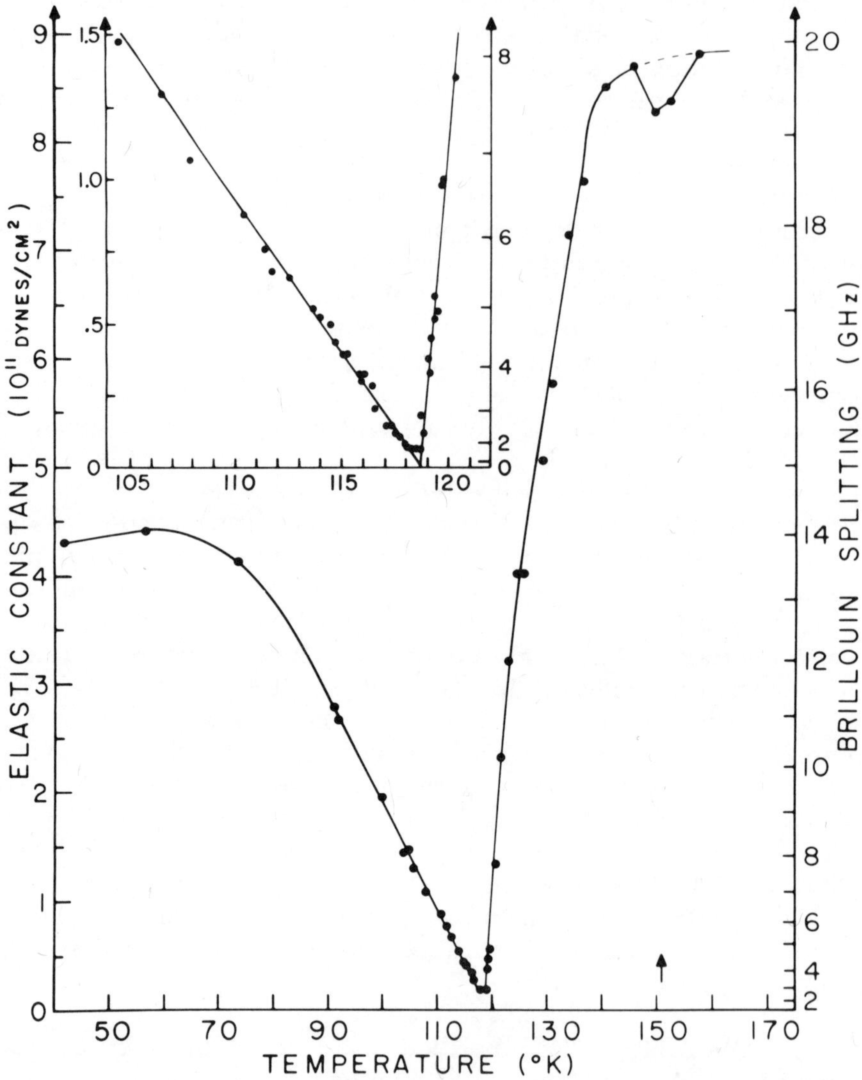

Fig. 4. Temperature dependence of soft mode frequency ν_B (right hand axis) and elastic constant $c_{ij} \sim \nu_B^2$ (left hand axis). Insert shows c_{ij} is linear in $|T - T_c|$, thereby confirming simple soft mode behavior and the mean field static exponent. The soft mode propagates in the [110] direction and is polarized in the [001] direction, referred to the $PrAlO_3$ orthorhombic axes. (See Ref. 25).

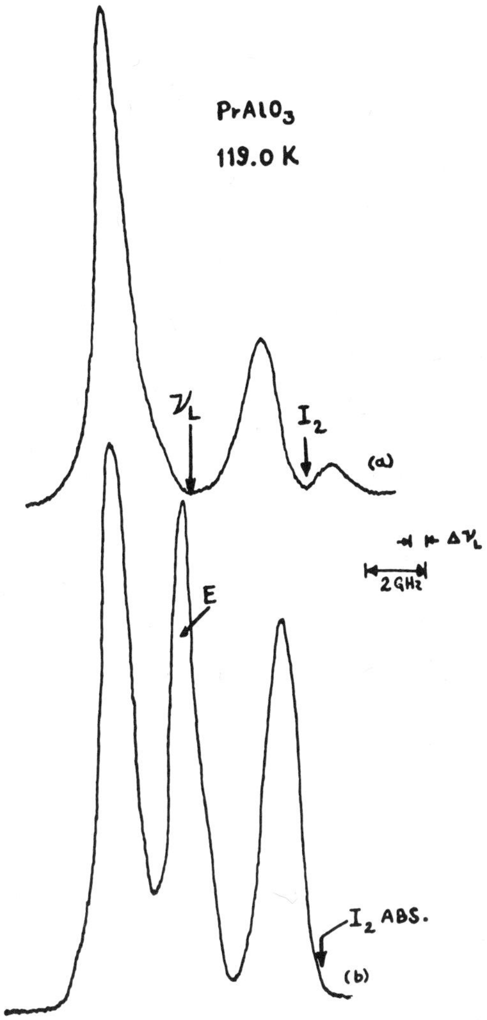

Fig. 5. High resolution spectra near the 118K transition showing the effect of slight detuning (b) of laser frequency (by 400 MHz) from exact resonance with iodine absorption. In (b) parasitic elastic light (E) is partially transmitted; whereas in (a) on-resonance there is no signal at the laser frequency, ν_L, thus demonstrating there is no dynamic central peak whose width exceeds that of the I_2 absorption line ($\lesssim .01\ cm^{-1}$). The two spectra can be registered in absolute frequency by matching the positions of the weak subsidiary I_2 absorption indicated by the arrows.

believe that the cause of the saturation lies elsewhere. To eliminate some of the other candidates (effects of finite correlation length ("$q\xi$" effects); or a small first order nature to the transition) would require different types of experiment to be done. In our opinion the most probable cause of the "rounding" observed near T_c is residual internal strains in the sample. By deliberate imposition of uncalibrated external stress (clamping the crystal) we were able to increase the amount of rounding observed. But even when we mounted our samples in the most strain-free manner consistent with adequate thermal contact, the smallest residual rounding observed was as shown in Fig. 6.

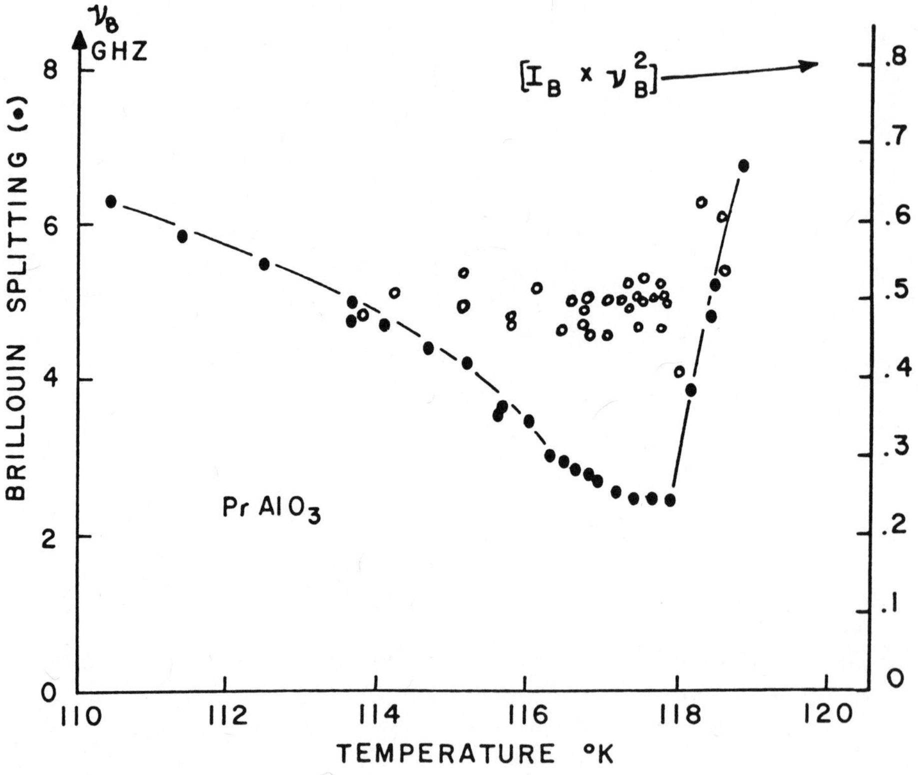

Fig. 6. Observed correlation between soft mode frequency (solid circles) and scattered intensity in the vicinity of the 118K pure strain transition in $PrAlO_3$. Open circles represent the product $I_B \times \nu_B^2$ (refer to right hand axis); and verify that within experimental error ($\pm 10\%$) $I_B \sim \nu_B^{-2}$ as expected for critical opalescence.

Finally regarding the critical opalescence question; we have carefully measured the intensity as well as the spectrum of the scattered light very near the 118K transition. Because the shear strain is the sole order parameter, the TA mode Brillouin scattering is precisely that feature which should exhibit critical opalescence as $T \to T_c$. In Fig. 6 we show the temperature dependence of the soft TA mode frequency ν_B (solid circles) together with the soft mode scattered intensity I_B multiplied by ν_B^2. For the expected correlation between scattered intensity and spectral content appropriate to critical opalescence, the product $I_B \nu_B^2$ should be a constant, independent of temperature. The open circles in Fig. 6 verify that this behavior is displayed in these experiments.

We may conclude then from the experimental results discussed here that critical opalescence is observed in structural phase transitions, and that for those systems studied with sufficient care thus far no departure from mean field behavior is indicated for static exponents. As far as dynamics is concerned, the simple quasi-harmonic approximation sufficies to describe the 118K transition in $PrAlO_3$; whereas a slight extension which accounts for coupling between the soft mode and the transverse acoustic phonons is required for both KDP and the 151K transition in $PrAlO_3$.

I am grateful to P. D. Lazay for experimental collaboration and to P. C. Hohenberg for constructive comments on this manuscript.

REFERENCES

1. See for example, H. Z. Cummins and H. L. Swinney in "Progress in Optics", Vol. 8 (edited by E. Wolf) North-Holland, Amsterdam (1970).
2. I. A. Iakovlev, T. S. Velichkina, and L. F. Mikheeva, Dokl. Akad. Nauk. SSSR 107, 675 (1956); Usp. Fiz. Nauk 69, 411 (1957).
3. W. D. Johnston and I. P. Kaminow, Phys. Rev. 168, 150 (1968).
4. P. D. Lazay, J. H. Lunacek, N. A. Clark and G. B. Benedek in "Light Scattering Spectra of Solids" (edited by G. W. Wright) Springer-Verlag, New York (1969) p. 593.

5. E. F. Steigmeier, G. Harbeke and R. K. Wehner in "Light Scattering in Solids" (edited by M. Balkanski) Flammarion Sciences, Paris (1971) p.396.
6. E. F. Steigmeier, H. Auderset and G. Harbeke, Solid St. Comm. 12, 1077 (1973).
7. S. M. Shapiro and H. Z. Cummins, Phys. Rev. Letters 21, 1578 (1968).
8. Without the evidence from spectral resolution, even the observation of temperature dependent angular variation of the scattered intensity (i.e. scattered wave vector dependence) cannot be said to be definitive.[6] For example, characteristic domain sizes will often depend on $(T - T_c)$ and could account for such observations.
9. V. L. Ginzburg, Usp. Fiz. Nauk. 77, 621 (1962); transl. Sov. Phys. Uspekhi 5, 649 (1963).
10. P. A. Fleury and J. M. Worlock, Phys. Rev. Letters 18, 665 (1967).
11. For a review see: J. F. Scott, Rev. Mod. Phys. 46, 83 (1974).
12. T. Schneider, G. Srinivasan and C. P. Enz, Phys. Rev. A5, 1528 (1972).
13. P. A. Fleury, Comments on Solid State Phys. IV, 149 and 167 (1972).
14. V. L. Ginzburg and A. P. Levanyuk, Zh. Eksp. Teor. Fiz. 39, 192 (1960); translation: Sov. Phys. JETP 12, 138 (1961).
15. P. W. Anderson, in Fizika Dielektrekov (G. Shansky, ed.) Acad. Sci. USSR, Moscow (1960).
16. W. Cochrane, Adv. Phys. 9, 387 (1960).
17. P. A. Fleury and P. D. Lazay, Phys. Rev. Letters 26, 1331 (1971); and in "Light Scattering in Solids" (edited by M. Balkanski), Flammarion Sciences, Paris (1971), p.406.
18. T. Riste, E. J. Samuelsen, K. Otnes and J. Feder, Solid State Comm. 9, 1455 (1971).
19. S. M. Rytov, Zh. Eksp. Teor. Fiz. 33, 671 (1958); transl. Sov. Phys. JETP 6, 513 (1958).
20. A. P. Levanyuk and A. A. Sobyanin Zh. Eksp. Teor. Fiz. 53, 1024 (1967); transl. Sov. Phys. JETP 26, 612 (1968).
21. See "Anharmonic Lattices, Structural Transitions and Melting" (edited by T. Riste) Noordhoff, Leiden (1974).
22. N. Lakagos and H. Z. Cummins, Phys. Rev. B10, 1063 (1974).
23. R. T. Harley, W. Hayes, A. M. Perry and S. R. P. Smith, J. Phys. C6, 2382 (1972).

24. R. J. Birgeneau, J. K. Kjems, G. Shirane and L. G. Van Uitert, Phys. Rev. B10, 2512 (1974).
25. P. A. Fleury, P. D. Lazay and L. G. Van Uitert, Phys. Rev. Letters 33, 492 (1974).
26. R. J. Elliott, R. T. Harley, W. Hayes, and S. R. P. Smith, Proc. Roy. Soc. A328, 217 (1972).
27. G. E. Devlin et al., Appl. Phys. Lett. 19, 138 (1971).

RAMAN SPECTRA AND STRUCTURAL PHASE TRANSITIONS IN IMPROPER FERROELASTICS Hg_2Cl_2 AND Hg_2Br_2

A. A. Kaplyanskii

A. F. Ioffe Physico-Technical Institute
USSR Academy of Sciences
Leningrad, USSR

INTRODUCTION

A new interesting group of materials has recently been synthesized in the form of artificial single crystals: halogenides of monovalent mercury Hg_2X_2, X = Cl, Br, I [1]. These isomorphous compounds at 20°C have a peculiar crystal structure comprised of parallel chains of linear molecules Hg_2X_2, which are relatively weakly bound to each other (Fig. 1). These molecules form a body-centered tetragonal lattice D_{4h}^{17} with one molecule in a primitive cell [2]. The chain-like structure of Hg_2X_2 crystals leads to extremely strong anisotropy of their physical properties. Thus, calomel crystals, Hg_2Cl_2, possess very high elastic anisotropy, one of the sound velocities being minimal from among those known in condensed media; it is lower than the sound velocity in air [3]. Crystals of Hg_2Cl_2 also have uniquely high birefrigence ($\Delta n = +0.65$ [4]). In [5,6] the spectroscopic properties of Hg_2X_2 single crystals were investigated via infrared (IR) spectra and Raman spectra (RS).

In studying the temperature dependence of RS for Hg_2X_2 it has been found that below $T_c = 185°K$ (Hg_2Cl_2) and $T_c = 143°K$ (Hg_2Br_2) there occurs in scattering spectra a number of qualitative changes which are indicative of a phase transition. The main effect of this transition consists in the appearance of additional weak lines in the first-order RS at $T \leq T_c$, such lines being absent in RS at $T > T_c$ [7]. This phase transition is directly

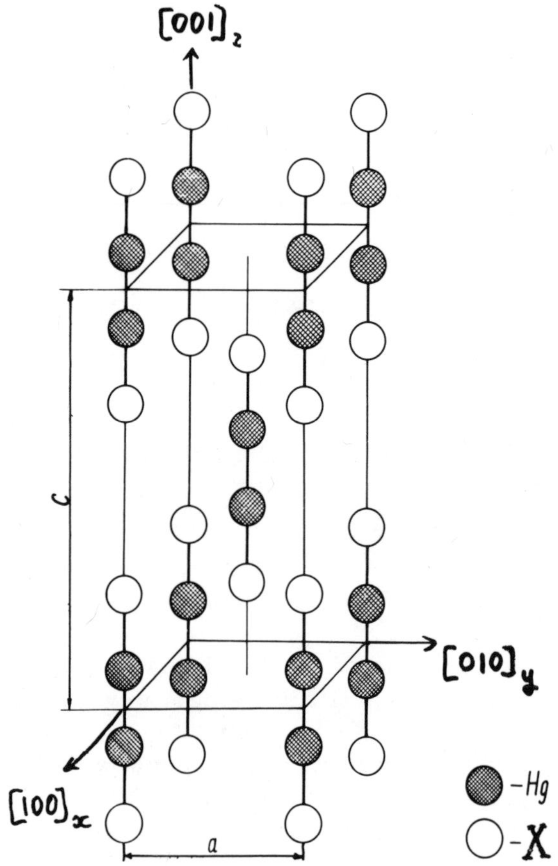

Fig. 1. Elementary unit cell of Hg_2X_2 crystals.

confirmed by the observation of domain structure in Hg_2Cl_2 and Hg_2Br_2 at $T < T_c$ [8]. According to [8] at $T \leq T_c$ the tetragonal point group D_{4h} of the crystal lowers down to the centro-symmetrical orthorhombic group D_{2h}, its principal axes being directed along [001], [110], [1$\bar{1}$0] (Fig. 1). At $T < T_c$ there also appears a spontaneous strain, and a monodomainization of the samples is possible by means of their uniaxial compression ("pure ferroelastics").

The distinct observation of phase transition effects in RS, in particular, the revelation of a soft mode [9] provided, on the basis of spectroscopic data, a detailed interpretation of the nature of the transition in Hg_2X_2. This transition has been found to be a structural one, $D_{4h}^{17} \rightarrow D_{2h}^{17}$, with unit cell doubling. This transition is induced by phonon instability at $T \leq T_c$, of a transverse acoustic branch of the vibrational spectrum at the X point at the Brillouin zone boundary of the tetragonal phase ("improper ferroelastic").

EXPERIMENTAL

Raman spectra were measured by a double monochromator (Coderg-PHO) with a He-Ne laser (Spectra Physics) with 60 mW power. For the determination of the scattering tensor, measurements were carried out for the 90° geometry of observation with polarized light, on oriented single crystals grown by the method described in [1]. The orientation of the samples was determined by the cleavage planes of the tetragonal crystals along [110], [110]. At $T > T_c$ the tensor was determined for the tetragonal axes x, y, z parallel to [100], [010], [001] (Fig. 1).

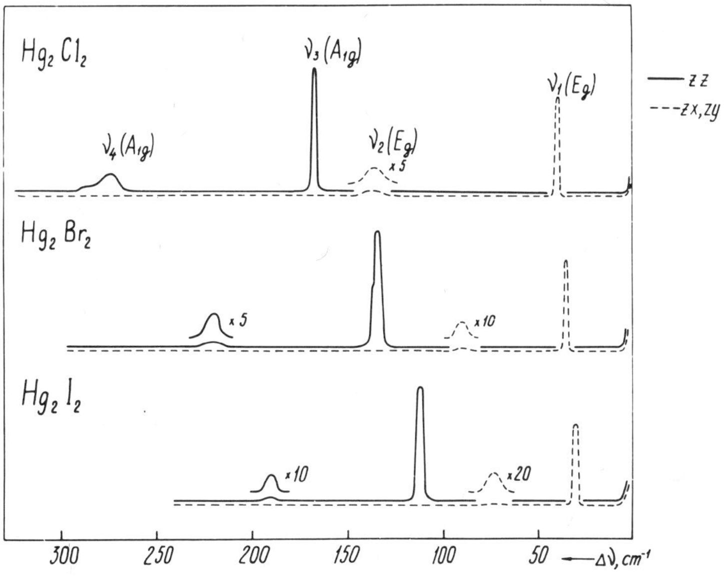

Fig. 2. Raman scattering spectra of Hg_2X_2 at 20°C in polarized light.

Figure 2 shows the first-order RS for crystals at 20°C. The spectra of the three crystals Hg_2Cl_2, Hg_2Br_2, Hg_2I_2 are seen to be completely analogous with respect to the number, relative position and polarization of the lines. In all, four lines are observed in the spectrum: two of them have polarization which correspond to doubly degenerate vibrations and two other (ν_3 and ν_4) that correspond to fully symmetric vibrations.

Experimental spectra completely agree with the results of a group-theoretical analysis of the fundamental vibrations at the Γ point of a tetragonal lattice D_{4h}, containing one linear molecule Hg_2X_2 in a primitive cell. Column 3 in Table 1 gives the symmetry of vibrations at the Γ point, and column 2 the selection rule for them. It may be seen that 12 fundamental lattice vibrations classify as (see columns 3, 4): 1) 6 odd vibrations: three acoustic and three-dipole-optic ones active in IR-spectra (A_{2u} and double generate E_u) and 2) 6 even vibrations active in RS (two fully symmetric A_{1g} stretching modes and two doubly generate E_g vibrations, bending and low-frequency librational modes). In column 1 of Table 1 for Hg_2Br_2 and Hg_2Cl_2, observed first order RS frequencies as well as IR frequencies known from [5], are identified.

The temperature dependence of the first-order RS has been investigated while cooling crystals from 300°K down to ~10°K. This interval includes the phase transition point T_c where the symmetry of the crystals lowered to a rhombic one. At $T \leq T_c$ the spectra of the rhombic phase have been studied on single domain samples obtained under uniaxial pressure in a basal plane along [110]. The scattering tensor was determined in the rhombic axes x ‖ [110], y ‖ [$1\bar{1}0$], z ‖ [001] (Fig. 1).

The results of a study of first-order RS are summarized in Fig. 3. On the ordinate axis spectral intervals in the range of fundamental frequencies are given, which are active in RS for tetragonal crystals at $T > T_c$ (Table 1), as well as intervals where at $T \leq T_c$ new RS lines of the first order appear. Solid lines denote the temperature dependence of the position of four fundamental frequencies ($2A_{1g}$ and $2E_g$) observed at $T > T_c$ in RS for tetragonal crystals (Table 1). Cooling from 300°K causes a small short-wave shift of these frequencies. At the point T_c all curves have a distinct anomaly, while the frequency of a doubly degenerate bending vibration $\nu_2(E_g)$ is split into a polarized **(zx, zy)** doublet, its width growing upon further crystal cooling below T_c.

Table 1. Fundamental Frequencies and Symmetry of Hg_2X_2 Crystals Vibrations

Phase D_{4h}^{17} $(T > T_c)$								Phase D_{2h}^{17} $(T < T_c)$	
1		2*	3	4	5	6	7*	8	
(300°K) Hg_2Cl_2	(300°K) Hg_2Br_2	α_{ik}, M_i	$\Gamma(D_{4h})$	$X_1(D_{2h})$	$\Gamma(D_{2h})$	$\Gamma(D_{2h})$	α_{ik}, M_i	Hg_2Cl_2	Hg_2Br_2
RS frequencies (cm^{-1}) [6]									
ν_1 = 40	35.6	zx, zy	E_g	$B_{2g}+B_{3g}$	$B_{2g}+B_{3g}$	A_u+B_{1u}	Z		
ν_2 = 137	91	zx, zy	E_g	$B_{2g}+B_{3g}$	$B_{2g}+B_{3g}$	A_u+B_{1u}	Z		
ν_3 = 167	135	zz, xx+yy	A_{1g}	A_g	A_g	B_{3u}	X		
ν_4 = 275	221	zz, xx+yy	A_{1g}	A_g	A_g	B_{3u}	X		
IR frequencies (cm^{-1}) [5]								RS frequencies (cm^{-1}), T = 90°K	
ν_5^T = 67	47	x, y	E_uTO	B_{3u}	B_{3u}	A_g	XX, YY, ZZ	$\nu_5^{L'}$ = 72	52
ν_5^L = 135	94		E_uLO	B_{2u}	B_{2u}	B_{1g}	XY	$\nu_5^{L'}$ = 144	97
ν_6^T = 254	168	z	A_{2u}TO	B_{1u}	B_{1u}	B_{2g}	ZX	$\nu_6^{T'}$ = 265	176
ν_6^L = 299	196		A_{2u}TO	-	-	-		-	-
Sound velocity [3] cm/sec·10^{-5}									
v [1$\bar{1}$0] [110]** = 0.347			E_uTA	B_{3u}	B_{3u}	A_g	XX, YY, ZZ	ν_{sm} = 13.6	8.7
v [1$\bar{1}$0] [001] = 1.084			A_{2u}TA	B_{1u}	B_{1u}	B_{2g}	ZX	ν'_A = 39	35
v [110] [110] = 2.08			E_uLA	B_{2u}	B_{2u}	B_{1g}	XY	-	-

*Single indices - nonzero components of dipole moment vector, double indices - scattering tensors. Column 7 gives selection rules only for lines flaring at $T < T_c$.

**The first index - direction of propagation, the second one - polarization of the wave.

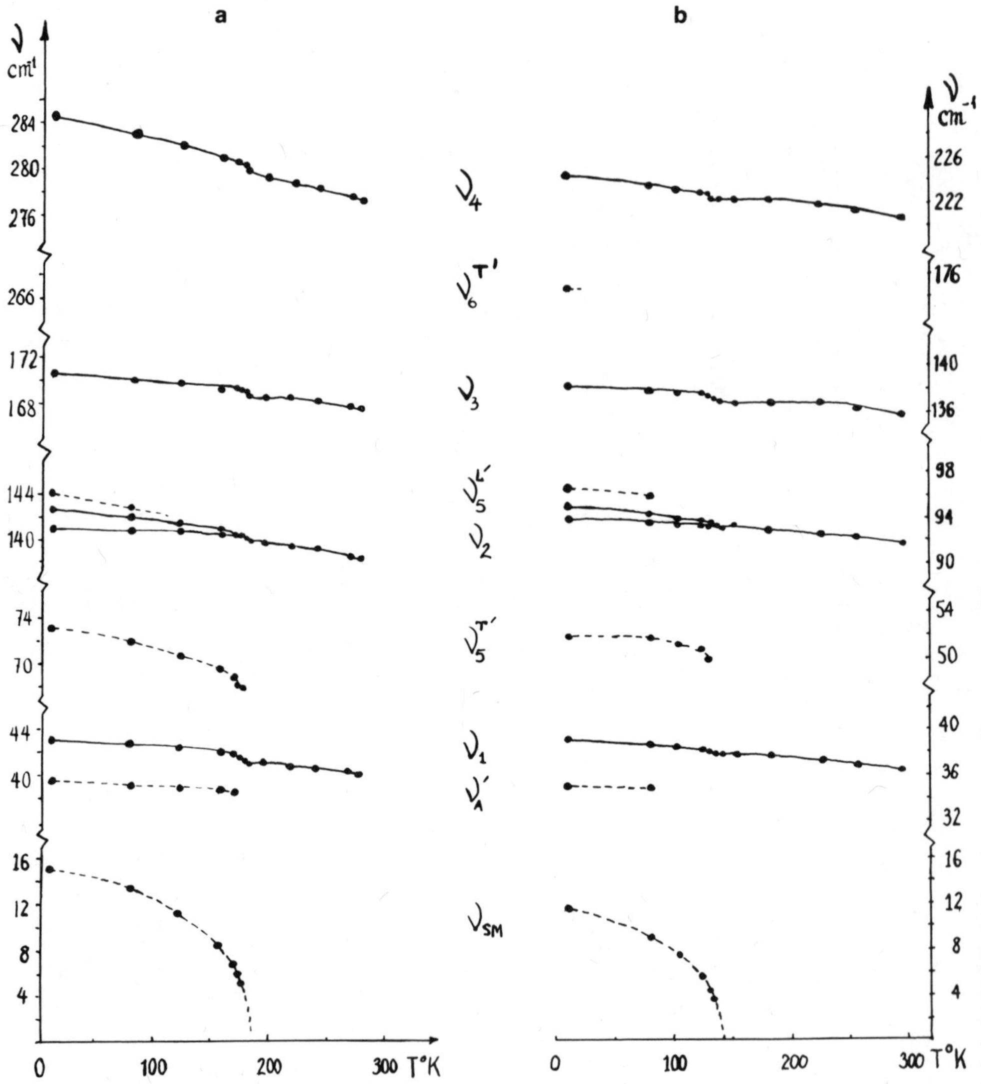

Fig. 3. Temperature dependence of line positions for (a) Hg_2Cl_2 and (b) Hg_2Br_2.

Broken lines denote the position of new first-order lines appearing in RS at the point $T = T_c$ and attenuating with further temperature decrease below T_c. The new lines are polarized. In all, five new lines have been observed in each spectra for Hg_2Cl_2 and Hg_2Br_2, their properties being similar in RS for both crystals:

1) The line ν_A', at a distance 2-3 cm^{-1} on the longwave side of the frequency ν_1 of the librational E_g vibration (Fig. 4). The line is fully polarized in the non-diagonal component zx. The intensity ν_A' grows with increasing $T_c - T$ (Fig. 5).

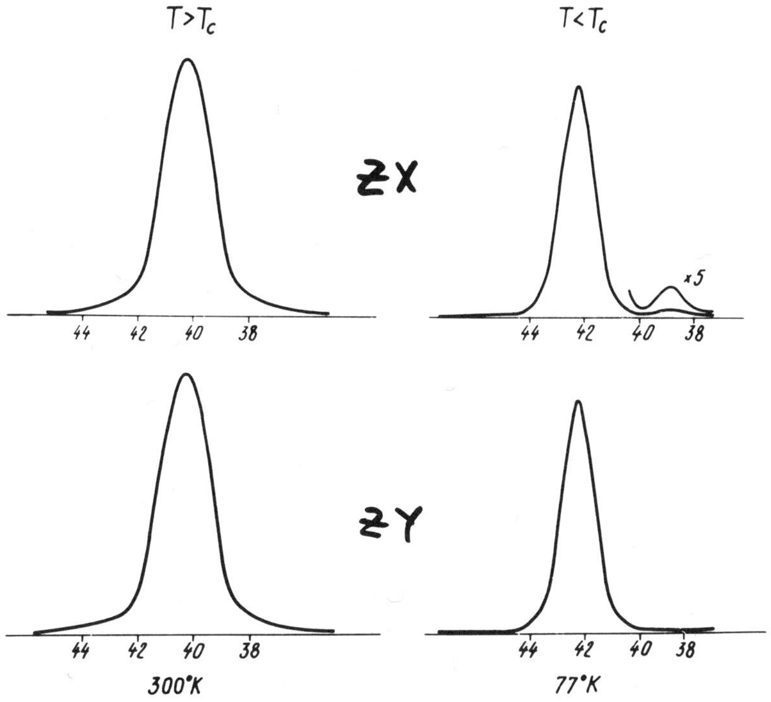

Fig. 4. "Flaring" at $T < T_c$ of a longwave polarized satellite near the line ν_1 of the librational vibration in RS of Hg_2Cl_2.

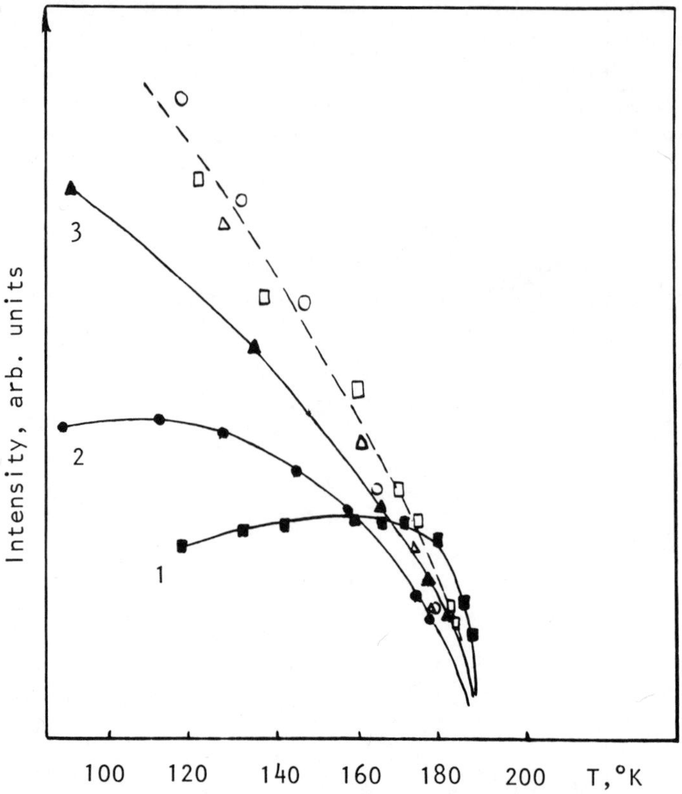

Fig. 5. Temperature dependence of the intensity of flaring lines in the RS of Hg_2Cl_2. 1 - ν_{sm} ; 2 - ν_A' ; 3 - $\nu_5^{T'}$. Broken line - reduced intensity.

2) The line $\nu_5^{T'}$, in the range of the fundamental vibration frequency $\nu_5^T(E_u, TO)$ active in IR-absorption. The line is observed in the diagonal polarization, yy, its intensity growing with increasing $T_c - T$ (Figs. 5, 6). It may be seen from Fig. 6 that the growth in intensity of $\nu_5^{T'}$ with crystal cooling contrasts with the behavior of the overtone $2\nu_1$, which is frozen out upon cooling. Upon cooling the crystal from T_c down to 10^oK a relatively large ($\sim 5 - 7\ cm^{-1}$) short-wave line shift becomes characteristic.

3) The line $\nu_5^{L'}$, in the range of the bending ν_2 vibration frequency E_g and longitudinal ν_5^L vibration frequency E_u, LO. A line

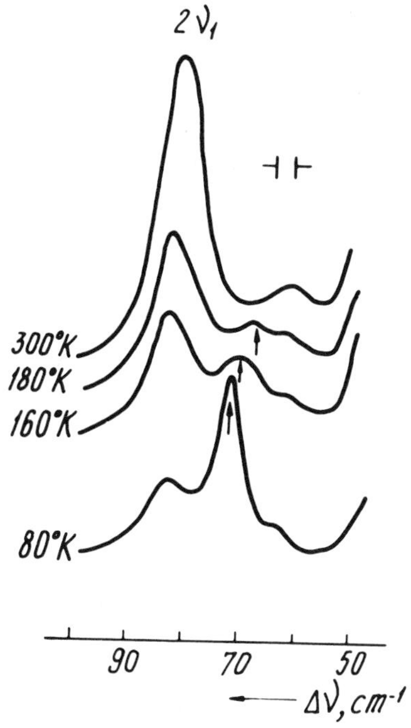

Fig. 6. "Flaring at $T < T_c$ of the line $\nu_5^{T'}$ in RS of Hg_2Cl_2.

of very weak intensity $\nu_5^{L'}$ is superimposed on the short-wave length tail of the line ν_2 (E_g) and it may be clearly distinguished in the spectrum only with polarized light. The polarization of $\nu_5^{L'}$ corresponds to xy.

4) The line $\nu_6^{T'}$, in the frequency range ν_6^{T} of the transverse A_{2u}, TO vibration, which is active in IR spectra. The line has nondiagonal polarization zx and is very weak (it is observed only upon cooling down to 10°K, whence the RS spectrum of the second order is frozen out [6]).

5) The line ν_{sm}, in the low-frequency spectrum range is the

most important line of additional RS appearing at $T \le T_c$ (Fig. 7). The frequency of the line is highly temperature dependent (at $T \to T_c^-$ the frequency $\nu_{sm} \to 0$), which doubtlessly is evidence of this line corresponding to a soft mode. The line is polarized in diagonal components, predominantly in yy. The line is very narrow, and its half-width ($\Delta\nu \approx 1.5$ cm^{-1}) undergoes a small change at $T \to T_c^-$ (Fig. 8), so that the increase of damping $\Delta\nu / \nu_{sm}$ at $T \to T_c^-$ (Fig. 8) is due mainly to the decrease of ν_{sm}. The intensity first grows rapidly with the increase of ν_{sm} and then, after passing the maximum, falls down, which is its difference in behavior from other new ("flaring") lines (Fig. 5).

There has been found to be a very strong dependence of the soft mode in RS on uniaxial stress in single domain samples along the axis [110]. The frequency and intensity of the line ν_{sm} increase considerably with strain (Fig. 9). This effect is greater, the nearer the sample temperature is to T_c. Figure 9 shows the dependence of the relative strain shift of the line

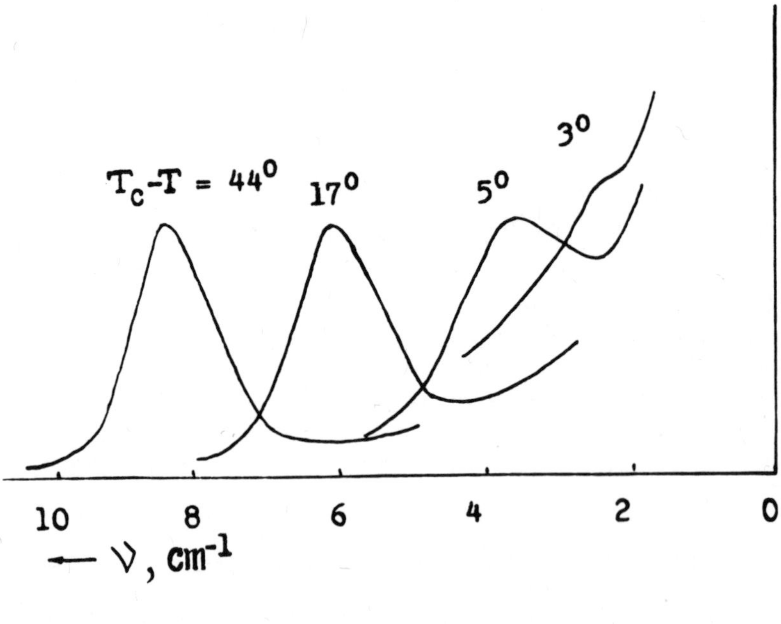

Fig. 7. Soft mode line in RS of Hg_2Br_2.

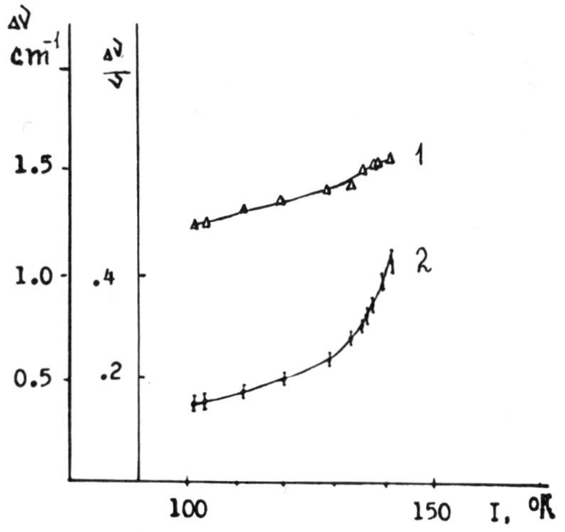

Fig. 8. Temperature dependence of half-width (1) and quenching (2) of the soft mode line in RS of Hg_2Br_2.

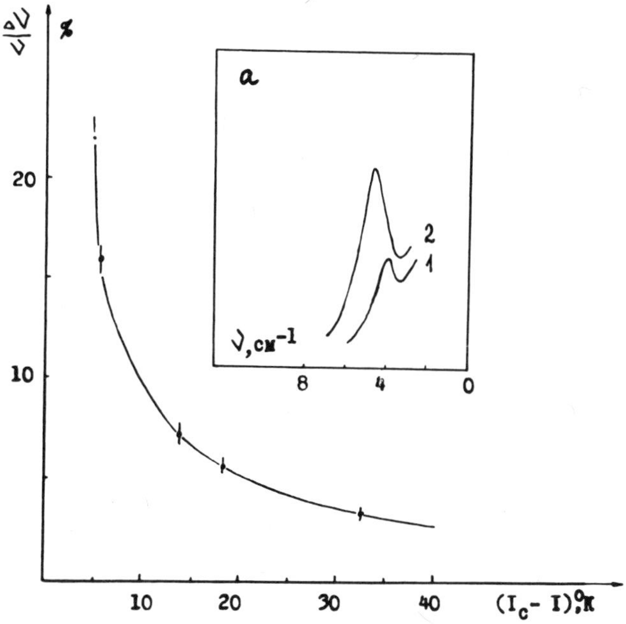

Fig. 9. Temperature dependence of a relative frequency shift in ν_{sm} RS upon deformation of Hg_2Br_2. a - soft mode spectrum of Hg_2Br_2 at T_cT = 10°K, when (1). σ = 0 and (2) σ = 0.5 kg/mm^2.

$$\frac{\nu_{sm}(\sigma) - \nu_{sm}(0)}{\nu_{sm}(0)} \text{ on } (T_c - T) .$$

It was possible to observe a relative increase of more than 30% in the soft mode frequency (at $T_c - T = 10^oK$, $\sigma = 0.5$ kg/mm^2).

The appearance of new lines in the RS of crystals at phase transitions was first properly explained in [10] when studying similar effects in RS of $SrTiO_3$ (structural phase transition at $T_c = 110^oK$). The increase in the number of fundamental vibrations appearing in the first-order RS at $T \leq T_c$ indicates that the number of molecules in the cell is doubled at the transition. Such a transition is induced by the soft mode, corresponding to lattice vibrations at the Brillouin zone (BZ) boundary. As a result, at $T \leq T_c$ the instability point of the BZ is transferred to the centre (Γ point) of a new halved-volume BZ for the low temperature phase and, accordingly, vibrations from this point become active at $T \leq T_c$ in first order optical processes. Based on this idea and on experimentally observed RS properties at $T \leq T_c$, a concrete model for phase transition may be proposed, which explains all known experimental data on phase transitions in Hg_2X_2.

PHASE TRANSITION MODEL AND OPTICAL SPECTRA

The body-centered tetragonal lattice (BCTL), corresponding to the structure Hg_2X_2 at $T > T_c$, has in (x, y, z) coordinates the basis vectors

$$\vec{a}_1 \tfrac{1}{2}(a, -a, c), \quad \vec{a}_2 \tfrac{1}{2}(-a, a, c), \quad \vec{a}_3 \tfrac{1}{2}(a, a, -c) \tag{1}$$

where a and c are parameters of the unit cell (Figs. 1, 10). The vectors of the reciprocal lattice are:

$$\vec{b}_1(\frac{1}{a}, 0, \frac{1}{c}), \quad \vec{b}_2(0, \frac{1}{a}, \frac{1}{c}), \quad \vec{b}_3(\frac{1}{a}, \frac{1}{a}, 0) \tag{2}$$

The first BCTL Brillouin zone shown by the solid line in Fig. 10b has some special points at the boundary: z (0, 0, 1/c) and two nonequivalent X points: X_1 (1/2a, 1/2a, 0) and X_2(1/2a, - 1/2a, 0).

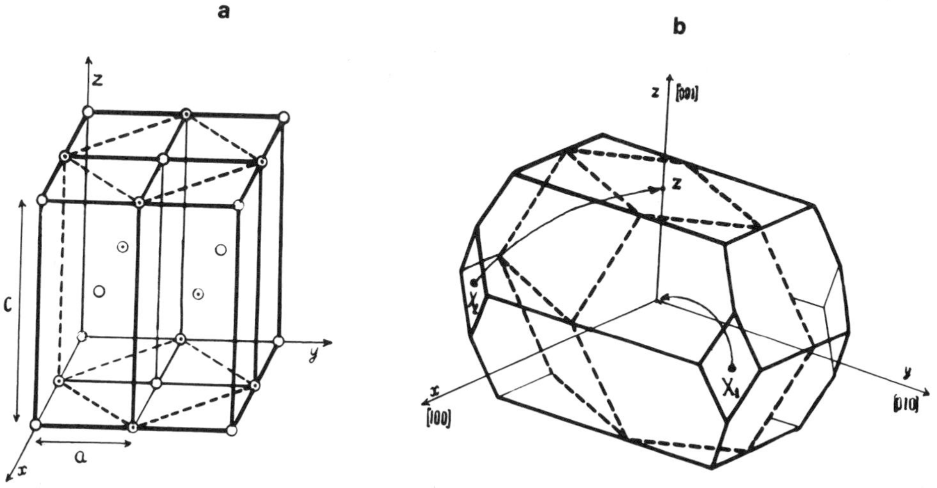

Fig. 10. Straight Bravais lattice (a) and Brillouin zone (b) of Hg_2X_2 crystals in tetragonal (solid lines) and orthorhombic (broken lines) phases.

We assume that the transition is induced by a phonon instability at the X points of the BCTL Brillouin zone at $T \le T_c$. The observed low frequency of the soft mode ($\nu_{sm}^{max} \approx$ 10-15 cm^{-1}) indicates that the appropriate vibration is connected energetically with the lowest acoustic branch of the Hg_2X_2 vibrational spectrum. As judged from acoustic measurements [3] this branch is a transverse TA branch with atom displacements in the crystal basal plane.

Let us consider the two nonequivalent points at the BZ boundary of BCTL, X_1 and X_2, with coordinates $\vec{k}(X_1) = \frac{1}{2}\vec{b}_3$ and $\vec{k}(X_2) = \frac{1}{2}(\vec{b}_2 - \vec{b}_1)$. The vectors $\vec{k}(X_1)$ and $\vec{k}(X_2)$ form a star of an irreducible representation of the crystal tetragonal group D_{4h}. The point group of the wave vector $\vec{k}(X_i)$ corresponds to D_{2h}. Acoustic transverse displacements in the basal plane at points X_1 and X_2 correspond to representations of B_{3u} and B_{2u} and the wave-vector group D_{2h}.

The basic vectors of the $B_{3u}[\vec{k}(X_1)]$ and $B_{2u}[\vec{k}(X_2)]$ representations,

$$\varphi_1 = e^{-i\vec{k}(X_1)\vec{r}}\,\psi_{B_{3u}}(\vec{r}),\quad \varphi_2 = e^{-i\vec{k}(X_2)\vec{r}}\,\psi_{B_{2u}}(\vec{r}) \qquad (3)$$

form the basis of a two-dimensional representation of a crystal space group D_{4h}^{17} [11]. Since the representation of τ is two-dimensional, the transition parameter is a double-component one [11]. Therefore, the electronic density in the crystal may be written in the form

$$\rho(\vec{r}) = \rho_0 + c_1\varphi_1 + c_2\varphi_2$$

where ρ_0 is the density in a tetragonal crystal, and c_1, c_2 are transition (displacement) parameters, which transform as φ_1 and φ_2.

In [12] a thermodynamic potential was derived for the two-component transition under consideration. The analysis of the properties of this transition in the frame of the traditional Landau theory predicts two theoretically possible low-temperature phases. Below, only the (rhombic) phase which is experimentally realized is considered.

The rhombic phase corresponds to four solutions: $c_1 = \pm 1$, $c_2 = 0$; $c_1 = 0$, $c_2 = \pm 1$, which are typical of four domain types which differ only by the orientation of the molecular displacements in the lattice. It may be seen that the given solutions correspond (with due regard to (3), (4)) to the instability of acoustic vibrations at one of the X points.

Figure 11 shows schematically displacements in a single domain ($c_1 = +1$; $c_2 = 0$) formed by "freezing" of the acoustic B_{3u} vibrations (3) at the X_1 point. Parallel displacements of the centers of gravity of Hg_2X_2 molecules take place along [110] directions, the directions of these displacements being opposite in adjacent atomic planes [110]. It follows from calculations [13] that the molecules are bent into trapezia, the result being that molecular chains linear for $T > T_c$ acquire for $T \leq T_c$ the form of a "crankshaft" (Fig. 11b). The displacements in the domain $c_1 = -1$, $c_2 = 0$ differ only in their signs (antiphase domains [14]); in the domains $c_1 = 0$, $c_2 = \pm 1$ the appropriate displacements take place in a perpendicular plane system [110].

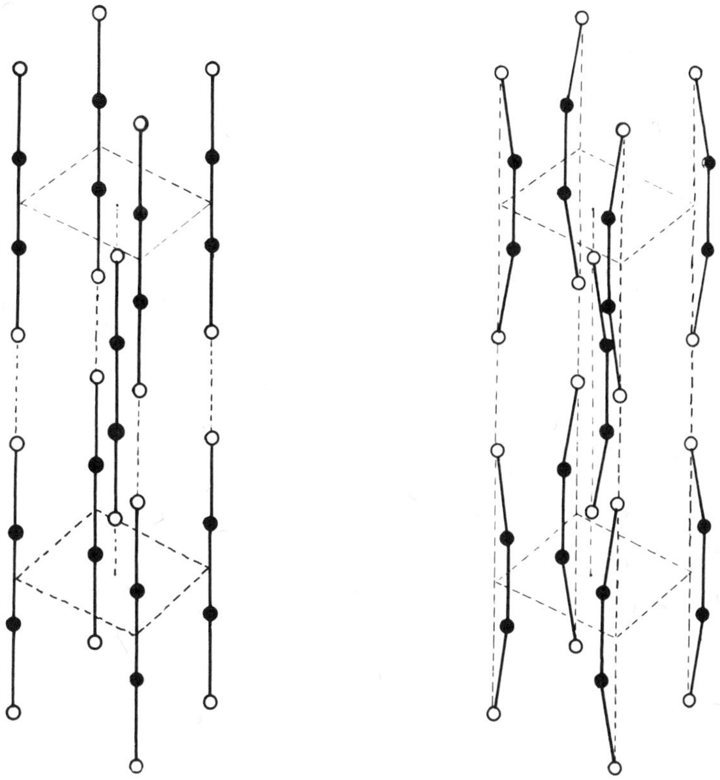

Fig. 11. Lattice scheme for Hg_2X_2 at $T > T_c$ (a) and (b) $T < T_c$.

A simple analysis of the structural geometry geometry in Fig. 11b shows that at the transition, the site symmetry of a Hg_2X_2 molecule becomes lowered from tetragonal D_{4h} to orthorhombic c_{2v}, the symmetry of the whole crystal lowers from tetragonal D_{4h} to rhombic D_{2h}. Molecules in adjacent planes [110] become noncongruent and the number of molecules in the cell is doubled. In Fig. 10 a broken line denotes a new orthorhombic cell (the broken line connects molecules which remain congruent), which correspond to a face centered rhombic lattice (BCRL). The three principal axes of the orthorhombic lattice, (z, x, y), are directed along the previous axes of the tetragonal crystal [001], [110], [110]. Such conclusions concerning the structure can also be drawn in a purely mathematical way from the condition of the invariance of the density $\rho(\vec{r})$ (4) at $c_1 = 0$ or $c_2 = 0$, related to approximate symmetry transformations. These transformations determine the space group of the low temperature phase as D_{2h}^{17}.

The presence of mixed terms in the thermodynamic potential, which include strain, explain the appearance of a spontaneous strain at $T \leq T_c$. The strain component ϵ_{xy} is the same for a pair of antiphase domains $c_1 = \pm 1$, $c_2 = 0$ and differs in sign for a pair $c_1 = 0$, $c_2 = \pm 1$. Thus, these two domain pairs differ macroscopically. Since a transition parameter is represented by a relative shift of molecular layers [110] Hg_2X_2 crystals should be referred to as "improper" ferroelastics.

Let us now consider the influence of the transition on RS. For this purpose it should be clarified how the Brillouin zone transforms at the transition. In the BZ of a tetragonal crystal, vibrations at points X_1 and X_2 are doubly degenerate for $T > T_c$. In the rhombic phase at $T < T_c$, the degeneracy is removed as X_1 and X_2 are transferred to different points of the BZ for the low-temperature BCRL. Reciprocal vectors for this lattice are of the form

$$\vec{b}'_1 = \vec{b}_1 - \tfrac{1}{2}\vec{b}_3,\ \vec{b}'_2 = \tfrac{1}{2}\vec{b}_3,\ \vec{b}'_3 = \vec{b}_3 - \vec{b}_1 - \vec{b}_2, \tag{5}$$

its Brillouin zone being denoted by a broken line in Fig. 10b. It may be seen that $X_1(\frac{1}{2}\vec{b}_3)$ corresponds to a reciprocal vector and, therefore, is transferred to the BZ center (Γ point) of the orthorhombic lattice. The point $X_2[\frac{1}{2}(\vec{b}_2 - \vec{b}_1)]$ falls on the new BZ boundary at the point z.

This transfer $X_1 - \Gamma$ results in all vibrations at X_1 including the soft mode ν_{sm}, which were optically inactive at $T > T_c$, becoming active at $T \leq T_c$ in first-order optics. Selection rules for the new fundamental vibrations at $T \leq T_c$ can be easily obtained with the help of group theory either by an analysis of vibration symmetry transformations at X_1 for the transfer $X_1 \rightarrow \Gamma$, or by a direct derivation of the symmetry of the fundamental crystal vibrations D_{2h} with two Hg_2X_2 molecules in a unit cell, which possess site symmetry c_{2v} (according to the method given in [15]). Table 1 gives the calculated results for vibrations from all branches of the vibrational spectra for tetragonal crystal Hg_2X_2 at the X point, which at $T \leq T_c$ are transferred to the Γ point of an orthorhombic crystal. In column 4 the symmetry of the vibrational branches at the X_1 point is given for a tetragonal crystal. In the right-hand part, data for an orthorhombic crystal D_{2h}^{17} are presented. Column 5 gives the symmetry of fundamental vibrations at the point $\Gamma(D_{2h})$, originating from fundamental vibrations at $\Gamma(D_{4h})$; the main effect of the $D_{4h} \rightarrow D_{2h}$ transition at the Γ point is the doublet splitting of the doubly degenerate E_g vibrations. In column 6 the symmetry at the Γ pount (D_{2h}) is given for vibrations originating from vibrations at the X_1 point of the BZ of a tetragonal crystal. For these new fundamental frequencies a selection rule is given in column 7 for IR and Raman spectra. It is seen that in RS at $T \leq T_c$ vibrations may appear from X, of all "odd" branches of the tetragonal crystal spectrum, three optical branches and three acoustic. Vibrations of "even" branches at X may give new lines in the IR spectrum of the orthorhombic phase.

It follows from general considerations that at $T < T_c$ both the value of the fundamental E_g vibration splitting and the intensity of the new lines in RS from the X point are proportional to the squared displacement parameter c^2. In the approximation of Landau theory $c^2 \sim (T_c - T)$. Therefore, the intensity of new RS Stokes lines is

$$I_s \sim (T_c - T)[n(T) + 1] \tag{6}$$

where

$$n(T) = [\exp(\frac{h\nu}{kT}) - 1]^{-1}$$

is the usual temperature factor which characterizes the population of vibrational modes.

COMPARISON OF EXPERIMENTAL AND CALCULATED RESULTS

The calculated data presented in Table 1 give an adequate explanation of the RS results at $T \leq T_c$. Particularly, the polarized splitting of the degenerate frequency of the bending vibration

$\nu_2(E_g)$ is explained at the transition $D_{4h} - D_{2h}(E_g = B_{2g} + B_{3g})$. The absence of a similar splitting for other degenerate librational vibrations $\nu_1(E_g)$ is most probably due to the smallness of this splitting (it equals $\sim 0.3\ cm^{-1}$, if the relative splitting ν_1 is assumed to be $\sim 1\%$, as is the case for ν_2).

The theory (columns 6, 7 in Table 1) also explains the properties of the new lines appearing in RS at $T \leq T_c$ and corresponding to fundamental vibrations of the D_{2h}^{17} phase, originating from the crystal X point D_{4h}^{17}. Five new lines from six theoretically feasible ones are experimentally observed. The position and polarization of the lines also agree with calculations. The frequencies of the new vibrations in RS from the X point for the three optical branches

$$\nu_5^{T'},\ \nu_5^{L'},\ \nu_6^{T'}$$

are close to the fundamental frequencies of the corresponding branches E_uTO, E_uLO and $A_{2u}TO$ known from IR. This is indicative of the small dispersion of the intermolecular dipole vibrations (an accurate comparison of the frequencies is difficult here since IR [5] and Raman spectra were measured at different temperatures).

In RS at $T \leq T_c$ there appear also vibrations (ν_{sm} and ν'_A) from the X point of two transverse acoustic branches; one of them is a soft mode. The values ν_{sm} and ν'_A agree with the evaluation of these branches at the X point obtained for Hg_2Cl_2 in the Debye approximation based on the known (column 1, Table 1) sound velocities:

$$\nu_{TA}(B_{3u}) = \nu_{[110]}^{[110]} \cdot q_x c^{-1} = 18\ cm^{-1},$$

$$\nu_{TA}(B_{1u}) = \nu_{[110]}^{[001]} \cdot q_x c^{-1} = 54\ cm^{-1},$$

where $q_x = (2a)^{-\frac{1}{2}}$ is the inverse wavelength of a phonon at X ($a \approx 4.5$ Å), and c is the light velocity. Just as should be expected, experimental frequencies are less than Debye ones due to dispersion. The vibrations ν_{sm} and ν'_A at $T \leq T_c$ refer, evidently, to vibrations of the translational type. It may be seen from column 6 in Table 1 that the soft mode ν_{sm} at $T \leq T_c$ is fully symmetrical (A_g), as is the $\nu_5^{T'}$ vibration related to the branch E_uTO. Due to the same symmetry these vibrations interact with each other. Hence, a considerable decrease in frequency is manifested at $T \to T_c^-$ (Fig. 3). It is a consequence of interaction with a soft mode of frequency $\nu_{sm} \to 0$.

A detailed agreement of experimental and calculated data on RS is convincing proof in favor of the transition model assumed in the third section. Consequences of this model also explain all the other known experimental facts concerning macrosymmetry of the orthorhombic phase (with a center of inversion), the directions of optical centers in it, orientations of domain boundaries, and ferroelastic properties [8]. The spatial group of the low-temperature phase D_{2h}^{17}, obtained from the analysis of purely spectroscopic data, is confirmed by direct x-raying of Hg_2Cl_2 at low temperatures.* It should be noted also that an instability of the acoustic transverse branch at the X point (imaginary value of the appropriate frequency) is obtained by microscopic calculation of the full vibrational spectrum of Hg_2Cl_2 from the data for the Γ point (elastic and dielectric constants, fundamental frequencies) [13].

A quantitative comparison of experimental results with the conclusions of phenomenological theory is also of interest. Such a comparison is impeded by the fact that the main temperature dependence of this theory for the displacement parameter $c^2 \sim (T_c - T)$ and soft mode frequencies $\nu_{sm}^2 \sim (T_c - T)$ near T_c are usually not realized because of correlation effects and large fluctuations of the displacement parameter [14]. Precision measurements of $\nu_{sm}(T)$ have not been performed; however, the curves obtained (Fig. 3), within the accuracy limits of the experiment, are definitely better described by the dependence $\nu_{sm}^3 \sim (T_c - T)$ in a certain region near T_c. A dependence of such a type has been obtained before, e.g., for structural phase transitions in $SrTiO_3$ [16]. In [16] it was shown also by the example of $SrTiO_3$ that the proportionality $c^2(T) \sim \nu_{sm}^2(T)$ is realized in the experiment in a sufficiently wide range $(T_c - T)$, though the dependences $c^2(T)$ and ν_{sm}^2 (T) taken separately deviate considerably from Landau theory predictions. This fact allows one to use for Hg_2X_2 experimentally obtained values $\nu_{sm}(T)$ as an argument when checking theoretical dependences of different values of the (unknown) displacement parameter c(T).

*Measurements have been performed by M. E. Boiko and A. A. Vaipolyn at the Physico-Technical Institute.

In Fig. 12 (the curve "a") the experimental width of the doublet splitting (at $T \leq T_c$) of the frequency ν_2 of the bending vibration $E_g(B_{2g} + B_{3g})$ is plotted as a function of ν_{sm}^2. A theoretically predicted linear dependence takes place. When analyzing the intensities of the "flaring" lines in RS, one should proceed from the values I, which account for the population of phonon modes n(T)(6). Figure 5 gives normalized "reduced" values of I (T) for the frequencies ν_{sm}, ν'_A and $\nu_5^{T'}$ obtained by dividing experimental values (curves 1, 2, 3) by the value n(T) + 1 (for the soft mode ν_{sm} in n(T), a temperature shift of the frequency is taken into account). It is seen that the temperature dependence of the intensity follows practically the same law for different lines, thus being described by a universal curve

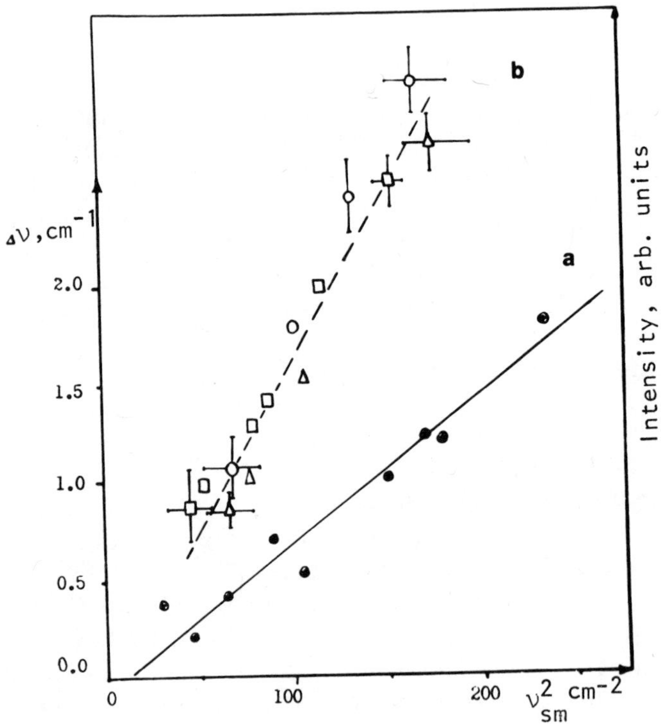

Fig. 12. The dependence of the splitting value $\nu_2(E_g)$ (a) and the intensity of flaring lines (b) on the square frequency of the soft mode in Hg_2Cl_2.

$I_0(T)$ (broken line in Fig. 5). In Fig. 12 the curve $I_0(T)$ (Fig. 5) is recalculated with the function $\nu_{sm}(T)$ to yield the dependence $I_0(\nu^2_{sm})$, which is linear (the curve "b") according to theory.

It should be noted in conclusion that Hg_2Cl_2 and Hg_2Br_2 are pure improper ferroelastics with a two component transition parameter, which is connected with the soft mode at a point on the Brillouin zone boundary. It is important that double degeneracy of the soft mode at $T > T_c$ is related to the existence of two non-equivalent points X on the zone boundary. Herein lies the difference between the case of Hg_2X_2 and the classical one, i.e., the transition of $SrTiO_3$ at 110°K, where non-one-dimensionality of the transition parameter is a consequence of a (threefold) degeneracy of proper vibrations at one (R) point in the BZ.

The simplicity of the crystallographic structure and the fundamental spectrum of Hg_2X_2, the clarity of spectrographic evidence of the phase transition (in particular, the narrowness of the soft mode line which allows one to follow its frequency decrease for $T \to T_c^-$ down to the minimal value of a few cm^{-1}, as reported in the literature) make Hg_2X_2 a convenient object for spectroscopic study of general problems of structural phase transitions in crystals.

V. V. Kulakov and Yu. F. Markov participated in the studies described in the present paper. The measurements were performed on single-crystals of Hg_2X_2 grown by Ch. Barta at IFTT Institute of the Czechoslovak Academy of Sciences (Prague).

REFERENCES

1. C. Barta, Kristal. u. Technik 5, 541 (1970).
2. H. Mark, I. Steinbach, Zs. f. Kristallogr. 64, 79 (1926).
3. I. M. Sil'vestrova, C. Barta, G. F. Dobrzhanskii, L. M. Belyaev, Yu. V. Pisarevskii, Kristallographya 20, 359 (1975).
4. M. H. Dufet, Bull. Soc. Franc. Mineralogie 21, 90 (1898).
5. I. Petzelt, I. Mayerova, C. Barta, L. Kislovskii, Czech. J. Phys., B23, 845 (1973).
6. C. Barta, A. A. Kaplyanskii, V. V. Kulakov, Yu. F. Markov, Optika and Spectroskopiya 37, 95 (1974).
7. C. Barta, A. A. Kaplyanskii, V. V. Kulakov, Yu. F. Markov, Fizika Tverd. Tela 16, 3125 (1974).

8. C. Barta, A. A. Kaplyanskii, V. V. Kulakov, Yu. F. Markov, Fizika Tverd. Tela 17, 1129 (1975).
9. C. Barta, A. A. Kaplyanskii, V. V. Kulakov, Yu. F. Markov, Pis'ma v ZhETF, 21, 121 (1975).
10. P. A. Fleury, J. F. Scott, J. E. Worlock, Phys. Rev. Lett. 21, 16 (1968).
11. G. Ya. L'ubarskii, Teoriya grupp i eje primenenije v fizike. Fizmatgiz, 1957; O. V. Kovalev, Neprivodimije predstavleniya prostranstvennykh grupp, Izd. AN USSR, Kiev, 1961.
12. C. Barta, A. A. Kaplyanskii, V. V. Kulakov, B. Z. Malkin, Yu. F. Markov, Zhurn. Eksper. Teoret. Phys. (to be published).
13. B. Z. Malkin, Materialy 8th Vsesoyuznogo soveshchaniya po teorii poluprovodnikov, Kiev, 1975 .
14. A. P. Levanyuk, D. G. Sannikov, Uspekhi Phys. Nauk, 112, 561 (1974).
15. A. Pule, G. P. Mathieu, Kolebatel'nije spectri i symmetriya kristallov, M., "MIR", 1973.
16. E. F. Steigmeier, H. Auderset, Sol. St. Comm. 12, 565 (1973).

MEASUREMENT OF LIQUID CRYSTAL ORIENTATIONAL STATISTICS BY RAMAN SCATTERING*

P. S. Pershan

Division of Engineering and Applied Physics
Harvard University
Cambridge, Massachusetts 02138 USA

Polarized Raman spectra have been used to obtain measures of $P_2 = 2^{-1} < 3\cos^2\theta - 1 >$ and $P_4 = 8^{-1} < 35\cos^4\theta - 30\cos^2\theta + 3 >$ of different liquid crystals. The experimental results disagree with existing theoretical predictions regarding the orientational statistics of nematic liquid crystals.

The results described in this manuscript are the collaborative works of a number of colleagues. Preliminary accounts have been published by Priestley et al,[1] Jen et al,[2] and a more extensive paper is currently under preparation.[3]

Statistical theories of nematic liquid crystals usually begin with the assumption that the molecules are rigid and cylindrical.[4] Taking $\cos\theta$ to be the direction cosine between the principal axis of an individual molecule and the nematic axis, theories have been developed to predict the probability distribution $f(\cos\theta)$ for individual molecules, $(\int_{-1}^{1} f(\cos\theta) d\cos\theta = 1)$.[5-9] All previous experimental efforts to characterize the nematic order consist of

---

*Work supported in part by the Joint Services Electronics Program (U.S. Army, U.S. Navy, and U.S. Air Force) under Contract No. N00014-75-C-0648 and by the National Science Foundation under Grant No. DMR72-03020-A05 and DMR72-02088.

measurements of the anisotropy in tensors of the second rank which, under the assumptions mentioned above, are equivalent to measurements of $S = \langle P_2(\cos\theta)\rangle = \int_{-1}^{1} P_2(x)f(x)dx$ where $P_2(x)$ is the Legendre polynomial of second order.[6,10-14] We discuss here simultaneous measurements of both $\langle P_2(\cos\theta)\rangle$ and $\langle P_4(\cos\theta)\rangle$ in nematic and smectic liquid crystals. Disagreement between the experimental results and existing theories of nematic order will be demonstrated. We argue that, in view of the number of adjustable parameters, the excellent agreement that has been reported between recent self-consistent mean field theories and measured values of $\langle P_2(\cos\theta)\rangle$,[6] over a range of temperature, is neither a meaningful test of the theories, nor of the assumption of rigid rod-like molecules.

Firstly, we have used a Raman scattering technique[1] to measure both $\langle P_2\rangle$ and $\langle P_4\rangle$ for a probe molecule, N-(p'-butoxybenzylidene)-p-cyanoaniline (BBCA) dissolved in N-(p'-methoxybenzylidene)-p-n-butylaniline (MBBA). BBCA was chosen as a probe because its structure is similar to that of MBBA, it readily dissolves in MBBA, and the physical properties of the solution are not very different from pure MBBA. Furthermore, the bond axis of the C ≡ N group at the end of the BBCA molecule is nearly parallel to the major axis of the molecule and there is a strong, narrow, anisotropic Raman line associated with the C ≡ N stretching vibration that can be spectroscopically isolated from all of the other Raman lines of the solution. The Raman polarizability tensor for the localized C ≡ N vibration has the uniaxial form $\underline{\alpha}_{CN} = a\delta_{ij} + (b-a)\delta_{i3}\delta_{j3}$ where the 3-direction coincides with the uniaxial direction of the C ≡ N bond. Since $a \neq b$, the Raman cross section for polarized light scattered by a BBCA molecule is dependent on the orientation of the molecule relative to polarization directions. For example, we denote the integrated intensity per unit solid angle for scattered light polarized along z with incident light polarized along x as I_{xz}. Since there is no phase coherence between C ≡ N vibrations on different molecules, the quantity I_{xz} is a statistical average for individual BBCA molecules. Thus in the nematic phase (take the nematic axis along z) the depolarization ratios $\rho_1 \equiv I_{zx}/I_{zz}$ and $\rho_2 = I_{xz}/I_{xx}$ contain information on the statistical averages $\langle P_2(\cos\theta)\rangle$ and $\langle P_4(\cos\theta)\rangle$ for a BBCA molecule. In the isotropic phase of BBCA-MBBA solutions at different concentrations, measurements of $\rho_1 = \rho_2$ determine the ratio $a/b = .06 \pm .02$ to be independent of temperature and concentration. Raman spectra in the nematic phase were observed using monodomain, planar samples, with thicknesses varying from 25μ to 250μ, homogeneously aligned between glass slides. Several alignment techniques (e.g., rubbing) gave identical results.[15] The back scattering geometry was used with incident light (λ = 5145 Å or 6471 Å) normal to the sample (x, z) plane.

In order to extract $<P_2>$ and $<P_4>$ from the measured values of ρ_1 and ρ_2, several factors must be taken into account. In samples of finite thickness multiple scattering by the director fluctuations causes spurious depolarization of incident and scattered light. This effect was eliminated by measuring ρ_1 and ρ_2 as a function of thickness and extrapolating to zero thickness. Secondly, the measured ratios must be corrected for the differences in solid angle outside and inside the sample, anisotropic demagnification of the scattering volume, and anisotropic reflection effects at the liquid crystal glass interface; i.e., the so-called surface coupling factors.[16] The corrected ratios are $\rho_1^c = \{(n_g + n_o)^2/(n_g + n_e)^2\}\rho_1$ and $\rho_2^c = \{(n_g + n_e)^2/(n_g + n_o)^2\}\rho_2$, where n_g, n_o, and n_e are the refractive indices of the bounding glass plates and of the liquid crystal sample perpendicular and parallel to the optic axis, respectively.[17] These refractive indices were measured as a function of temperature through the entire nematic range with a Pulfrich refractometer using 5145 Å and 6328 Å light. Values of n_g, n_o, and n_e at 5811 Å (wavelength of scattered light, λ_s, for incident 5145 Å radiation) were determined by interpolation; those at 7563 Å (λ_s for incident 6471 Å radiation) were calculated from the measured values using a one-term Sellmeier equation. Finally, a small correction must be made for the fact that the C ≡ N bond is not co-linear with the long axis of BBCA, there being an angle β between them. Taking $10^o \pm 2^o$ as a reasonable estimate for β and using the experimentally determined value for the ratio a/b together with the zero thickness, corrected ρ_1^c and ρ_2^c one can obtain $<\cos^2\theta>$ and $<\cos^4\theta>$ and, hence, $<P_2(\cos\theta)>$ and $<P_4(\cos\theta)>$ for BBCA in MBBA. These results are shown in Fig. 1.

In order to confirm that the statistics of the probe molecule are also characteristic of MBBA, we compare, in Fig. 1, our Raman measurements of $<P_2>$ for the BBCA probe with five independent measurements of $<P_2>$ for MBBA. Firstly, there are absolute NMR measurements of partially deuterated MBBA.[18] Secondly, there are measurements of the optical $(\Delta\alpha)$[19] anisotropy and three separate measurements of the magnetic $(\Delta\chi)$[9-11]

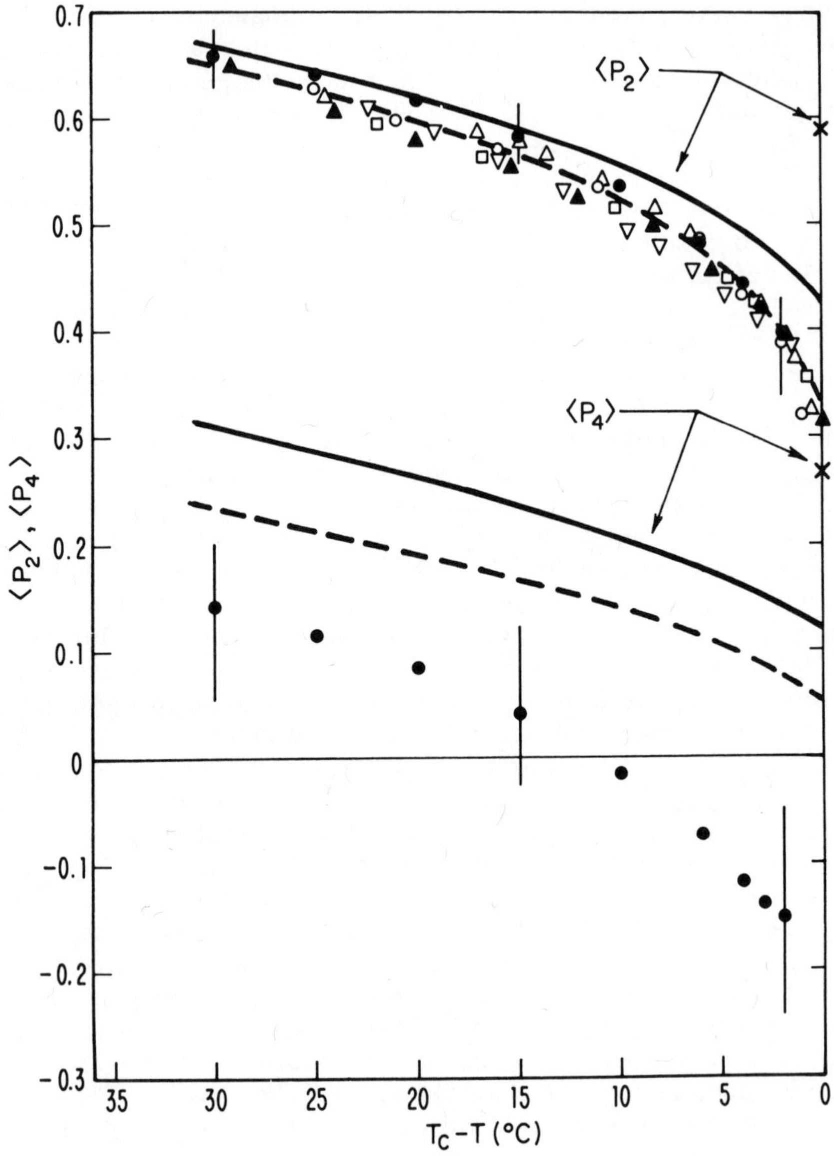

Fig. 1. Theoretical and experimental values of the nematic order parameters $<P_2(\cos\theta)>$ and $<P_4(\cos\theta)>$: theoretical results of Maier-Saupe (——)[5]; Humphries-James-Luckhurst[6] (-----) with $\lambda = -0.55$ and $\gamma = 10$; Onsager-Lakatos[9] (X); the present Raman measurements (●); the NMR results on partially deuterated MBBA (O)[18]; the relative values obtained from measurements of the optical anisotropy, $\Delta\alpha$, (□)[19]; and of the magnetic anisotropy, $\Delta\chi$, (▽),[14] (△),[13] and (▲).[12]

anisotropy of MBBA. Relative values of $<P_2>$ from the latter measurements were obtained by assuming $<P_2>$ is proportional to $\Delta\alpha$ and $\Delta\chi$ and fixing the proportionality factor to obtain agreement at $T_c - T = 2^oC$. The temperature dependence of $<P_2>$ calculated from the different measured anisotropies of MBBA are identical to the temperature dependence obtained from the Raman data on the BBCA molecule. This result, along with the structural similarity of BBCA and MBBA supports our contention that the orientational statistics of the BBCA guest are similar to those of the MBBA host.

The results of the Maier-Saupe theory (MS) for $<P_2>$ and $<P_4>$ are also included in Fig. 1.[5] Although they do show the qualitative behavior of the measured quantities, there are significant quantitative discrepancies. Extensions of the MS theory, including molecular field terms varying as $P_4(\cos\theta)$ have been proposed.[6,7] These results are also shown in Fig. 1 for the choice of parameters[6] $[\lambda = -0.55$ and $\gamma = 10]$ that obtain the best fit to the experimental values of $<P_2>$. Although with the extra free parameters of the Humphries-James-Luckhurst[6] (HJL) theory the fit to the $<P_2>$ data is perfect, the fit to the $<P_4>$ data is not significantly improved. Furthermore the choice of $\gamma = 10 \pm 2$ that is required to fit the $<P_2>$ data implies an unusually strong dependence of order parameter on density. With more reasonable values of γ [i.e., $\gamma = 4$], the fit of the HJL theory to $<P_2>$, although no longer perfect, improves upon the MS theory. The predictions for $<P_4>$, however, still disagree significantly with the data. Thus the most important discrepancy between theory and experiment concerns the values of $<P_4>$. To examine this difference more carefully let us expand

$$f(\cos\theta) = \sum_{\ell=0} A_\ell P_{2\ell}(\cos\theta)$$

where $A_\ell = 2^{-1}(4\ell+1)<P_{2\ell}(\cos\theta)>$, and consider the first three terms

$$f^{III}(\cos\theta) = 2^{-1}[1+5<P_2>P_2(\cos\theta)+9<P_4>P_4(\cos\theta)] .$$

In Fig. 2 we plot f^{III} for two temperatures using experimental values of $<P_2>$, $<P_4>$ and values calculated from the HJL model after the parameters were adjusted to obtain good agreement with

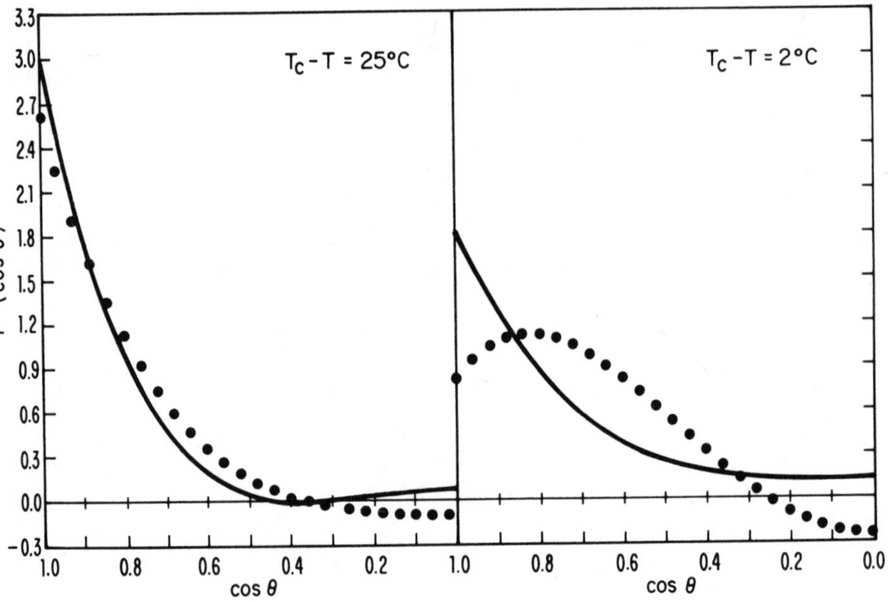

Fig. 2. Plot of the theoretical and experimental truncated angular distribution functions $f^{III}(\cos\theta)$: Humphries-James-Luckhurst[6] (——) and Raman measurements (.....).

$\langle P_2 \rangle$. Note that, although $f(\cos\theta)$ is positive definite, $f^{III}(\cos\theta)$ need not be. The principal result to see in Fig. 2 is that the BBCA (and presumably also the MBBA) molecules have a stronger tendency to be tipped away from the nematic axis than predicted by mean field theory. This tendency is strongest near the nematic-isotropic phase transition. Aside from the mean field theories, other statistical mechanical models of nematic ordering are available. Specifically there is the work of Onsager[8] for the ordering of a system of hard rods as a function of density. Using the Onsager model, Lakatos[9] has recently calculated values of $\langle P_n(\cos\theta) \rangle$ for all n. In Fig. 1 we also show these results for $\langle P_2 \rangle$ and $\langle P_4 \rangle$ at the transition density. The ratio of $\langle P_4 \rangle / \langle P_2 \rangle$ as a function of $\langle P_2 \rangle$ is basically the same as the results of mean field theory and also disagrees with our results.

Alben[4a] pointed out that for molecules which are cylindrically symmetric the angular distribution function f(cos θ) must be replaced by a more general function of two Eulerian angles. In view of the fact that the C ≡ N axis is not precisely parallel to the major axis of the BBCA molecule, neglect of this effect could, in principle, lower the value of $<P_4>$ deduced from the measured values of ρ_1 and ρ_2. However, we have found in practice that the value of $<P_4>$ obtained from the Raman data is changed only by a small fraction of the overall experimental uncertainty (see the error bars in Fig. 1) even when very large anisotropies of the type suggested by Alben are included. We believe this results from the near parallelism of the CN bond and the principal molecular axis and justifies the neglect of possible effects due to cylindrical asymmetries.

On the other hand, mean field theory assumed rigid molecules, an assumption that is clearly questionable.[4b, c] The butyl tail of MBBA is quite flexible and we have no knowledge of the statistical distribution of different conformers in MBBA. To the extent molecules are not rigid there is no unique choice for a nematic order parameter. In the present case the agreement between the NMR results and all the other measurements of $<P_2>$ suggest that $<P_2(\cos\theta)>$ may be suitable if cos θ is the direction cosine of the rigid part of MBBA with respect to the nematic axis. In that case there is no a priori reason to expect that the mean field theory should apply to only one part of a complex molecule. For example, if "L" shaped isomers were relatively probable, the observed behavior of $<P_4>$ could be rationalized.

Of course it is also possible that the approximation of a rigid molecule is acceptable, but that the mean field theory fails because it neglects interactions that induce splay on a microscopic scale, i.e., pair correlations in which neighboring molecular axes are not parallel. For example, dipole-dipole interactions could do this.

A slightly different way to interpret the discrepancy is in terms of

$$\sigma = \frac{(\cos^2\beta - <\cos^2\beta>)}{<\cos^2\beta>} .$$

By Schwartz inequality we have

$$\langle\sigma^2\rangle = \frac{(\langle\cos^4\beta\rangle - \langle\cos^2\beta\rangle^2)}{\langle\cos^2\beta\rangle^2} \geq 0 .$$

Note that in the isotropic phase $\langle\sigma^2\rangle = 0.8$ and in the perfectly aligned state $\langle\sigma^2\rangle = 0$. However $\langle\sigma^2\rangle = 0$ does not imply perfect alignment. Rather a state in which the principal axis of every molecule makes some definite angle β or $\pi-\beta$ with respect to the z-axis will have $\langle\sigma^2\rangle = 0$ even if all azimuthal angles are equally probable. Taking

$$P_2 = \tfrac{1}{2}\langle 3\cos^2\beta - 1\rangle$$

$$P_4 = \frac{1}{8}\langle 35\cos^4\beta - 30\cos^2\beta + 3\rangle ,$$

we can write

$$P_4 = \frac{1}{72}\left[140(1+\langle\sigma^2\rangle)(P_2)^2 - 20(2-7\langle\sigma^2\rangle)P_4 - 7(4-5\langle\sigma^2\rangle)\right].$$

Figure 3 is a plot of P_4 versus P_2 obtained from the above by assuming several fixed values of $\langle\sigma^2\rangle$. The experimental points, as well as those predicted by the mean field theories, are shown. From this plot one can read off the experimental values of $\langle\sigma^2\rangle$ and see that they satisfy both the Schwartz inequality and the expectation that $\langle\sigma^2\rangle_{nematic}$ is greater than $\langle\sigma^2\rangle_{isotropic} = 0.8$. One can see further that the increase in the nematic order accompanying a decrease in temperature also decreases the empirical value of $\langle\sigma^2\rangle$. Viewed this way there is nothing unphysical about negative values of P_4.

Although the analysis is much more complicated, it has been possible to obtain similar results from some of the intrinsic vibrations of pure MBBA. In Fig. 4 we repeat the Raman data shown in Fig. 1 for the cyano vibration and also include data extracted from the 1576 cm^{-1} (squares), 1597 cm^{-1} (triangles), and the 1625 cm^{-1} (circles) of the MBBA molecule. The open data points are for the 20% mixture and they agree with the cyano data. The solid points are for pure MBBA and although they are closer to the theoretical curves, the discrepancies persist.

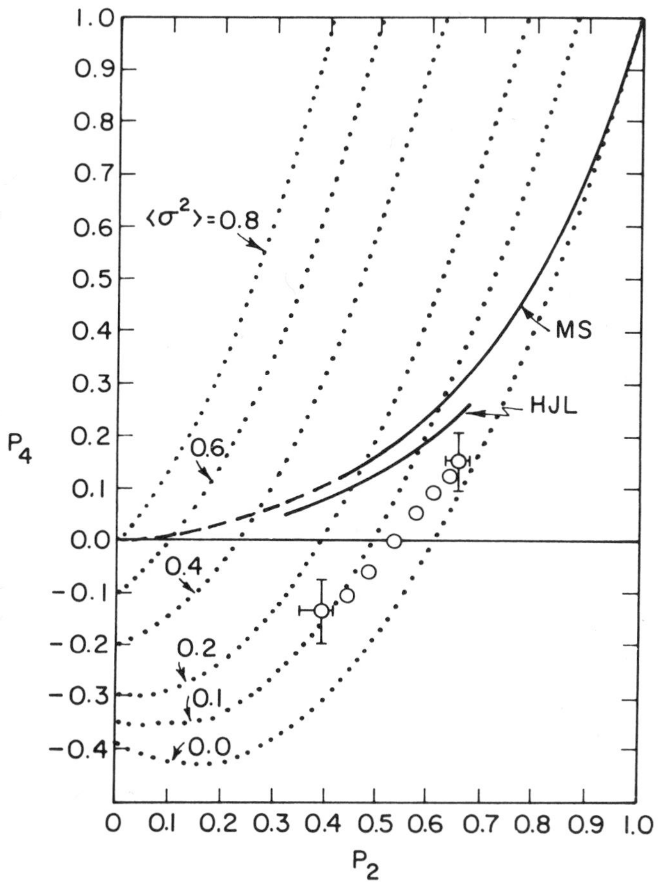

Fig. 3. Plot of P_4 vs. P_2 assuming σ^2 (see text) is constant. The open circles are the experimental values and the solid lines are the Maier-Saupe (MS) and the Humphries-James-Luckhurst (HJL) theories.

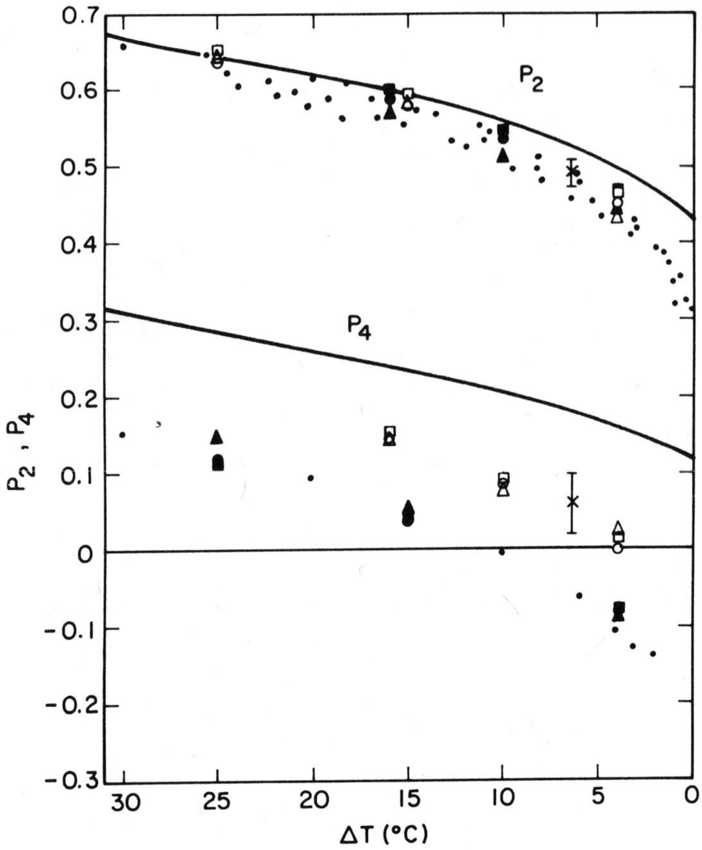

Fig. 4. Plot of P_2 and P_4 vs. temperature. The small points repeat the experimental results displayed in Fig. 1. The solid line is the Maier-Saupe theoretical result. The dashed lines are other theoretical curves that will be discussed elsewhere.[3] The remaining data points are discussed in the text.

A similar study was also done on the compound 40.8 [N-(p'-butoxybenzylidene)-p-n-octylaniline] that has smectic B, smectic A, and nematic phases. The data for P_2 and P_4 vs. temperature is shown in Fig. 5. The plot of P_4 vs. P_2 for constant values of $<\sigma^2>$ is shown in Fig. 6.

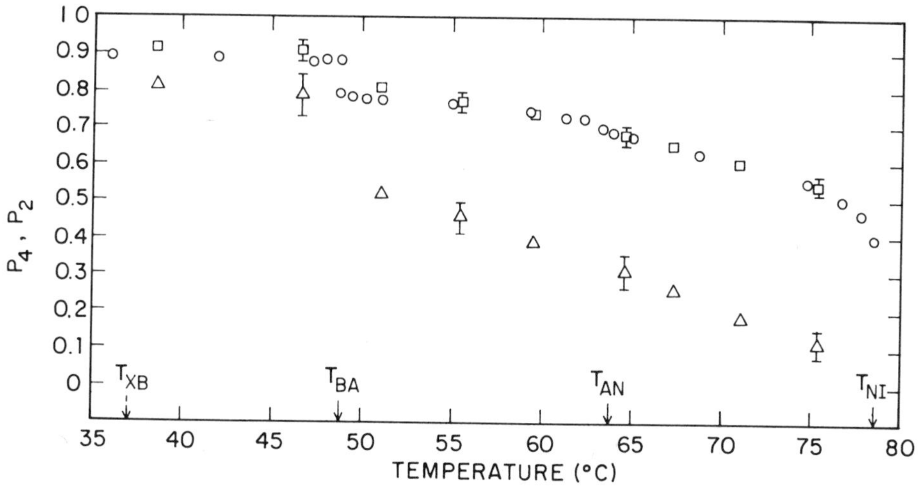

Fig. 5. Plot of P_2 and P_4 vs. temperature for 40.8 (see text).

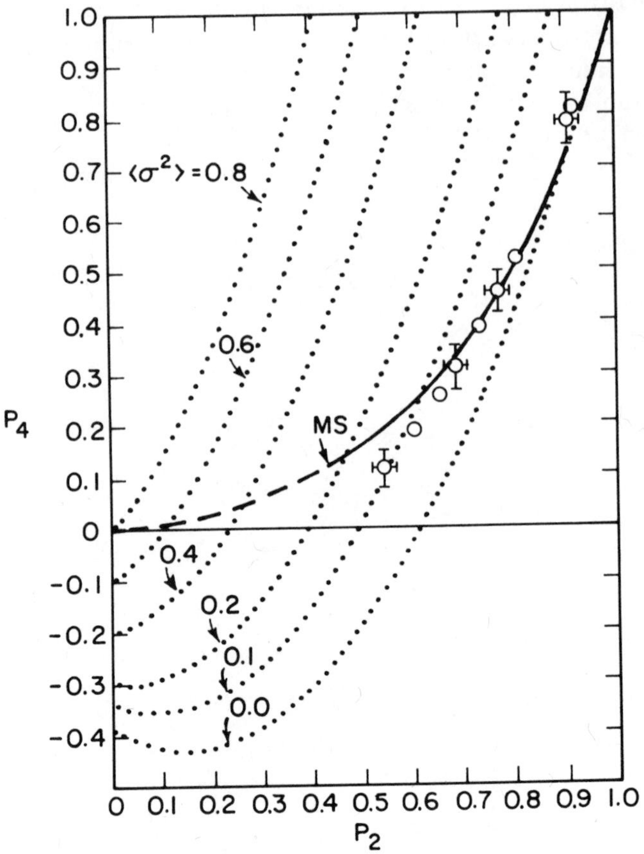

Fig. 6. Plot of P_4 vs. P_2 assuming σ^2 (see text) is constant for 40.8. The solid line is the Maier-Saupe theory.

This presentation briefly displays experimental results that are described more fully in other publications. The principal point illustrated by these results is that polarized Raman spectra can be used to obtain information that has not previously been available on the orientational statistics of anisotropic molecules. In particular for nematic liquid crystals the information thus obtained is in conflict with existing theories.

REFERENCES

1. E. B. Priestley, P. S. Pershan, R. B. Meyer, and D. H. Dolphin, Raman Memorial Volume, Vijnana Parishad Anusandhan Patrika 14, 93 (1971).
2. Shen Jen, Noel A. Clark, P. S. Pershan, and E. B. Priestley, Phys. Rev. Lett. 31, 1552 (1973).
3. Shen Jen, Noel Clark, P. S. Pershan, and E. B. Priestley (in preparation). A preliminary version of this work is available as Shen Jen, Raman Scattering from Liquid Crystals: A Study of Orientational Order, Ph.D. Thesis, Harvard University, 1975 (unpublished).
4. For exceptions to this see: a) R. Alben, J. R. McColl, and C. S. Shih, Solid State Commun. 11, 1081 (1972); b) F. Jähnig, to be published; and c) H. Stenschke, Solid State Commun. 10, 653 (1972).
5. W. Maier and A. Saupe, Z. Naturforsch. 13a, 564 (1958); 14a, 882 (1959); and 15a, 287 (1960).
6. R. L. Humphries, P. G. James, and G. R. Luckhurst, J. Chem. Soc. Faraday Trans. II, 68, 1031 (1972).
7. S. Chandrasekhar and N. V. Madhusudana, Mol. Cryst. and Liq. Cryst. 10, 151 (1970).
8. L. Onsager, Ann. New York Acad. of Sci. 51, 627 (1949).
9. K. Lakatos, J. Stat. Phys. 2, 121 (1970).
10. A. Saupe and W. Maier, Z. Naturforsch. 16a, 816 (1971). P. Pincus, J. Phys. (Paris) 30, C4-8 (1969).
11. N. V. Madhusudana, R. Shashidhar, and S. Chandrasekhar, Mol. Cryst. and Liq. Cryst. 13, 61 (1971). S. Chandrasekhar and N. V. Madhusudana, J. de Phys. 30, C4-24 (1969).
12. G. Sigaud and H. Gasparoux, J. de Chim. Phys. 70, 699 (1973).
13. I. Haller, J. Chem. Phys. 57, 1400 (1972).
14. P. I. Rose, presented at the Fourth International Liquid Crystal Conference, Kent State University, Kent, Ohio (1972).
15. P. Chatelain, Bull. Soc. France Miner. Cryst. 66, 105 (1943). J. L. Janning, Appl. Phys. Lett. 21, 173 (1972). E. Guyon, P. Pieranski, and M. Boix, Letters in Appl. and Eng. Sci., to be published.
16. M. Lax and D. F. Nelson in Coherence and Quantum Optics, Edited by L. Mandel and E. Wolf (Plenum Press, New York) 1973, p.415.
17. N. V. Madhusudana, R. Shashidhar, and S. Chandrasekhar (Ref. 11) first drew attention to the empirical fact that the temperature dependence of the anisotropy in the optical

frequency polarizability tensor, $\Delta\alpha$, scales with other measures of $<P_2>$. The proportionality between $\Delta\alpha$ and nematic order is direct with no need for local field corrections. Similarly, the Raman data for $<P_2>$ is in excellent agreement with other data for $<P_2>$ without introduction of local field effects. The empirical fact that local field corrections are not necessary for either of these measurements can be rationalized as follows: local field corrections depend on dipole-dipole summations over neighboring molecules. Although the problem can be solved in a number of ways, the answers are ultimately dependent on near neighbor correlations. Near neighbor correlations appear to be relatively insensitive to temperature implying the same about the local field corrections. In the present case the ratio of a/b in the isotropic phase, which should also reflect local field effects, was measured to be independent of temperature and we assume the same value applies to the nematic phase. The $<P_2>$ data support this assumption.

18. Y. S. Lee, Y. Y. Hsu, and D. Dolphin, presented at the American Chemical Society Symposium on Ordered Fluids and Liquid Crystals, Chicago, Illinois (1973).
19. $\Delta\alpha$ was obtained from our refractive index measurements and the "Vuks equation". See Ref. 11.

THE VAN DER WAALS FORCES AND LIGHT SCATTERING IN LIQUID CRYSTALS

I. E. Dzyaloshinsky

L. A. Landau Institute for Theoretical Physics
Academy of Sciences of the USSR
Moscow, USSR

INTRODUCTION

Recent experiments on light scattering in liquid crystals performed by McMillan [1] demonstrated essential deviations from the theoretical law predicted [2] earlier for the intensity $I = 1/q^2$ ($\vec{q}$ is the transferred momentum). One of the reasons for these deviations may be the influence of the Van der Waals (VdW) interaction between molecules at distances of the order $1/q$. The contribution of the VdW forces may be not small due to their strength, although they have an exponential decrease with distance : actually (see below), the law for the scattering intensity acquiı es the form:

$$I \sim (\vec{q}^{\,2} + \text{const}\, |\vec{q}|^3)^{-1}$$

The first time the influence of the VdW forces on light scattering near the critical point in an ordinary liquid was taken into account was by Kemoklidze and Pitaevsky [3]. They calculated the VdW interaction between the density fluctuations and showed that the usual Ornstein-Zernicke formula $I \sim 1/(a + b\vec{q}^{\,2})$ would be replaced by the law $I \sim 1/(a + b\vec{q}^{\,2} + c|\vec{q}|^3)$. The difference between light scattering in an isotropic liquid and in liquid crystals is accounted for by the fact that in liquid crystals scattering takes place mainly not from the density fluctuations but from the fluctuations of the optical axis direction, i.e., fluctuations of the director $\delta\vec{n}$. Therefore, in liquid crystals critical scattering [2] always occur.

Thus, our task is to take into account along with the Oceen-Franck energy

$$F_{O.F.} = \frac{1}{2}\int dV\,\{k_{11}(\operatorname{div}\vec{n})^2 + k_{22}(\vec{n}\operatorname{rot}\vec{n})^2 + k_{33}[\vec{n}\operatorname{rot}\vec{n}]^2\} \quad (1)$$

corresponding to the inhomogeneous director distribution $\vec{n}(\vec{r})$, the interaction energy of VdW molecules oriented differently in various points of space. Therefore, we shall make use of the general theory of VdW forces in condensed solid and liquid bodies [4] (also see [5]). According to this theory the VdW energy in an inhomogeneous condensed system is equal to the change in the fluctuating electromagnetic field energy due to inhomogeneity. If inhomogeneities are macroscopic, the fluctuating field is determined by the local value of the dielectric function $\epsilon_{ik}(\omega,\vec{r})$. In an isotropic liquid $\epsilon_{ik} = \epsilon\delta_{ik}$ and all inhomogeneities reduce to the inhomogeneous density $\rho(\vec{r})$:

$$\epsilon(\omega,\vec{r}) = \epsilon(\omega,\rho(\vec{r})).$$

In a liquid crystal the relevant tensor ϵ_{ik} becomes

$$\epsilon_{ik}(\omega,\vec{r}) = \epsilon(\omega)\delta_{ik} + \epsilon_a(\omega)n_i(\vec{r})n_k(\vec{r}) \quad (2)$$

where $\epsilon = \epsilon_\perp$ is the transverse dielectric function and $\epsilon_a(\epsilon_\parallel - \epsilon_\perp)$ is the difference between the transverse and longitudinal dielectric functions. We shall ignore possible inhomogeneities of the functions ϵ, ϵ_a, which was done for the isotropic case by Kemoklidze and Pitaevsky [3] and confine ourselves to the characteristics of the liquid crystal case with essential inhomogeniety of the director $\vec{n}(\vec{r})$.

The corresponding calculations are performed in the next section. The third section contains formulae for the scattered light intensity.

ENERGY OF THE VdW INTERACTION IN A LIQUID CRYSTAL

So, let us calculate the fluctuating electromagnetic field energy. It is completely determined by the temperature Green function of the electric field $D_{ik}(\vec{r},\vec{r}',\omega)$ (see e.g., [4,5]). This function satisfies the equation [4,5]

$$[\epsilon_{i\ell}(\vec{r},i|\omega|)\omega^2 + (\mathrm{rot}^2)_{i\ell}]\, D_{\ell k}(\vec{r},\vec{r}',\omega) = 4\pi\omega^2\delta(\vec{r}-\vec{r}')\delta_{ik} \tag{3}$$

in which the dielectric function ϵ_{ik} is evaluated at the imaginary (Matsubara) frequency value $\epsilon_{ik} \equiv \epsilon_{ik}(i|\omega|)$. Here and elsewhere we shall use a set of units with $\hbar = c = 1$.

Naturally, it is not possible to solve the equation for D for an arbitrary distribution of the vector in Eq. (2) for ϵ_{ik}. We shall therefore employ the smallness of the dielectric anisotropy ϵ_a in real liquid crystals (usually $\epsilon_a/\epsilon \approx 0.1 - 0.01$) and shall seek the Green function and the fluctuating field energy in the form of a power series expansion in ϵ_a. Then for the energy we can immediately apply the expression for the first variant with respect to $\delta\epsilon_{ik}$ [4,5]:

$$\delta F = \frac{1}{8\pi^2}\int_0^\infty d\omega \int d\vec{r}\, D_{ik}(\vec{r},\vec{r}',\omega)\delta\epsilon_{ik}(\vec{r},i\omega) \tag{4}$$

where in our case of small ϵ_a

$$\delta\epsilon_{ik} = \epsilon_a(i|\omega|)n_i(\vec{r})n_k(\vec{r}) \tag{5}$$

Note that Eq. (4) takes into account only quantum fluctuations and in fact is not satisfied at very high temperatures. It is obviously valid at room temperature (for details, see [3,5]).

Confining oneself to quadratic accuracy in ϵ_a in the energy, it is sufficient to find D to linear accuracy in ϵ_a. We have:

$$D_{ik}(\vec{r},\vec{r}') = D^{(0)}_{ik}(\vec{r}-\vec{r}') - \frac{1}{4\pi}\omega^2\int d\vec{r}''D^{(0)}_{i\ell}(\vec{r}-\vec{r}'')D^{(0)}_{mk}(\vec{r}''-\vec{r}')\delta\epsilon_{\ell m}(\vec{r}'') \tag{6}$$

where $D^{(0)}$ is the solution of Eq. (3) at $\epsilon_a = 0$, i.e., the Green function of radiation in the homogeneous space of the dielectric function $\epsilon\delta_{ik}$. For the Fourier components,

$$D_{ik}^{(0)}(\vec{q}, \omega) = \frac{4\pi\omega^2}{\epsilon(i|\omega|)\omega^2 + q^2}\left(\delta_{ik} + \frac{q_i q_k}{\epsilon(i|\omega|)\omega^2}\right). \tag{7}$$

The substitution of the first term of (6) into the formula for the energy gives a term principally linear in ϵ_a. However, this is of no interest to us as the spacial integral in it has the structure

$$\int D_{ik}^{(0)}(\vec{r} - \vec{r} \equiv 0; \omega) n_i(\vec{r}) n_k(\vec{r}) d\vec{r}$$

and evidently describes the energy of a homogeneous director distribution. Using the concrete form (7) it is easy to show that all inhomogeneity effects reduce to surface integrals.

The quadratic contribution in ϵ_a is just the energy of the VdW attraction. It has the form:

$$F_{vdw} = \frac{1}{16\pi^2} \frac{1}{(2\pi)^7} \int_0^\infty \omega^4 d\omega \int d\vec{p} d\vec{q} D_{ik}^{(0)}(\vec{p} + \frac{\vec{q}}{2}) \cdot \times D_{m\ell}^{(0)}(\vec{p} - \frac{\vec{q}}{2}) \delta\epsilon_{im}(-\vec{q}) \delta\epsilon_{k\ell}(\vec{q}) \tag{8}$$

Here $\delta\epsilon_{ik}(\vec{q})$ is the Fourier component from $\epsilon_a n_i(\vec{r}) n_k(\vec{r})$,

$$\delta\epsilon_{ik}(\vec{q}) = \epsilon_a(i|\omega|) N_{ik}(\vec{q}), \quad N_{ik}(\vec{r}) = n_i(\vec{r}) n_k(\vec{r}) \tag{9}$$

Integration over $\vec{p}$ in Eq. (8), with $D^{(0)}$ from (7) in general form, leads to complicated formulae. Simple expressions are obtained in the limits of large and small q. These limits correspond to large and small distances respectively in the theory of VdW forces (see [4, 5]). The criterion as usual is a certain characteristic wavelength λ_0 in the absorption spectrum (see a more accurate definition below). The calculations practically completely coincide with the corresponding calculations in the work by Kemoklidze and Pitaevsky [3], and therefore we shall cite the results without any discussion.

At large distances $R \gg \lambda_0$ ($q\lambda_0 \ll 1$), the principal role is played by the delayed part of the VdW interaction. The corresponding energy is quadratic in q and (in a standard system of units) has the form

$$F_{vdw} = \frac{L}{(2\pi)^3} \int d\vec{q}\, \{4q_k q_\ell N_{i\ell}(\vec{q}) N^*_{ki}(\vec{q}) - q^2 N_{i\ell}(\vec{q}) N^*_{i\ell}(\vec{q})\} \tag{10}$$

$$L = \frac{\hbar}{192\pi^2 c} \int_0^\infty \frac{\epsilon_a^2(i\omega)}{\epsilon^{3/2}(i\omega)} \omega d\omega \tag{11}$$

In the coordinate representation (10) with (9) taken into account has the form of the Oceen-Franck energy:

$$F_{vdw} = \tfrac{1}{2} \int dv\{8L(\mathrm{div}\,\vec{n})^2 - 8L(\vec{n}\,\mathrm{rot}\,\vec{n})^2 + 8L[\vec{n}\,\mathrm{rot}\,\vec{n}]^2\} \tag{12}$$

and represents the renormalization of the bare coefficient k in the energy (1). It is of interest to note that the VdW forces tend to decrease k_{22} and increase k_{11} and k_{33}, which correctly describes the situation in real liquid crystals where in all experimentally known cases $k_{22} < k_{11}, k_{33}$.

Small distances $R \ll \lambda_0$ ($q\lambda_0 \gg 1$) represent a region of action of ordinary VdW forces. The corresponding energy is cubic in $|\vec{q}|$ and has the form

$$F_{vdw} = \frac{M}{(2\pi)^3} \int d\vec{q}\{2|\vec{q}|^3 N_{ik}(\vec{q}) N^*_{ik}(\vec{q}) - 4|\vec{q}|\, q_i q_k N_{i\ell}(\vec{q}) N^*_{k\ell}(\vec{q}) + 3\frac{q_i q_k q_\ell q_m}{|\vec{q}|} N_{i\ell}(\vec{q}) N^*_{km}(\vec{q})\} \tag{13}$$

$$M = \frac{\hbar}{2048\pi} \int_0^\infty \frac{\epsilon_a^2(i\omega)}{\epsilon^2(i\omega)} d\omega \tag{14}$$

It is easy to observe that in contrast to (10) the energy (13) is positive definite. In the coordinate representation the energy (13) is not local and is given by the formula:

$$F_{vdw} = \frac{M}{2\pi^2} \int d\vec{r}d\vec{r}'\{24 \frac{(n(\vec{r})n(\vec{r}'))^2}{|\vec{r} - \vec{r}'|^6}$$

$$+ \frac{8}{|\vec{r} - \vec{r}'|^4} \frac{\partial}{\partial x_i} (n_i(\vec{r})n_\ell(\vec{r})) \frac{\partial}{\partial x'_k} (n_k(\vec{r}')n_\ell(\vec{r}'))$$

$$- \frac{3}{|\vec{r} - \vec{r}'|^2} \frac{\partial^2 n_i(\vec{r})n_\ell(\vec{r})}{\partial x_i \partial x_\ell} \frac{\partial^2 n_k(\vec{r}')n_m(\vec{r}')}{\partial x'_k \partial x'_m} \quad (15)$$

The numeric estimate of the coefficients L and M is fairly difficult since it is necessary to know $\epsilon(i\omega)$ and $\epsilon_a(i\omega)$ over a wide frequency range. It is in fact necessary to know [4] the dispersion and anisotropy of absorption over the whole spectral interval. A qualitative comparison of Eqs. (10) and (13) enables us to find the above mentioned wavelength λ_0. Assuming $q \sim 1/\lambda_0$ and $L/\lambda_0^2 \sim M/\lambda_0^3$, we find $\lambda_0 \sim c/\omega_0$

$$\omega_0 \sim \int_0^\infty \frac{\epsilon_a^2(i\omega)}{\epsilon^{3/2}(i\omega)} \omega d\omega \Bigg/ \int_0^\infty \frac{\epsilon_a^2(i\omega)}{\epsilon^2(i\omega)} d\omega$$

Thus ω_0 is the frequency at which the essential dispersion of the dielectric function anisotropy begins (recall that at $\omega \to \infty$ the dielectric function ϵ tends to 1, and ϵ_a tends to 0 no slower than ω^{-2}).

LIGHT SCATTERING

Equations (13) and (15) for the energy may lead in principle to a number of new qualitative effects. These effects will be especially clear, if the renormalization of the coefficients k in the quadratic region, i.e., the value L, will be of the same order as the coefficients k. This situation a priori is quite possible since the values L and M are determined only by the optical properties of a liquid crystal and are independent, at least at first glance, of the non-renormalized coefficients k, which are determined by the interaction of the neighboring molecules. The above mentioned smallness of the coefficient k_{22} also supports the assumption of a large value of L.

In the case of large L for sufficiently small distances $R \lesssim \lambda_0$ and correspondingly large $q\lambda_0 \gtrsim 1$, the VdW energy will exceed the Oceen-Franck energy. Accordingly, the director distribution $\vec{n}$ in thin films $d \lesssim \lambda_0$ of liquid crystals and their behavior in sufficiently high magnetic fields H will change. In particular, the dependence of the coherence length ξ_H introduced by De Gennes will transform from the law $\xi_H \sim H^{-1}$ in low fields to the law $\xi_H \sim H^{-2/3}$ in high fields.

Here we shall only study the influence of the VdW forces on light scattering by thermal fluctuations of the director direction in the vicinity of a constant equilibrium value $\vec{n}_0$. Introducing small deviations $\vec{V}$

$$\vec{n}(\vec{r}) = \vec{n}_0 + \vec{V}(\vec{r}) \ , \ \vec{n}_0 \vec{V}(\vec{r}) = 0$$

we shall write the fluctuation energy to quadratic accuracy in $\vec{V}$.

At small $q\lambda_0 << 1$ the energy is mainly given by the Oceen-Franck expression (1) with renormalized values of the coefficients:

$$k_{11} \to k_{11} + 8L, \ k_{22} \to k_{22} - 8L, \ k_{33} \to k_{33} + 8L$$

The additions are of the fourth order in q (actually they are proportional to $q^4 \ln q$, see [3]), and can be omitted. Let us as usual introduce for the Fourier component $\vec{V}(\vec{q})$ a system of coordinates (V_1, V_2, V_z) with the z axis along $\vec{n}_0$, i. e., $V_z = 0$; axis 2 (unit vector $\vec{e}_2(\vec{q})$) perpendicular to $\vec{q}$ and $\vec{n}_0$, and axis 1 (unit vector $(\vec{e}_1(\vec{q}))$ orthogonal to $\vec{n}_0$ and axis 2. In these coordinates the fluctuation energy has the form [2]

$$F_{O.F.} = \tfrac{1}{2}\int d\vec{q} \ \{(k_{33}q_z^2 + k_{11}q_\perp^2)|V_1(\vec{q})|^2 + (k_{33}q_z^2 + k_{22}q_\perp^2|V_2(\vec{q})|^2\} \tag{16}$$

At large $q\lambda_0 >> 1$ the Oceen-Franck energy contribution will be given by the same formula (16) but with "bare" values of the coefficients k, i.e., with $k_{11} - 8L$, $k_{22} + 8L$, $k_{33} - 8L$ instead of k_{11}, k_{22}, k_{33}. There will be an addition to the VdW energy resulting from the linearization of Eq. (13). In the same axes 1, 2, z we have:*

$$F_{vdw} = \frac{4M}{(2\pi)^3} \int d\vec{q} \, \{ q q_\perp^2 |V_2(\vec{q})|^2 + 3 \frac{q_\perp^2 q_z^2}{q} |V_1(\vec{q})|^2 \} \tag{17}$$

Now let us write a formula for the total intensity of the scattered light [2]. Let $\vec{i}$ and $\vec{f}$ be polarization vectors of the incident and scattered light and $\vec{q}$ be the transferred momentum. The scattering per unit solid angle (the extinction coefficient dh) is the sum of scattering from the director fluctuations $V_1(\vec{q})$, $V_2(\vec{q})$ along axes $\vec{e}_1$ and $\vec{e}_2$ (the axis z is along n_0):

$$\frac{dh}{d\Omega} = \frac{\omega^4}{16\pi^2 c^4} |\epsilon_a(\omega)|^2 \{ I_1(\vec{q})(f_z i_1 + f_1 i_z)^2 + I_2(\vec{q})(f_z i_2 + f_2 i_z)^2 \} \tag{18}$$

Let us stress that here, in contrast with all the remaining formulae, the anisotropy of the dielectric function ϵ_a is evaluated at the real (!) frequency ω, the frequency of the incident and scattered light.

For small $q\lambda_0 \ll 1$ the De Gennes formula may be presented [2]:

$$I_1(\vec{q}) = \frac{T}{k_{33} q_z^2 + k_{11} q_\perp^2} \; ; \; I_2(\vec{q}) = \frac{T}{k_{33} q_z^2 + k_{22} q_\perp^2} \tag{19}$$

At large $q\lambda_0 \gtrsim 1$ we should take into account the VdW energy. Taking into consideration the renormalization of the coefficients k, we have with L from (11) and M from (14)

$$I_1(\vec{q}) = T / [k_{33} - 8L) q_z^2 + (k_{11} - 8L) q_\perp^2 + 12M \frac{q_z^2 q_\perp^2}{q}]$$

---

*In our preliminary communication [6] the formula for the VdW energy and the corresponding expressions for the intensity were not correct.

$$I_2(\vec{q}) = T/[(k_{33} - 8L)q_z^2 + (k_{22} + 8L)q_\perp^2 + 4Mqq_\perp^2] \qquad (20)$$

In the intermediate region $q \sim 1/\lambda_0$ the expressions for the intensity are too complicated and are given by the integrals in Eq. (8).

Unfortunately, a direct comparison of Eqs. (19) and (20) with experiment is at present not possible since, on the one hand, the [1] McMillan experiments are performed in the vicinity of the nematic-smectic transition point A, and, on the other hand, it is not easy to obtain numeric values of the coefficients L and M.

REFERENCES

1. W. L. McMillan, Phys. Rev. 7A, 1673 (1973); 8A, 328 (1973).
2. Groupe d'etude des crystals liquides, J. Chem. Phys. 51, 816 (1968).
3. M. P. Kemoklidze, L. P. Pitaevsky, JETP 59, 2187 (1970).
4. I. E. Szyaloshinsky, E. M. Lifshits, L. P. Pitaevsky, UFN, 73, 381 (1961); Adv. in Phys. 10, 165 (1961).
5. A. A. Abrikosov, L. P. Gor'kov, I. E. Dzyaloshinsky, "Methods of the Quantum Field Theory in Statistical Physics", Physmatguiz (1962) VI.
6. I. E. Dzyaloshinsky, S. G. Dmitriev, E. I. Katz, JETP Lett. 19, 586 (1974); JETP 68, 2335 (1975).

Section II

Strong Anharmonicity Effects and Electron–Phonon Interactions in Crystals

QUASIPARTICLE INTERACTIONS IN MANY-BODY SYSTEMS*

J. Ruvalds

Physics Department, University of Virginia
Charlottesville, Virginia 22901 USA

A survey of recent developments in the study of interacting elementary excitations is presented. The general features of the formalism are discussed in the formation of bound states of quasiparticles, and their possible hybridization with other excitations. The relevant theoretical features and their relation to experimental data are briefly discussed for phonons in solids, rotons in liquid helium, spin waves, and plasmon excitations in an electron gas.

The concept of weakly interacting quasiparticles or elementary excitations of various many-body systems was developed extensively by Landau.[1] In this paper we survey some aspects of quasiparticle interactions which have been the subject of considerable study and controversy in recent years. In particular we shall focus on the possible strong coupling between excitations which may yield bound states or resonances in the two-excitation spectrum. These resonances may couple to other modes of the system and thus modify the single quasiparticle spectrum.

---

*Work supported by the National Science Foundation Grant No. NSF-GH-32747.

First we describe the general formulation of the theory in terms of a simple model Hamiltonian. Then we proceed to discuss the criterion for resonant scattering of two excitations in terms of the one-excitation energy spectrum. Finally we mention briefly the microscopic origin of the interaction parameters, and their estimated values in the following cases: Excitations in superfluid helium; Phonons in solids; Spin waves in magnetic systems; and Plasmon modes in an electron gas.

The theoretical discussion is greatly simplified by neglecting the momentum dependence of the quasiparticle coupling terms. Thus, for convenience, we consider the model Hamiltonian

$$H = \sum_{k} \epsilon_k b_k^+ b_q + g_3 \sum_{k,q} (b_k^+ b_q b_{k-q} + h.c.) \qquad (1)$$
$$+ q_4 \sum_{k,p,q} b_{k+q}^+ b_{p-q}^+ b_p b_k + \ldots ,$$

where ϵ_k denotes the excitation energy, $b_k (b_k^+)$ define the quasiparticle destruction (creation) operators. The three-quasiparticle coupling g_3 has been studied primarily in connection with modifications in the one-excitation spectrum, which may influence the renormalization of the excitation energy spectrum. The quasiparticle scattering is represented by the g_4 coupling parameter; by using a point interaction model for the coupling in real space (g_4 = constant) we limit the discussion to s-wave scattering. The general situation of scattering in other angular momentum channels may be treated by a straightforward extension of the theory, as exemplified in the theory of two-roton resonances in helium.[2]

With the above model, the Bethe-Salpeter equation for the multiple scattering of two quasiparticles becomes a simple series with the solution

$$F(Q, \omega) = \frac{F^{(o)}(Q, \omega)}{1 - g_4 F^{(o)}(Q, \omega)} , \qquad (2)$$

where the propagator for two non-interacting excitations is given by[2]

$$F^{(0)}(Q,\omega) \cong \int \frac{\rho_2^{(0)}(Q,\omega')d\omega'}{\omega - \omega' + 2i\Gamma} , \tag{3}$$

and $\rho_2^{(0)}(Q,\omega)$ is the joint density of states for two quasiparticles which is determined by the energy spectrum ϵ_k. A bound state is split off from the two-phonon continuum providing that

$$1 = g_4 F^{(0)}(Q,\omega) . \tag{4}$$

Now the formalism indicates that a bound state is formed by arbitrarily weak coupling providing that there is a discontinuity or singularity in the two-excitation density of states.[2] In other cases a resonance may appear for intermediate values of the coupling, and the coupling must exceed a critical value to form a bound state.

A direct probe of the two-excitation spectrum is provided by second-order Raman scattering experiments. Although these light scattering experiments sample only the region of very small total momentum of the excitation pair, they have provided very accurate information on the lineshapes and peak positions which are sensitive to the excitation interactions as discussed below. These studies sample the density of states for interacting excitations which follows from Eq. 2 and is given by

$$\rho_2(Q,\omega) = -\frac{1}{\pi} \mathrm{Im} F(Q,\omega) . \tag{5}$$

The coupling between excitations may also influence the single excitation spectrum as a consequence of level mixing or hybridization. This process was originally suggested by Fermi[3] to explain an anomalous peak in the Raman spectrum of CO_2 which was attributed to the overtone of a vibrational mode. In molecules the vibrational energy levels have negligible dispersion so that the overtone appears as a sharp peak in the two-excitation density of states. By contrast the excitations in liquids and solids generally exhibit a broad spectrum of quasiparticle pair states. The hybridized single excitation spectrum can be expressed as

$$\rho_1(Q,\omega) = -\frac{1}{\pi} \mathrm{Im} \frac{1}{\omega - \epsilon_Q + i\Gamma_Q - g_3^2 F(Q,\omega)} , \tag{6}$$

where the density of states $\rho_1(Q, \omega)$ yields one peak at the renormalized excitation energy $\omega \cong \epsilon_Q$, with a width Γ_Q. Hybridization results in a transfer of intensity from this peak to the pair quasiparticle manifold with increasing coupling g_3.

The combined effects of the above interactions can give rise to two distinct peaks in the dynamic structure factor of the system. This point is evident from Eqs. 2-6, since the formation of a bound state is manifested by a pole in the propagator in $F(Q, \omega)$, which in turn yields a secondary pole structure in $\rho_1(Q, \omega)$. Experimentally the cross-section is generally dominated by $\rho_1(Q, \omega)$. Hence the hybridization provides an interesting mechanism for probing the excitation pair states, especially since the intensity transfer may be varied by changing the excitation energy ϵ_k as a function of pressure, temperature, or other conditions.

EXCITATIONS IN SUPERFLUID HELIUM

Liquid helium at very low temperatures exhibits well-defined excitations in the superfluid phase. These consist of acoustic phonons at long wavelengths and rotons at intermediate momenta ($k_0 \sim 2$ Å^{-1}).[1] The latter states exhibit an energy minimum and a dispersion of the type $\epsilon_k \cong \Delta + (k-k_0)^2/2\mu$, where μ denotes an effective mass for these modes. These states are intrinsically stable at zero temperature, but become overdamped at the λ phase transition.

For the present discussion we shall focus on these roton states, in part because of their correspondingly large density of states. As a result of their vanishing group velocity (for $k = k_0$), the two-roton density of states exhibits a singularity $\rho_2^{(0)}(Q = 0, \omega) \propto (2\Delta - \omega)^{-\frac{1}{2}}$ at zero total momentum and a discontinuity at other values of Q.[2] Consequently a two-roton bound state is split off from the continuum for arbitrarily weak attractive roton coupling.[2]

The zero total momentum spectrum has been probed by second-order Raman scattering, and provides evidence for a bound roton pair with angular momentum $\ell = 2$ symmetry.[4] Raman scattering in helium appears to be limited to the d-wave channel,[5] so that the possibility of bound states in other channels remains an intriguing question, as discussed in this Seminar by Pitaevski and others. It may be interesting to probe the $\ell = 0$ states by light scattering from polarization modes[6] formed by the mixing of light and excitation pairs in helium in the presence of an external electric field.

Neutron scattering provides an interesting probe of the hybridization processes discussed here. However, these experiments are limited to relatively large momentum transfers and poor resolution in the case of liquid helium. Nevertheless, these techniques have verified the strong hybridization of the phonon-roton spectrum with two-roton states as predicted by theory.[7,2]

A further consequence of the resonance formation and hybridization can be seen in the theory of the roton lifetime. At finite temperatures the lifetime is limited by collision broadening. Thus the final state interactions have a profound influence on the density of states (as in Eqs. 2, 5), and the Born approximation[8] is inadequate for these excitations. Recent theoretical studies conclude that strong roton scattering in several angular momentum channels is required to explain the observed lifetime.[9]

It would be most interesting to develop a microscopic theoretical basis for the quasiparticle interactions. The original efforts[10] along these lines were directed to the excitation spectrum and yield results in qualitative agreement with experiment, but substantially deficient in the roton region.

Recently there has been progress in the theoretical understanding of the renormalized excitation spectrum including the hybridization interactions from first principles.[11] One promising avenue relies on a self-consistent solution for the liquid structure factor including quasiparticle interactions; this method eliminates the need for a helium pseudopotential and consequently requires no adjustable parameters in principle. Nevertheless, the search for a microscopic theory of roton interactions[12] poses several interesting challenges.

SPIN WAVES IN MAGNETIC SYSTEMS

The general theory of spin-wave interactions in a Heisenberg ferromagnet was analyzed in some detail by Dyson.[13] Because the magnons have a parabolic dispersion with a relatively small effective mass, the magnon coupling required to form a bound pair is large. Nevertheless it has been shown[13] that the Heisenberg model yields interactions strong enough (and attractive) to split a bound state off the two-magnon continuum for values of the pair momentum exceeding a critical value. For this reason the bound states are inaccessible to light scattering experiments.

The hybridization of one and two-magnon states is found to be substantial,[13] however, and thereby admit the analysis of the bound states by neutron scattering.

From the experimental point of view, the evidence for bound states of two magnons is best illustrated by infrared measurements of metamagnetic systems[14] (for example, $FeCl_2 \cdot 2H_2O$). Here the magnons show very little dispersion and therefore require very weak interactions to bind. Also the spin waves in these systems are highly sensitive to temperature and external magnetic fields. These properties provide interesting testing grounds for theories of magnon-phonon coupling[15] as well.

PHONONS IN SOLIDS

Stimulated by an anomalous sharp peak in the two-phonon Raman spectrum of diamond, Cohen and Ruvalds[16] proposed the existence of a bound state of two optic phonons which is formed by a repulsive fourth-order anharmonic interaction. This proposal provides a simple physical explanation for the peak width, which is about twice the single phonon width, and the lineshape.[17]

In solids the occurrence of bound optic phonon pairs is favored as a result of their small dispersion. In this regard diamond is a more favorable situation than either germanium or silicon. Recently it has been noted that diamond may exhibit an extermely flat optic spectrum, which may account for the anomalout Raman peak without invoking phonon interactions.[18] Further accurate experiments are needed to resolve the issue for diamond.

As in the other systems mentioned above, phonon hybridization in solids may elucidate the two-phonon spectrum. One such example is the Raman spectrum of quartz, which demonstrates the level crossing of a soft phonon mode with a two-phonon excitation.[19] Another coupling mechanism of interest is the electron hybridization with a pair of phonons. The latter case has been observed in photoconductivity measurements on n-type InSb,[20] and suggests the existence of a bound state of two phonons in InSb. It would be of great interest to check this prediction by Raman scattering experiments in similar semiconductors.

PLASMONS IN AN ELECTRON GAS

Traditional treatments of plasmon excitations in an electron gas, such as the random phase approximation (RPA), yield reasonable results for the plasmon energy and dispersion. These theories neglect plasmon interactions which are related to electron correlations.

According to the above considerations, the observed large dispersion of plasmon excitations in most metals would make them unlikely candidates for binding. However, it turns out that the effective interaction between two plasmons is sufficiently attractive to form a bound state in low density systems.[21] The physical origin of the plasmon-plasmon interaction is the exchange of an electron-hole pair. A crude estimate of this process suggests resonant enhancement of plasmons in Al and a strongly bound plasmon state in Cs and in various degenerate semiconductors.

Finally we mention the case of interacting plasmon excitations in highly anisotropic materials, which may exhibit an effective strong coupling due to the small dispersion of the excitations. Thus it may be of interest to extend the recent analysis of quasi-one-dimensional systems[22] (e.g., TCNQ), with a view toward the quasiparticle interactions.

It is a pleasure to acknowledge the hospitality of the physics department of the University of California, La Jolla, where this manuscript was completed.

REFERENCES

1. Collected Papers of L. D. Landau, (ed. D. Ter Haar, Gordon and Breach, New York 1947). L. D. Landau, J. Phys. USSR 5, 71 (1947).
2. J. Ruvalds and A. Zawadowski, Phys. Rev. Lett. 25, 333 (1970); F. Iwamoto, Prog. Theor. Phys. 44, 1135 (1970); A. Zawadowski, J. Ruvalds and J. Solana, Phys. Rev. A5, 399 (1972).
3. E. Fermi, Z. Phys. 71, 250 (1931).

4. T. J. Greytak and James Yan, Phys. Rev. Lett. 22, 987 (1969); T. J. Greytak, R. Woerner, J. Yan, and R. Benjamin, Phys. Rev. Lett. 25, 1547 (1970).
5. M. Stephen, Phys. Rev. 187, 279 (1969).
6. A. Bagchi and J. Ruvalds, Phys. Rev. Lett. 32, 209 (1974).
7. L. P. Pitaevski, Zh. Eksp. Teor. Fiz. 36, 1168 (1959); Sov. Phys. JETP 9, 830 (1966).
8. Theory of Superfluidity, I. M. Khalatnikov, (Benjamin, New York, 1965).
9. J. Yan and M. Stephen, Phys. Rev. Lett. 27, 482 (1971); I. A. Fomin, Sov. Phys. JETP 33, 637 (1971); J. Solana, V. Celli, J. Ruvalds, I. Tüttö and A. Zawadowski, Phys. Rev. A6, 1665 (1972); K. Nagai, K. Nojima, and A. Hatano, Prog. Theor. Phys. 46, 355 (1972).
10. N. N. Bogoliubov, J. Phys. 11, 23 (1947); R. P. Feynman and M. Cohen, Phys. Rev. 102, 1189 (1956); E. Feenberg, Theory of Quantum Fluids (Academic Press, New York 1969).
11. T. Nishiyama, Prog. Theor. Phys. 6, 366 (1951); 7, 417 (1952); 8, 655 (1952); 9, 245 (1953); A. K. Rajagopal and G. S. Grest, Phys. Rev. A10, 1837 (1974); G. S. Grest and A. K. Rajagopal, Phys. Rev. A10, 1395 (1974); J. A. Carballo and J. Ruvalds, Phys. Rev. B11, 4278 (1975). Much of this theoretical work is based on the formalism of: N. N. Bogoliubov and D. N. Zubarev, Sov. Phys. JETP 1, 83 (1955).
12. A. K. Rajagopal, A. Bagchi, and J. Ruvalds, Phys. Rev. A9, 2707 (1974).
13. F. J. Dyson, Phys. Rev. 102, 1217, 1230 (1956); M. Wortis, Phys. Rev. 132, 85 (1963); J. Hanus, Phys. Rev. Lett. 11, 336 (1963).
14. J. B. Torrance and M. Tinkham, Phys. Rev. 187, 587; 595 (1969).
15. K. L. Ngai, J. Ruvalds and E. N. Economou, Phys. Rev. Lett. 31, 166 (1973).
16. M. H. Cohen and J. Ruvalds, Phys. Rev. Lett. 23, 1378 (1969): The Raman data has been obtained by R. S. Krishnan, Proc. Indian Acad. Sci. 24, 25 (1946); S. Solin and A. K. Ramdas, Phys. Rev. B1, 1687 (1970).
17. J. Ruvalds and A. Zawadowski, Phys. Rev. B2, 1172 (1970).
18. K. Uchinokura, T. Sekini, and E. Matsuura, J. Phys. Chem. Solids 35, 171 (1974); S. Go, H. Biltz and M. Cardona, Phys. Rev. Lett. 34, 580 (1975); R. Tubino and J. L. Birman, Phys. Rev. Lett. 35, 670 (1975).

19. J. F. Scott, Phys. Rev. Lett. 21, 907 (1968), and references cited therein.
20. R. Kaplan and R. F. Wallis, Phys. Rev. Lett. 20, 1499 (1968); J. Ruvalds, E. N. Economou and K. L. Ngai, Phys. Rev. Lett. 27, 417 (1971).
21. J. Ruvalds, A. K. Rajagopal, and J. Carballo, (submitted to Phys. Rev. Lett. May, 1975).
22. I. E. Dzyaloshinskii and E. I. Kats, Sov. Phys. JETP 28, 178 (1969); recent data on the plasmon spectrum of TTF-TCNQ may be found in J. J. Ritsko, D. J. Sandman, A. J. Epstein, P. C. Gibbons, S. E. Schnatterly, and J. Fields, Phys. Rev. Lett. 34, 1330 (1975).

WEAKLY-BOUND EXCITATION STATES IN CRYSTALS

L. P. Pitaevsky

Institute of Physical Problems
Academy of Sciences of the USSR
Moscow, USSR

INTRODUCTION

In the present paper it is shown that in crystals bound states of two elementary excitations (two phonons, a phonon and an electron) may be formed at arbitrarily weak interaction between them. This occurs in the vicinity of some special points in the quasi-momentum space of the excitations.

The formation of bound states of excitations has recently been a problem of considerable attention. Wortis [1] investigated bound states of two magnons. Cohen and Ruvalds [2], Ruvalds and Zavadovsky [3], and Agranovich [4] studied bound states of phonons. However, in the cases considered by these authors, the energy of the phonon interaction had to exceed a certain threshold value for bound states to be formed. The situation is known to be different in liquid He_4. There a bound state of two rotons is formed at arbitrarily weak attraction between them [5-7]. This is explained by the fact that the roton energy as a function of its momentum has its minimum on a whole sphere in momentum space. Thus, this phenomenon is closely related to the liquid isotropy. Kozhushner found that a bound state of two excitons is formed at arbitrarily weak interaction in a special model which takes account of interaction with only nearest "neighbors" [8].

Some examples of the formation of bound states at weak interaction in crystals have been considered by Rashba and Levinson

[9-10]. They deal, however, either with phenomena occuring when optical phonon dispersion is neglected, or with electrons in a magnetic field, which becomes a "one-dimensional" case.

Meanwhile, we shall see that weakly-bound excitation states may be formed in crystals with quite general properties, and this is a rule rather than an exception. We shall demonstrate this by considering the simpler case when two similar excitations are bound, i.e., phonons which belong to the same branch. Let the dispersion law of the binding excitations be of the form $\omega(\vec{k})$, where ω is the excitation energy, and $\vec{k}$ is the quasi-momentum. Hereafter, we shall always speak of the momentum instead of quasi-momentum. It will be shown in the next section of the paper that in this case the equation for the bound state energy is of the form

$$\frac{\lambda}{(2\pi)^3} \int \frac{d^3q}{\epsilon - \omega(\frac{\vec{p}}{2} - \vec{q}) - \omega(\frac{\vec{p}}{2} + \vec{q}) + i0} = 1. \tag{1}$$

Here λ represents an effective interaction constant, ϵ is the bound state energy, and $\vec{p}$ is its momentum. The function $\epsilon(\vec{p})$ determined by this equation is the dispersion law of this bound state. It may be easily seen from Eq. (1) that it can be solved at any arbitrarily small constant λ, if only the integral in the left-hand side diverges for a certain value of ϵ and $\vec{p}$. In fact, in the opposite case the left-hand side would be in any case less than the right-hand one; the divergence of the integral may compensate the negligibility of λ. Now we shall consider the case when such a divergence is deliberately made to take place. It should be noted that the expression

$$\omega(\frac{\vec{p}}{2} + \vec{q}) + \omega(\frac{\vec{p}}{2} - \vec{q})$$

is an even function of $\vec{q}$ and as such it may be expanded in even powers of the components of this vector. Let it now be assumed that the symmetry axis of the crystal is of order higher than second, and $\vec{p}$ is directed along this axis. Then for small values of q we have

$$\omega(\frac{\vec{p}}{2} + \vec{q}) + \omega(\frac{\vec{p}}{2} - \vec{q}) \approx 2\omega(\frac{\vec{p}}{2}) + a_{\parallel}(p)q_{\parallel}^2 + a_{\perp}(p)q_{\perp}^2 + \sim q^4 \tag{2}$$

where $\vec{q}_{\parallel}$ is the projection of vector $\vec{q}$ on the symmetry axis, and $\vec{q}_{\perp}$ is the projection of $\vec{q}$ on the plane perpendicular to this axis. The coefficients $a_{\parallel}$ and $a_{\perp}$ are functions of $\vec{p}$ and at a certain value of $\vec{p} = \vec{p}_0$ the function $a_{\perp}$ may go to zero. It can be easily seen that in this case the integral at $\vec{p} = \vec{p}_0$ and $\epsilon = 2\omega(\vec{p}_0/2)$ ($\vec{p}_0$ is a vector equal in magnitude to p_0 and directed along the symmetry axis) diverges at small values of q. When $a_{\perp} = 0$, in the right-hand side (2) the terms of the order of $q_{\perp}^4$ should be taken into account. To simplify further calculations it will be supposed that the symmetry axis is of the sixth order. This assumption makes no change in principle. Then the terms of the fourth order by $\vec{q}_{\perp}$ are of the same form as in the isotropic case, so that

$$\omega\left(\frac{\vec{p}_0}{2} + \vec{q}\right) + \omega\left(\frac{\vec{p}_0}{2} - \vec{q}\right) \approx 2\omega\left(\frac{\vec{p}_0}{2}\right) + a_{\parallel} q_{\parallel}^2 + b q_{\perp}^4 . \tag{3}$$

This expression represents the kinetic energy of relative motion of excitations (at a given value $\vec{p} = p_0$ of total momentum). The bound state at small values of $\vec{q}$ may be formed only in the case when this energy at $\vec{q} = 0$ has its maximum or minimum (and not a saddle point) or, in other words, if the constants $a_{\parallel}$ and b have the same sign. If $a_{\parallel} > 0$, $b > 0$, i.e., in the case of a minimum, the bound state energy should be lower than the minimum value $2\epsilon(\vec{p}_0/2)$ of the energy of the two excitations. In the opposite case, when $a_{\parallel} < 0$, $b < 0$, the bound state energy is higher than the maximum. Hereafter, for the sake of definiteness, we shall consider only the case of a bound state near a minimum (i.e., we assume $a_{\parallel} > 0$, $b > 0$).

Now it can be easily seen that the integral in (1) diverges at small $\vec{q}$ when $\vec{p} = \vec{p}_0$ and $\epsilon = 2\epsilon(\vec{p}_0/2)$. Substituting (3) into (1), let us represent d^3q in the form $q_{\perp} dq_{\perp} dq_{\parallel} d\varphi$ and integrate over the angle φ. Then the integral becomes

$$\Pi = -\frac{1}{(2\pi)^2} \int \frac{dq_{\parallel} q_{\perp} dq_{\perp}'}{\delta_0 + a_{\parallel} q_{\parallel}^2 + b q_{\perp}^4}$$

where the designation $\delta = 2\omega(\vec{p}_0/2) - \epsilon$ is introduced. Integration over $dq_{\parallel}$ gives

$$\Pi = -\frac{1}{4\pi a^{\frac{1}{2}}}\int\frac{q_\perp dq'_\perp}{(\delta_o + bq_\perp^4)^{\frac{1}{2}}}\ .$$

For $\delta_o = 0$ this integral diverges logarithmically for small $q_\perp$, divergence at large $q_\perp$ being connected with the unsuitability of expansion (2) at such $\vec{q}$. In calculations the integral should be cut at a certain $q_\perp = \Lambda$ of the order of the reciprocal lattice period. With due regard to this the calculation of the integral at $\delta_o << b\Lambda^4$ and substitution into (1) gives

$$-\lambda/[16\pi(a_{\parallel}b)^{\frac{1}{2}}]\,\ln\frac{4b\Lambda^4}{\delta_o} = 1$$

from which the bound state energy at $\vec{p} = \vec{p}_o$ is equal to

$$\epsilon = 2\omega\left(\frac{\vec{p}_o}{2}\right) - 4b\Lambda^4 \exp\left(-\frac{2}{g}\right)\ .$$

where

$$g = -\lambda/[8\pi(a_{\parallel}b)^{\frac{1}{2}}]$$

The value

$$\delta_o = 4b\Lambda^4 \exp\left(-\frac{2}{g}\right) \tag{4}$$

represents the binding energy of excitations at $\vec{p} = \vec{p}_o$. We see that a bound state is formed in the vicinity of the point $\vec{p} = \vec{p}_o$ for any arbitrarily weak interaction of the excitations. Only the constant g is required to be positive, this corresponding to the attraction of excitations.* However, the binding energy turns out to be exponentially small in the interaction constant. This is evidence of the obvious similarity of the phenomenon considered and the Cooper effect in superconductors. The dispersion law of

---

*This refers to the case when $a_{\parallel} > 0$, $b > 0$. In the opposite case, when $a_{\parallel} < 0$, $b < 0$, the value g must be negative, so that a bound state near the maximum is formed by the repulsion of the excitations.

a bound state, i.e., the dependence $\epsilon(\vec{p})$ at $\vec{p} \neq \vec{p}_0$ will be determined in the third section of the present paper.

Now we shall consider in more detail the problem of the existence of such a point $\vec{p}_0$ on the symmetry axis where the coefficient $a_\perp$ becomes zero. For this purpose the dispersion curve $\omega(\vec{k})$ of the binding excitations will be considered for $\vec{k}$ directed along the axis. Let $\omega(\vec{k}_\parallel)$ be of the form shown in Fig. 1 (it may be the case, e.g., of optical phonons). If now we consider $\vec{k}$ not only on the axis, then the function $\omega(\vec{k})$ may obviously have at the point A either a minimum or a saddle point, and at the point B a maximum or a saddle point. Let the point A be the minimum. Then at $p = 2k_A$ we have $a_\parallel > 0$ and $a_\perp > 0$. If now the point B is the maximum, then $a_\parallel(2k_B) < 0$, $a_\perp(2k_B) < 0$ and the coefficient $a_\perp$ should become zero somewhere at $2k_A < p_\xi < 2k_B$ ($a_\parallel$ becomes zero at some other point; this being unimportant for us). The same situation occurs, if both A and B are saddle points. It is clear from above that $a_\perp$ going to zero and, consequently, the formation of a bound state when constants are of the appropriate sign, should take place in many crystals with a symmetry axis of order higher than the second. Other possible cases of the formation of bound states for equal excitations will be briefly discussed at the end of the third section.

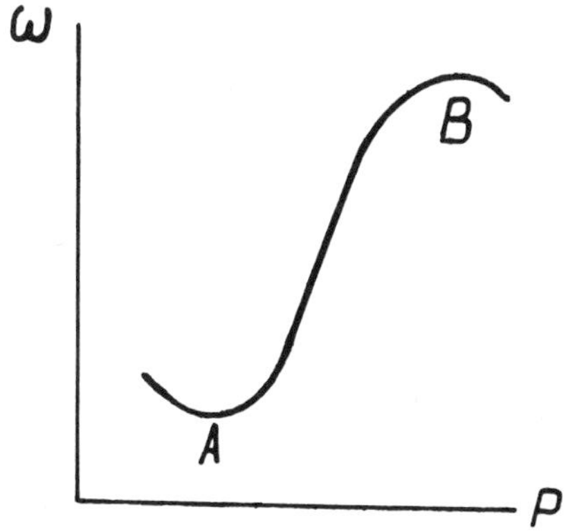

Fig. 1

BASIC EQUATIONS

The investigation of excitation bound states, as a rule, amounts to solving equations of the type (1). This is not surprising as (1) is analogous to the Schrödinger equation in the momentum representation for a pair of interacting excitations. In our case, when we deal with weakly-bound excitations, a closed form, general equation is obtained without any model representations. The point is that for low binding energy the process of the bound state "decay" into its constituent excitations becomes "almost possible". We shall describe a bound state as a spectral branch determined by a pole in the appropriate Green's function.* Accordingly, in an equation for this function those diagrams should be isolated which describe the above mentioned decay. Such a system of equations has already been obtained by the author [11]. In diagrams it looks like this:

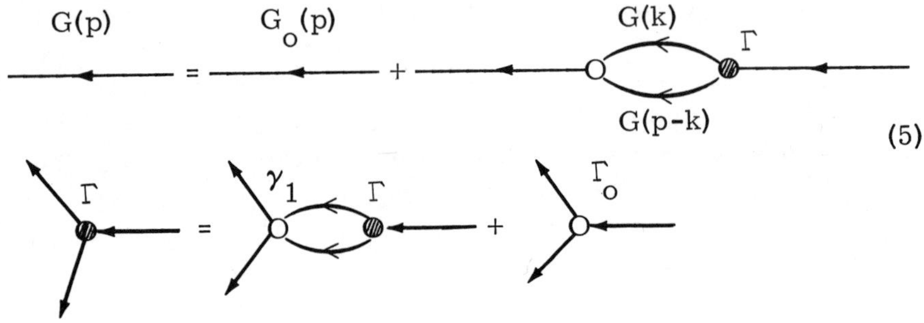

(5)

The vertex parts Γ_0 and γ_1 given by nonshaded circles are the sums of diagrams that cannot be divided between input and output ends into parts connected by only one or two lines; and the shaded total vertex part Γ contains all diagrams, except those connected by only one line. Equation (5) will not be written in an analytical form (see [11], formulae (3) and (27); in formula (3) the wrong sign on the right-hand side should be changed). The bound-state energy is determined by the pole of the Green function G(p). The equation for finding this energy is thus obtained from (5) upon

*It may seem that a bound state should be sought as a pole of the vertex part and not of a Green function. However, it can be easily seen that in a system where one excitation can transform into two excitations, the poles of both functions coincide.

neglect of the "free term" G_0. Omitting arguments of vertex parts this system can be written in the form

$$G_0^{-1}(p) = i\int \Gamma_0 G(k)G(p-k)\Gamma d^4k/(2\pi)^4$$
$$\Gamma = \Gamma_0 + i\int \gamma_1 G(k)G(p-k)\Gamma d^4k/(2\pi)^4 \qquad (6)$$

Equations (6) are exact. They can be simplified using the supposed negligibility of the binding energy. Due to this negligibility the indefiniteness in the energy and momenta of binding excitations is small; one may say that the excitations are connected by definite momentum values. From a mathematical point of view this is manifested in that only a small integration region is of importance, namely, where the product of Green functions is anomalously large. Therefore, the vertex parts may be removed from under the integration sign. After this the vertex part Γ may be excluded from the equations. The bound-state energy is finally determined by the following equation:

$$\frac{i\lambda}{(2\pi)^4}\int G(k)G(p-k)d^4k = 1 \qquad (7)$$

where

$$\lambda = [\gamma_1 + \Gamma_0^2 G_0(p)]. \qquad (8)$$

In (7) the arguments k and p-k of the vertex parts which correspond to inner lines in (5) should be assumed equal to the energies and momenta of binding excitations. While it is possible to take into account the dependence on the bound state momentum p, it is not of considerable importance as we are interested only in momenta in the vicinity of a singular value p_0 (see the introduction and next section). Therefore, the coefficient λ may be considered as constant. The increase of the interaction λ as the pole of the Green function G_0 is approached, should also be noted, i.e., in the case when the spectrum of a bound state is close to some branch of ordinary excitations. This problem, however, will not be touched on in any more detail.

The Green function G of the binding excitations may be written in the usual form

$$G(k) = [k_4 - \omega(\vec{k}) + i0]^{-1}.$$

Integration over the energy component of 4-momentum $k(k_4, \vec{k})$ finally gives:

$$\lambda\Pi(p) \equiv \lambda \int \frac{d^3k/(2\pi)^3}{\epsilon - \omega(\vec{k}) - \omega(\vec{p} - \vec{k}) + i0} = 1 \qquad [p = p(\epsilon, \vec{p})]. \tag{9}$$

This equation determines the dependence $\epsilon(p)$, i.e., the dispersion law of a bound state.

DISPERSION LAW FOR BOUND STATES OF EQUIVALENT EXCITATIONS

The value δ_0 calculated in the introduction is the binding energy of excitations with total momentum $\vec{p} = \vec{p}_0$. The binding energy varies with $\vec{p}$. The dispersion curve for a bound state can be also obtained from Eq. (9). In this section we shall consider again the case of a bound state of similar excitations which are binding at the same momentum values. In this case it is convenient to replace the variable $\vec{k} = \vec{p}/2 + \vec{q}$, whereupon Eq. (9) reduces to Eq. (1) given above. Let us consider this equation in more detail assuming, as has been done in the introduction, that the crystal has an axis of the sixth order, with the coefficient $a_\perp$ going to zero at the point $\vec{p}_0$ on the axis. First we shall expand the expression $\omega(\vec{p}_0/2 + \vec{q})$ in powers of $\vec{q}$

$$\omega(\vec{p}_0/2 + \vec{q}) = \omega(\vec{p}_0/2) + v_0 q_\parallel + \frac{a_\parallel}{2} q_\parallel^2 + \alpha q_\parallel q_\perp^2 + \frac{b}{2} q_\perp^4 . \tag{10}$$

The coefficients in (10) are designated in accordance with (3). The term with just $p_\perp^2$ is absent providing that $a_\perp(\vec{p}_0)$ goes to zero. Using (10), after simple transformations we find, to the accuracy required:

$$\begin{aligned} \omega(\vec{p}/_2 + \vec{q}) + (\vec{p}/_2 - \vec{q}) = 2\omega(\vec{p}/2) + \alpha\Delta p_\parallel q_\perp^2 + a_\parallel q_\parallel^2 + bq_\perp^4 \\ + b(\vec{p}_\perp \vec{q}_\perp)^2 + b/2\, p_\perp^2 q_\perp^2 \end{aligned} \tag{11}$$

where $\Delta\vec{p} = \vec{p} - \vec{p}_0$ and $\vec{p}_\perp$ is the projection of $\Delta\vec{p}$ on the plane perpendicular to the symmetry axis.

First, the value of $\epsilon(\vec{p})$ on the symmetry axis, i.e., at $\vec{p}_\perp = 0$ will be determined. Calculations for this case are again very simple:

$$\Pi \equiv \int \frac{d^3q/(2\pi)^3}{\epsilon - \epsilon(\vec{p}/2 - \vec{q}) - \epsilon(\vec{p}/2 + \vec{q})} = -\frac{1}{4\pi}\int \frac{q_\perp dq_\perp}{\sqrt{a_\parallel(\epsilon' + \alpha\Delta p_\parallel q_\perp^2 + bq_\perp^4}}$$

$$= -\frac{1}{8\pi(a_\parallel b)^{\frac{1}{2}}}\ln\frac{4b\Lambda^2}{2(b\epsilon')^{\frac{1}{2}} + \alpha\Delta p_\parallel}. \tag{12}$$

Here the designation $\epsilon' = 2\omega(\vec{p}/2) - \epsilon$ is introduced. It should be kept in mind that according to the assumption made $a_\parallel > 0$, $b > 0$. Substitution of (12) into (1) and the solution of the equation obtained for ϵ lead to

$$\epsilon(p) = 2\omega(\vec{p}/2) - [2(b\delta_0)^{\frac{1}{2}} - \alpha\Delta p_\parallel]^2/4b \tag{13}$$

where δ_0 has the same meaning as in (4). It should be noted that Eq. (1) can be easily shown to have a solution only on the condition that $2(b\delta_0)^{\frac{1}{2}} > \alpha\Delta p_\parallel$, so that formula (13) has meaning only within this region.

Formula (13) gives the absolute value of the bound state energy. However, a value of greater interest is that of the "binding energy", i.e., the spacing of the bound state level from the boundary of a continuum. For excitations formed near the maximum of expression (12) this binding energy is equal to

$$\delta(\vec{p}) = \epsilon_m(\vec{p}) - \epsilon(\vec{p})$$

where $\epsilon_m(\vec{p})$ is the minimum value of (11) at a given $\vec{p}$. It can be easily checked that for $\vec{p}$ lying on the axis $\epsilon_m = 2\omega(\vec{p}/2)$ for $\alpha\Delta p_\parallel > 0$ and

$$\epsilon_m = 2\omega(\vec{p}/2) - \frac{(\alpha\Delta p_\parallel)^2}{4b}$$

for $\alpha \Delta p_{\parallel} < 0$. With regard to this, Eq. (13) can be finally re-written as

$$\delta = \delta_0 \begin{cases} (1 - \Delta p_{\parallel}/r_{\parallel})^2 & \Delta p_{\parallel} > 0 \\ 1 - 2\Delta p_{\parallel}/r_{\parallel} & \Delta p_{\parallel} < 0 \end{cases} \tag{14}$$

where the following designation is introduced

$$r_{\parallel} = 2(b\delta_0)^{\frac{1}{2}}/\alpha \tag{15}$$

(for the sake of definiteness we assume $\alpha > 0$).

At $\Delta p_{\parallel} = r_{\parallel}$ the binding energy goes to zero and the bound state level at $\Delta p_{\parallel} > r_{\parallel}$ goes into the continuum. The point $\Delta p_{\parallel} = r_{\parallel}$ is the end point of the bound state spectrum in the sense of the work [11]. According to the classification given in that work, it corresponds to the second way of the spectrum ending. At negative values of $\Delta p_{\parallel}$ the binding energy increases. It should be borne in mind, however, that $|\Delta p_{\parallel}|$ in any case must be much less than the period of the reciprocal lattice. Accordingly, it can be easily checked, that the binding energy in any case is much less than $[\delta_0 \omega_0]^{\frac{1}{2}}$, where $\omega_0 \sim \epsilon(\vec{p}_0/2)$ is the characteristic frequency of an optical phonon.

Figure 2a shows graphically the binding energy $\delta(p_{\parallel})$. On the vertical axis δ is scaled in units of δ_0, and on the horizontal axis $\Delta p_{\parallel}$ is plotted in terms of $r_{\parallel}$.

Now let the total momentum $\vec{p}$ be no longer directed strictly along the symmetry axis, so that $\vec{p} - \vec{p}_0$ has a component $\vec{p}_{\perp}$ perpendicular to the axis. Substituting (11) into the expression for Π and calculating integrals over $dq_{\perp}$ and $dq_{\parallel}$ we obtain

$$\Pi = -\frac{1}{16\pi^2 (a_{\parallel} b)^{\frac{1}{2}}} \int_0^{2\pi} \ln \frac{4b\Lambda^2}{2(b\epsilon')^{\frac{1}{2}} + \alpha \Delta p_{\parallel} + b p_{\perp}^2 \cos\varphi} \, d\varphi$$

where φ is the angle between $\vec{q}_{\perp}$ and $\vec{p}_{\perp}$. The integral over $d\varphi$ is calculated with the use of the formula

$$\int_0^{\pi} \ln (1 + a\cos^2\varphi) d\varphi = 2\pi \ln \frac{1 + \sqrt{1+a}}{2} . \tag{16}$$

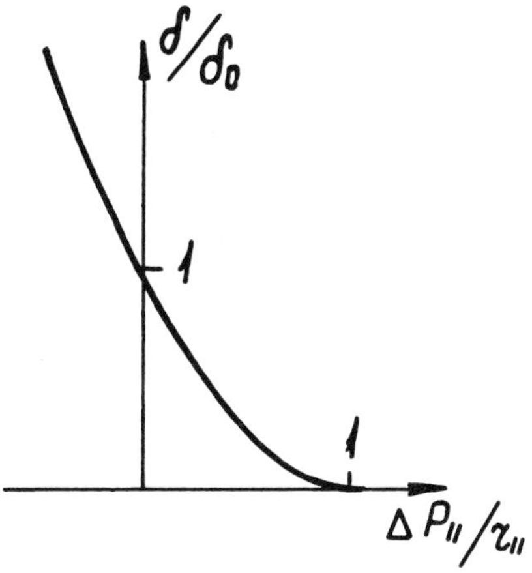

Fig. 2a

Finally:

$$\Pi = -\frac{1}{8\pi(a_\parallel b)^{\frac{1}{2}}}\,\ln\frac{8b\Lambda^2}{\{[2(b\epsilon')^{\frac{1}{2}}+\alpha\Delta p_\parallel+\frac{b}{2}p_\perp^2]^{\frac{1}{2}}+[2(b\epsilon')^{\frac{1}{2}}+\alpha\Delta p_\parallel+\frac{3}{2}bp_\perp^2]^{\frac{1}{2}}\}^2}\,. \tag{17}$$

Substituting (17) into (9) and solving the equation, the value $\epsilon(\vec{p})$ may be found. We may straightforwardly write the expression for binding energy $\delta(p)$, taking into account that the lower boundary of the continuum $\epsilon_m(\vec{p})$ is now of the form:

$$\epsilon_m(\vec{p}) = \begin{cases} 2\omega(\vec{p}/2), & \alpha\Delta p_\parallel + \frac{b}{2}p_\perp^2 > 0 \\ 2\omega(\vec{p}/2) - \dfrac{(\alpha\Delta p_\parallel + \frac{b}{2}p_\perp^2)^2}{4b}, & \alpha\Delta p_\parallel + \frac{b}{2}p_\perp^2 < 0\,. \end{cases}$$

The binding energy equals

$$\delta(\vec{p}) = \delta_0\{1 - 4(\frac{p_\perp}{r_\perp})^2[1 - \tfrac{1}{4}(\frac{p_\perp}{r_\perp})^2] - \frac{\Delta p_\parallel}{r_\parallel}\},\quad \alpha\Delta p_\parallel + \frac{b}{2}p_\perp^2 > 0 \tag{18}$$

and

$$\delta(\vec{p}) = \delta_0\{1 - \frac{3p_\perp^2}{r_\perp^2} + \frac{p_\perp^4}{r_\perp^4} - \frac{2\Delta p_\parallel}{r_\parallel}\}[1 - (\frac{p_\perp}{r_\perp})^2], \quad \alpha\Delta p_\parallel + \frac{b}{2}p_\perp^2 < 0$$

where $r_\perp = (2^{3/2}\delta_0^{\frac{1}{4}})/b^{\frac{1}{4}}$. The relationship between δ and $p_\perp$ at $\Delta p_\parallel = 0$ is given in Fig. 2b. It should be noted that the binding energy decreases with an increase of $p_\perp$, i.e., when moving away from the symmetry axis. The region where a bound state exists, i.e., the region where $\delta(\vec{p}) > 0$ is shaded in Fig. 3. On the vertical axis $\Delta p_\parallel / r_\parallel$ is plotted, and on the horizontal axis $p_\perp / r_\perp$ is plotted. On the boundary $\delta = 0$. It should be noted that the characteristic dimension of the existence region in the direction perpendicular to the axis is of the order of $r_\perp \sim \delta_0^{\frac{1}{4}}$, i.e., it highly exceeds $r_\parallel \sim \delta_0^{\frac{1}{2}}$.

Up to now we have considered the states formed in the vicinity of the symmetry axis in a crystal. Now we shall discuss the possibility of the formation of such states away from the axis, near a certain "nonsymmetrical" point in momentum space. In this case quadratic terms in the expression $\omega(\vec{p}/2 + \vec{q}) + \omega(\vec{p}/2 - \vec{q})$

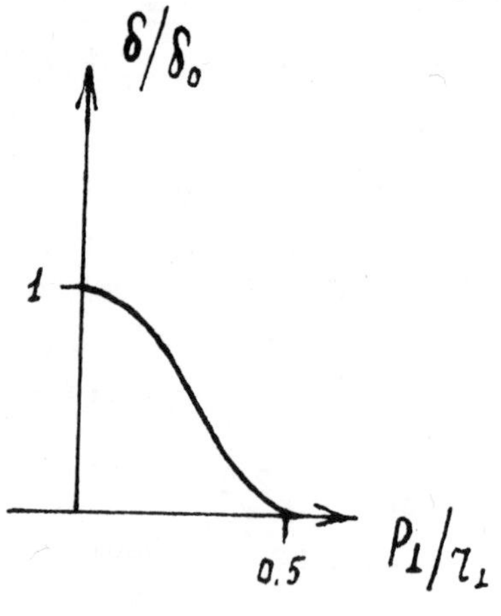

Fig. 2b

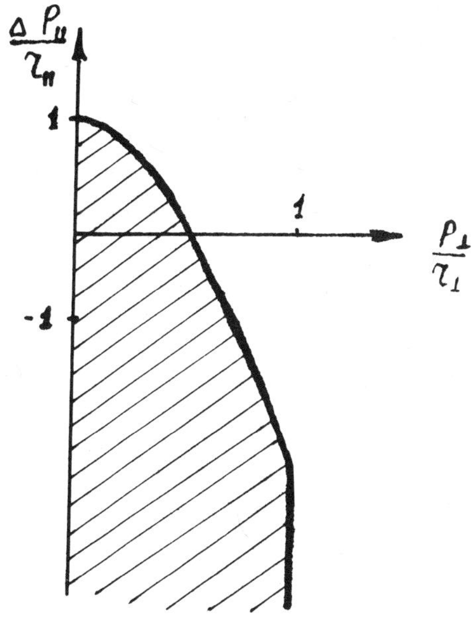

Fig. 3

are of the form $a_{ik}q_i q_k$. When directing the axis along the principal axes of the quadratic form $a_{ik}(\vec{p})$, we obtain

$$\omega(\vec{p}/2+\vec{q})+\omega(\vec{p}/2-\vec{q}) \approx 2\omega(\vec{p}/2)+a_1 q_1^2+a_2 q_2^2+a_3 q_3^2 \,.$$

It is evident that for the divergence of the integral (9) two of the coefficients a_i are required to go to zero at a certain point p_0. Let us see how many conditions should be satisfied for this purpose. When two eigenvalues are going to zero, it means that the matrix a_{ik} at the point p may be given as

$$a_{ik} = \ell_i \ell_k$$

where ℓ is an eigenvector of the matrix corresponding to the eigenvalue which does not go to zero. Excluding three components ℓ_i from these six equalities, we find the conditions:

$$a_{12}^2 = a_{11}a_{22}, \; a_{13}^2 = a_{11}a_{33}, \; a_{23}^2 = a_{22}a_{33} \, . \tag{19}$$

Equation (19) should be regarded as three equations for the determination of the three components of the vector p_0, near which a bound state is formed. The fact that the number of equations equals the number of unknowns means that such a point, generally speaking, does exist, so that this state may be formed away from the crystal axis. In this connection it should be emphasized that the above case of the state of the symmetry axis is an independent case, and not a particular case of the general one. In fact, an accidental occurrence of the point p_0 on the axis would be absolutely incredible. Meanwhile, the reasoning given in the introduction shows that p_0 could be placed strictly on the axis with the same probability as at a nonsymmetrical point. On the other hand, from the experimental point of view a search for the states in the vicinity of the axis is probably easier due to less arbitrariness. Bearing that circumstance in mind alongside with the great complexity of the formulae in the nonsymmetrical case, the latter will not be investigated here.

It should be noted also that a weakly-bound state can also be formed near the symmetry plane of a crystal. In fact, for p_0 on the symmetry plane the terms quadratic in q are of the form:

$$a_z q_z^2 + a_{\alpha\beta} \, q_\alpha \, q_\beta$$

where the axis z is normal to the plane and indices α, β enumerate the vector components in the plane. The coefficients a_z, $a_{\alpha,\beta}$ depend now on two components of the vector p in the plane. Generally speaking, these two components may be chosen in such a way that the value a_z and one of the eigenvalues of the matrix $a_{\alpha\beta}$ go to zero simultaneously. In this case the point p_0 will lie strictly on the plane.

BOUND STATES OF DIFFERENT EXCITATIONS

Let the binding excitations now be different. Of particular interest here are bound states of an electron and a phonon in a semiconductor, i. e., weakly bound polarons. We may also consider two phonons from different branches or two phonons which belong to one and the same branch, but are binding with different quasimomentum values.

The equation for the bound state energy in the case of two different excitations also has the form (9) with the only difference that $\omega(k) + \omega(p-k)$ is replaced with

$$\omega_1(k) + \omega_2(p-k) \tag{20}$$

where ω_1 and ω_2 are the energies of the binding excitations. In the previous case terms linear in q in the denominator (1) were absent due to the symmetry. Now they should be eliminated first of all. For this purpose we shall find the minimum of (20) as a function of k at a given p.* Let this minimum occur at $k = k_0$. (Naturally, k_0 is a function of p, $k_0 = k_0(p)$). We assume $k = k_0(p) + q$. Since the expression (20) has its minimum at $k = k_0$, its expansion in powers of q will not already contain linear terms.** Again we suppose that the order of a symmetry axis in a crystal is higher than the second. Let us direct p along the symmetry axis. Then we have:

$$\begin{aligned} \omega_1(p-k) + \omega_2(k) &= \omega_1(p-k_0-q) + \omega_2(k_0+q) \\ &\approx \omega_1(p-k_o) + \omega_2(k_0) + a_\perp q_\perp^2 + a_\parallel q_\parallel^2 \end{aligned} \tag{21}$$

and at a certain point $p = p_0$ on the axis, $a_\perp$ may go to zero. It should be noted that the previously considered case of equal excitations falls within this general scheme. However, there $k_0 = p/2$ due to the symmetry of the problem. But there is a significant difference, which lies in the fact that now the left-hand side of (21) is no more an even function of q. Meanwhile, for a bound state to exist it is required that in this expansion at $p = p_0$ the terms of the third order in q be absent. This condition will be met automatically, if the symmetry of the axis is not less than of the fourth order. In this case the presence of such an axis is essential for the existence of the effects under consideration. Weakly-bound states of two different excitations cannot be formed

*Here again the case of a bound state with energy lower than the boundary of a continuum is considered.
**Compare with similar reasoning in [12].

at a nonsymmetrical point and on the symmetry plane.

Specific calculations will again be carried out for an axis of the sixth order. It turns out that the expansion is more convenient not with respect to the point $k_o(p)$, but with respect to the point $k_o(p_{\parallel})$, where $p_{\parallel}$ is the projection of p on the axis. In other words, we assume $k = k_o(p_{\parallel}) + q$. We shall write the expansion of (20) with respect to powers of q straightforwardly on the basis of general considerations, without writing formulae which connect expansion coefficients with the derivatives of the functions ω_1 and ω_2. It is significant that when p is not directed strictly along the axis, the point $k_o(p_{\parallel})$ is not a minimum point and in the expansion there are always present terms linear in $q_{\perp}$ (and not in $q_{\parallel}$). With due regard to this we obtain:

$$\omega_1(p-k) + \omega_2(k) \approx \epsilon_o(p) + \alpha\Delta p_{\parallel} q_{\perp}^2 + 2\beta p_{\perp} q_{\perp} + a_{\parallel} q_{\parallel}^2 + bq_{\perp}^4 . \qquad (22)$$

The designation

$$\epsilon_o = \omega_1[p-k_o(p_{\parallel})] + \omega_2[k_o(p_{\parallel})]$$

is introduced. Due to the presence of a term linear in $p_{\perp}$ in (22) terms of higher order in this quantity need not be taken into account.

First of all it should be emphasized that on the axis, i.e., at $p_{\perp} = 0$ the expansions (22) and (14) coincide. This means that for a momentum directed along the axis, the binding energy for different excitations is also given by the same formula (14).

When deviating from the axis the situation is, naturally, different. Integration over $dq_{\parallel}$ gives an integral Π of the form

$$\Pi = \frac{1}{8\pi^2 (a_{\parallel})^{\frac{1}{2}}} \int \frac{q_{\perp} dq_{\perp} d\varphi}{(\epsilon' + \alpha\Delta p_{\parallel} q_{\perp}^2 - 2\beta q_{\perp} p_{\perp} \cos\varphi + bq_{\perp}^4)^{\frac{1}{2}}} \qquad (23)$$

where $\epsilon' = \epsilon_o - \epsilon$.

The integral (23) cannot be calculated in general. Only asymptotic formulae for small and large $p_{\perp}$ can be obtained. We restrict ourselves to the determination of a limit for the region of a bound state for small and large $p_{\perp}$, i.e., the equation of the surface $\delta(p) = 0$ in momentum space.

Let us begin with small values of $p_\perp$. First we shall determine the lower boundary of the continuum, i.e., the minimum value of (22) for a given p. The minimum is obtained when $\cos\varphi = 1$. (For the sake of definiteness we suppose that $\alpha > 0$; $\beta > 0$). For $\Delta p_\parallel > 0$ and small $p_\perp$ the term $bq_\perp^4$ near the minimum is insignificant; the minimum is reached at $q_\perp = \beta p_\perp/\alpha\Delta p_\parallel$ and the minimum energy value is equal to

$$\epsilon_m = \epsilon_o - \beta^2 p_\perp^2/\alpha\Delta p\ ; \text{ or } \epsilon'_m = \beta^2 p_\perp^2/\alpha\Delta p_\parallel\ .$$

Let the value of Π at $\epsilon' = \epsilon_m$ be designated by Π_m; then a differentiation is carried out twice with respect to $f = \beta p_\perp$ and a change of variables performed by defining $q_\perp = xf/a_\perp$, where $a_\perp = \alpha\Delta p_\parallel$ (good convergence of the integral permits us to change the upper integration limit to infinity). As a result we have:

$$\frac{\partial^2\Pi_m}{\partial f^2} = -\frac{3bf}{8\pi^2(a_\parallel^{\frac{1}{2}}a_\perp^{9/2})}\int_0^\infty\int_0^{2\pi}\frac{(1-x\cos\varphi)x\,dx\,d\varphi}{[1+x^2-2x\cos\varphi+(f^2b/a_\perp^3)x^4]^{5/2}}\ .$$

At small f large values of x are significant in the integral, so that the unity in the denominator may be neglected and only the first term in the expansion of this denominator in $2x\cos\varphi$ need be considered. Calculations give

$$\frac{\partial^2\Pi}{\partial f^2} = \frac{\sqrt{b}}{2\pi\, a_\parallel^{\frac{1}{2}}\, a_\perp^3}\ . \tag{24}$$

On the other hand when $f = 0$, we can easily obtain directly from (23):

$$\Pi_m(0) = -\frac{1}{8\pi(a_\parallel b)^{\frac{1}{2}}}\ln\frac{4b\Lambda}{a_\perp}\ ,\quad \frac{\partial\Pi_m}{\partial f} = 0\ .$$

Integration of (24) with these boundary conditions leads to

$$\Pi_m = -\frac{1}{8\pi(a_\parallel b)^{\frac{1}{2}}}\left[\ln\frac{4b\Lambda^2}{\alpha\Delta p_\parallel} - \frac{2b\beta^2p_\perp^2}{\alpha^3(\Delta p_\parallel)^3}\right].$$

If we substitute this expression into Eq. (9), we directly obtain the relation between $\Delta p_\parallel$ and $p_\perp$ on the surface $\delta(p) = 0$, i.e., at the boundary of existence of a bound state. Simple transformations give an equation for this boundary in the form:

$$1 - \Delta p_\parallel / r_\parallel = (p_\perp / S_\perp)^2 \tag{25}$$

where $r_\parallel$ has the same meaning as in (15), and

$$S_\perp = 2b^{\frac{1}{4}} \delta_o^{3/4} / \beta ; \tag{26}$$

formula (25) being true on condition that $p_\perp / S_\perp << 1$. The value $S_\perp$ characterizes a transverse dimension of the existence region of a bound state for different excitations. It should be noted that $S_\perp \sim \delta_0^{3/4}$ while for similar excitations the corresponding value $r_\perp \sim \delta_o^{\frac{1}{4}}$, so that if δ_o in both cases is of the same order, then

$$S_\perp << r_\perp .$$

Now we shall determine the boundary for large $p_\perp >> S$. It can be easily checked that large negative values $\alpha \Delta p_\parallel$ correspond large $p_\perp$ on the boundary. More specifically, $|\Delta p_\parallel| >> r_\parallel$, $\Delta p_\parallel < 0$ (we assume $\alpha > 0$, $\beta > 0$). In this case in the integral for Π_m the term with f is small compared to the rest and it is significant only near the denominator minimum. Therefore, q can be replaced there by its value at the minimum q_m. After this the integral with respect to $q_\perp$ can be calculated as has been done in the derivation of (12). Taking into account that

$$\epsilon'_m = - a_\perp q_m^2 + 2fg_m - bq_m^4$$

and taking within the accuracy required

$$q_m \approx |a_\perp|^{\frac{1}{2}} / (2b)^{\frac{1}{2}}, \quad (a_\perp = \alpha \Delta p_\parallel)$$

after expansion by f we obtain:

$$\Pi_m = \frac{1}{16\pi^2 (a_\parallel b)^{\frac{1}{2}}} \int_0^{2\pi} \ln \frac{2\Lambda^2 |b\alpha\Delta p_\parallel|^{\frac{1}{2}}}{2^{\frac{1}{2}} \beta p_\perp (1 - \cos\varphi)} d\varphi = \frac{1}{8\pi (a_\parallel b)^{\frac{1}{2}}} \ln \frac{2\Lambda^2 |2b\alpha\Delta p_\parallel|^{\frac{1}{2}}}{\beta p_\perp} .$$

The integral over $d\varphi$ is performed according to the formula

$$\frac{1}{\pi}\int_0^{\pi} \ln(1-\cos\varphi)d\varphi = -\ln 2.$$

Substituting into (9) and solving the equation we find the boundary equation at large $p_\perp$

$$-\frac{\Delta p_\parallel}{r_\parallel} = \left(\frac{p_\perp}{S_\perp}\right)^2. \quad (27)$$

Formally this is the limiting form of the same parabola as for small $p_\perp$. The existence region of a bound state for different excitations is shaded in Fig. 4. On the abscissa $p_\perp/S_\perp$ is plotted and on the ordinate $\Delta p_\parallel$ is plotted.

Thus, it is seen that bound states of the type considered should, evidently, exist in crystals rather often. They should be

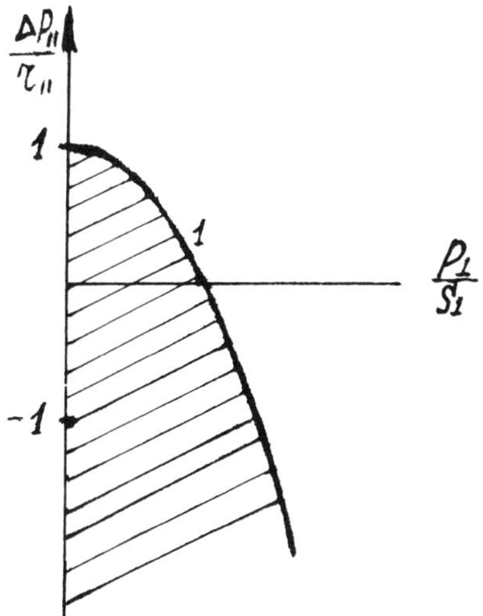

Fig. 4

searched for primarily in the vicinity of symmetry axes of high order in momentum space. The most straightforward method to detect them is the observation of inelastic neutron scattering. Exponential negligibility of the binding energy can hardly be a serious obstacle for their discovery, as in reality they are not very weak. On the other hand, the picture may "grow dim" due to the attenuation of excitations. From the form of the principal equation (9) it is clear that if ω_1 and ω_2 have imaginary parts j_1 and j_2, they can be neglected only for the rather strict condition that $j_1 + j_2 \ll \delta_0$. The level width of the binding excitation should be less than the binding energy. Therefore, it is necessary that the work should be carried out at low temperatures and those excitation branches be chosen for which decays are forbidden.

It is to be also noted that, even if the sign of the interaction is such that a bound state is not formed, nevertheless, the density of two-particle states in the vicinity of the point $\vec{p}_0$ has a certain singularity that can also be revealed experimentally.

The author is indebted to V. M. Agranovich, A. F. Andreev, I. E. Dzyaloshinsky, I. B. Levinson, and E. I. Rashba for useful discussion.

REFERENCES

1. M. Wortis, Phys. Rev. 132, 85 (1963).
2. M. H. Cohen, J. Ruvalds, Phys. Rev. Lett. 23, 1378 (1969).
3. J. Ruvalds, A. Zavadovsky, Phys. Rev., B2, 1172 (1970).
4. V. M. Agranovich, Fiz. Tverd. Tela. 12, 562 (1970).
5. F. Iwamoto, Progr. Theor. Phys. (Kyoto) 44, 1135 (1970).
6. J. Ruvalds, A. Zavadovsky, Phys. Rev. Lett. 25, 333 (1970).
7. L. P. Pitaevsky, Pys'ma v ZhTF, 12, 118 (1970).
8. M. A. Kozhushner, Zhurn. Exper. Theor. Phyz. 60, 220 (1971).
9. I. B. Levinson, E. I. Rashba, UFN 111, 683 (1973).
10. I. B. Levinson, E. I. Rashba, Rep. Progr. Phys. 36, 1499 (1973).
11. L. P. Pitaevsky, Zhurn. Exper. Theor. Phys. 36, 1168 (1959).
12. R. F. Kazarinov, O. V. Konstantinov, Zhurn. Exper. Theor. Phys. 40, 936 (1961).

POLARITON FERMI-RESONANCE IN THE RAMAN SPECTRUM OF AMMONIUM CHLORIDE

V. S. Gorelik, G. G. Mitin, and M. M. Sushchinskii

P. N. Lebedev Physical Institute
Academy of Sciences of the USSR
Moscow, USSR

To the present several theoretical papers [1-4] have been devoted to the investigation of bound states in the vibrational spectra of crystals. In these works the opportunity of investigating such states with the help of Raman scattering of light has been discussed. The theory predicts that vibrational bound states (biphonons or vibrational biexcitons for molecular crystals) arise as a result of anharmonicity. According to the results obtained [1-4] Fermi-resonance with strong anharmonicity is most favorable for the appearance of bound states.

The question of the appearance of bound states under conditions of polariton Fermi-resonance is of particular interest and it was investigated by Agranovich and Lalov [3, 4]. Such a resonance takes place, when a polariton branch intersects polar two-phonon bands or polar bound states (with $k \approx 0$), originating from the coupling of two phonons with opposite wave vectors ($\vec{k}_1 \approx -\vec{k}_2$). One of the main theoretical [3, 4] conclusions is that the presence of a polar bound state should give rise to a supplementary gap (alongside with the gaps caused by fundamental polar vibrations) in the polariton curve. Accordingly, a supplementary gap should arise in the polariton Raman spectrum.

The strong anharmonicity effects in Raman spectra for large k ($\sim 10^5$ cm^{-1}) have been discussed recently [5-7] for a number of crystals (SiO_2, $AlPO_4$, diamond). In spite of this the question of the existence of bound states for large k and for the polariton region is not clear up to now (see also [8]). It is interesting also to continue the investigation of polariton Fermi-resonance into the region of two-phonon zones for comparison with the first results [8] on this question.

In the present paper we report the results of Raman studies of intermolecular vibrational excitations of an ammonium ion for the low temperature NH_4Cl phase.

THE STRUCTURE AND FUNDAMENTAL VIBRATIONS

The NH_4Cl crystal has a very simple structure; at low temperature its elemental volume contains only one unit NH_4Cl (Fig. 1). The space group is T_{α}^{1}, the point group is T_d. Such a structure refers to a cubic system, but the point group has no inversion (due to the asymmetric $(NH_4)^+$ group). Thus, polaritons can be investigated with the help of Raman spectroscopy.

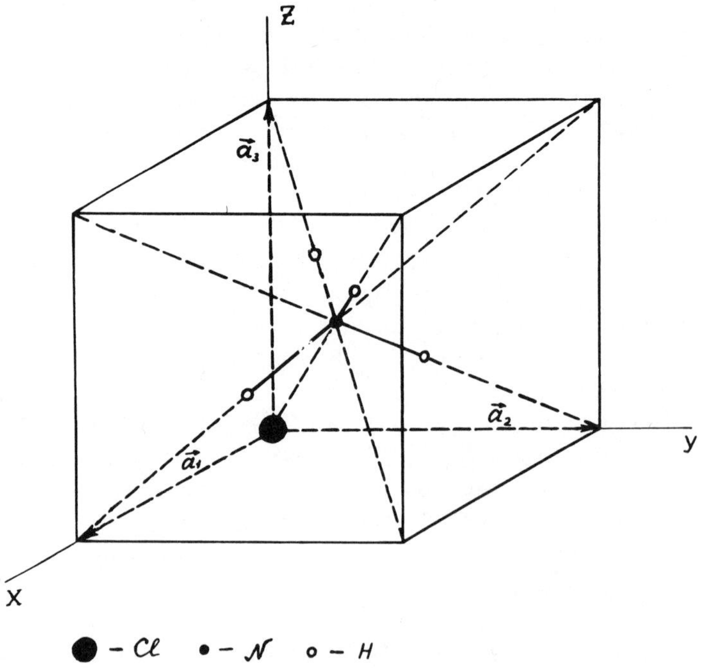

Fig. 1. The elemental volume in the low-temperature structure of the NH_4Cl crystal.

According to the conventional group theory analysis we have the vibrational representation for k = 0 in the form:

$$T_{vib} = (F_1 + F_2) + (A_1 + E + F_2 + F_2) \quad (1)$$

The first term in (1) corresponds to external vibrations. The Raman and infrared inactive F_1-type mode ($\nu_L(F_1)$ = 380 cm^{-1}) is due to librational motion of the $(NH_4)^+$ group; the polar F_2-type (ν_{2t} = 188 cm^{-1}, $\nu_{2\ell}$ = 275 cm^{-1}) corresponds to translational motion of Cl^{-1} and $(NH_4)^+$ ions and is Raman and infrared active. The second term in (1) describes internal fundamental vibrations (vibrational excitons) of the ammonium group (all types are Raman active; A_1 and E types are infrared inactive). The frequencies of the internal fundamental vibrations are: $\nu_1(A_1)$ = 3042 cm^{-1}, $\nu_2(E)$ = 1716 cm^{-1}, $\nu_{4t}(F_2)$ = 1400 cm^{-1}, $\nu_{4\ell}(F_2)$ = 1418 cm^{-1}, $\nu_{3t}(F_2)$ = 3122 cm^{-1}, $\nu_{3\ell}(F_2)$ = 3159 cm^{-1}.

RAMAN SPECTRA AT LARGE ANGLES

The shape of the 90°-Raman spectrum for the region of vibrational excitons recorded with the help of a laser source is presented in Figs. 2 and 3. Figure 2 shows a TO-LO doublet (1400, 1418 cm^{-1}), corresponding to the fundamental vibration $\nu_4(F_2)$, an intense sharp line ($\nu_2(E)$ = 1716 cm^{-1}) and two wide second order satellites. The frequencies of these satellites are close to $\nu_4(F_2) + \nu_L(F_1)$ and $\nu_2(E) + \nu_L(F_1)$, respectively. Polarization measurements show that these Raman maxima, in spite of their large intensity should be regarded as two-particle bands, but not as bound states.

Figure 3 gives a more complex picture of the spectrum. The most intense line (3042 cm^{-1}) and intense doublet (3122, 3159 cm^{-1}) correspond to the fundamental vibrations $\nu_1(A_1)$ and $\nu_3(F_2)$. In the region of the overtone $2\nu_4(F_2)$ there exists a wide band with a sharp peak 2805 cm^{-1} (halfwidth - 4 cm^{-1}) that exhibits the same polarization properties as $\nu_1(A_1)$. The total width of this band is 110 cm^{-1}, which is approximately twice the width of the fundamental $\nu_4(F_2)$ band. As it was noted in papers [9, 10], the sharp peak should be regarded as a non-polar (non-interacting with polaritons) vibrational exciton; we suppose [9] that the binding takes place due to the proximity of the $\nu_1(A_1)$ and $2\nu_4(F_2)$ frequencies ("weak" Fermi-resonance). Near $\nu_1(A_1)$ there are: a

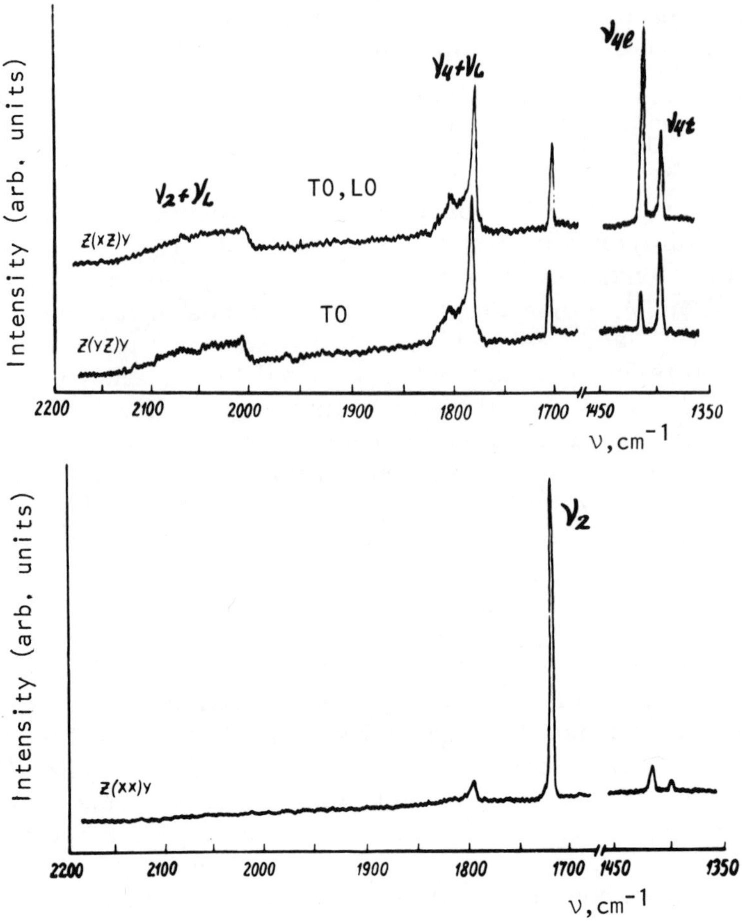

Fig. 2. The Raman spectrum of a NH_4Cl monocrystal for low-frequency vibrational excitons, recorded with different geometries.

polar TO-LO doublet (ν_t' = 3052 cm^{-1}, ν_ℓ' = 3070 cm^{-1}), and a polar two-particle band $\nu_4(F_2) + \nu_2(E)$ (maximum of this band corresponds to 3096 cm^{-1}). The peaks ν_t', ν_ℓ' cannot be attributed to fundamental vibrations. We believe that it is a polar bound state, originating as a result of a Fermi-resonance of the $\nu_4(F_2) + \nu_2(E)$ combinational tone with the fundamental vibration $\nu_3(F_2)$.

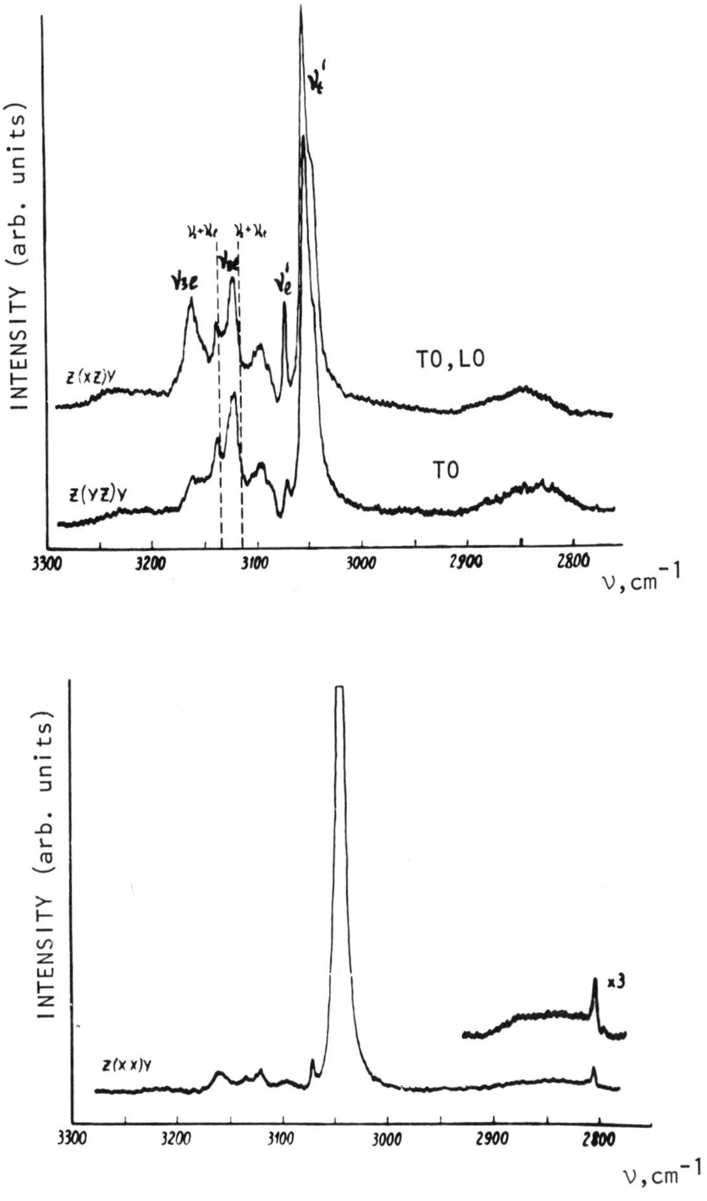

Fig. 3. The Raman spectrum in the high frequency region for different geometries.

Our explanation of the peculiarities of the second-order Raman spectra was confirmed by investigations of deuterium analogs, temperature measurements and angular-dependent measurements.

RAMAN SPECTRA AT SMALL ANGLES

Investigations of light scattering at small angles were realized with the help of a well known frequency-angle dependence scheme [8].

We have investigated polariton Raman spectra in the regions of $\nu_4(F_2)$ polar vibrations; $\nu_2(E)+\nu_L(F_2)$, $\nu_4(F_2)+\nu_L(F_1)$, $\nu_2(E)+\nu_4(F_2)$ bands; and polar bound states $\nu_t'(F_2)$, $\nu_\ell'(F_2)$, and we have also studied more carefully, as compared to the first investigation [8], the polar two-particle band $2\nu_4(F_2)$. The most striking feature of the polariton Raman spectra obtained is the existence of a number of gaps in the region of the polariton Fermi-resonance: 1) near the polar bound state ν_t', ν_ℓ' 2) near the band $\nu_2(E)+\nu_4(F_2)$ (in this case the gap is very difficult to observe) 3) near the bands $\nu_4(F_2)+\nu_L(F_1)$ and $2\nu_4(F_2)$.

We have also observed that the polariton curve in the Raman spectrum was very sharp below the two-particle band $2\nu_4(F_2)$. On "entering" this band the polariton curve has become diffused and then has disappeared completely (inside this band).

Besides, we have observed the increase of the polariton Raman intensity near the polar bound state ν_t', ν_ℓ'.

THE DISPERSION LAW FOR POLARITONS

From the frequency-angular dependence obtained we have calculated the dispersion law for polaritons, using the formula:

$$K = 2\pi \{ (\nu_0 n_0)^2 + [(\nu_0 - \nu)n']^2 - 2(\nu_0 - \nu)\nu_0 n_0 n' \cos\theta \}^{\frac{1}{2}} . \quad (2)$$

In (2) K and ν are the wave vector and wave number of polaritons; ν_0 is the wave number of the exciting photon; n_0, n' are corresponding refractive indices. Figure 4(a-c) shows the dispersion

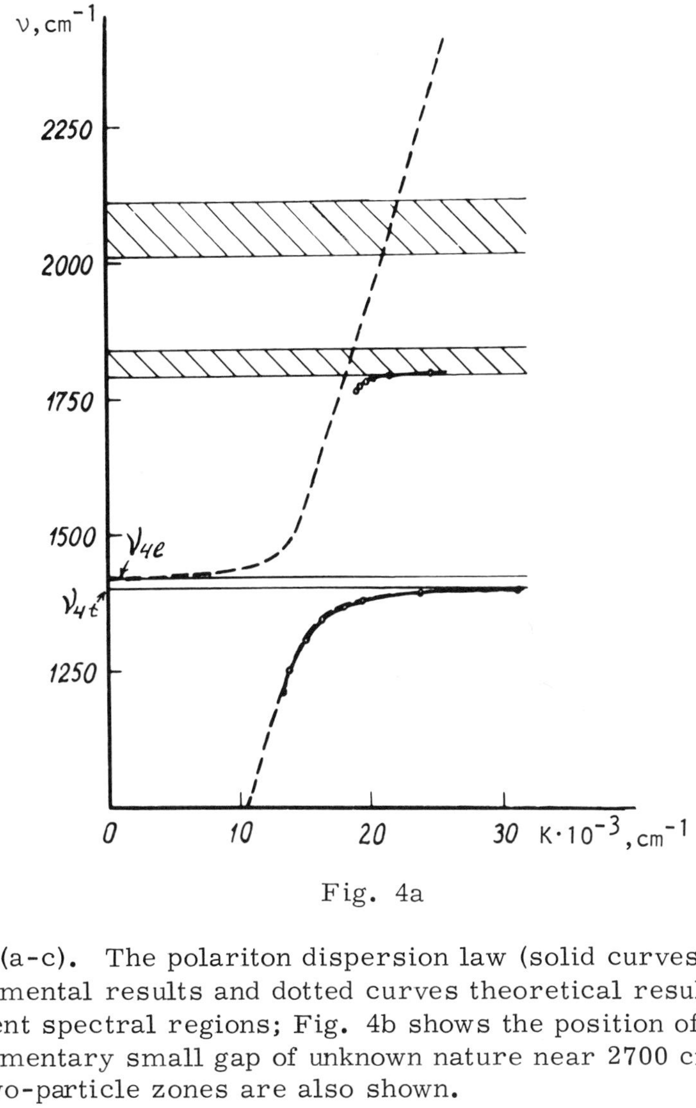

Fig. 4a

Fig. 4(a-c). The polariton dispersion law (solid curves are experimental results and dotted curves theoretical results) for different spectral regions; Fig. 4b shows the position of the supplementary small gap of unknown nature near 2700 cm^{-1}. The two-particle zones are also shown.

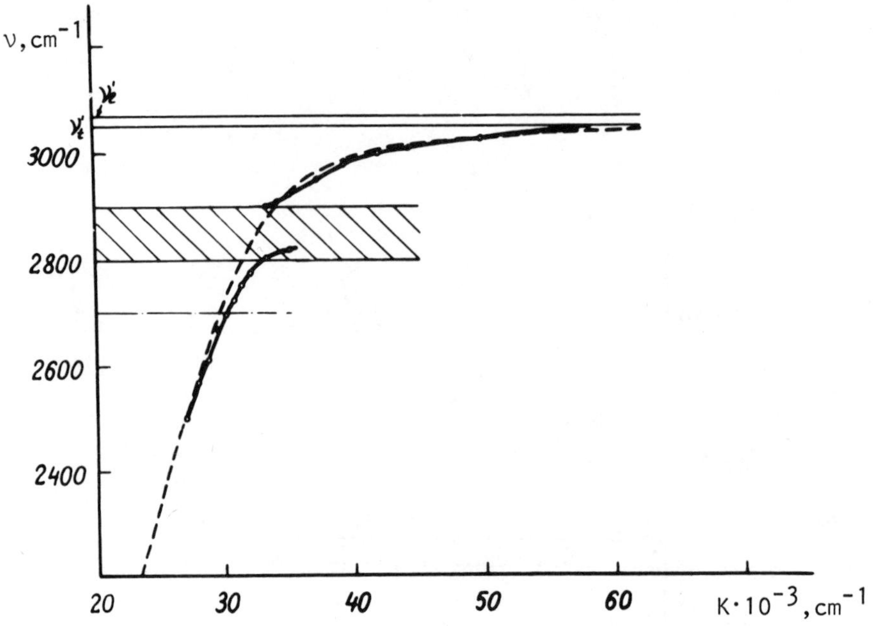
ν, cm⁻¹
3000
2800
2600
2400
20
30
40
50
60
K·10⁻³, cm⁻¹

Fig. 4b

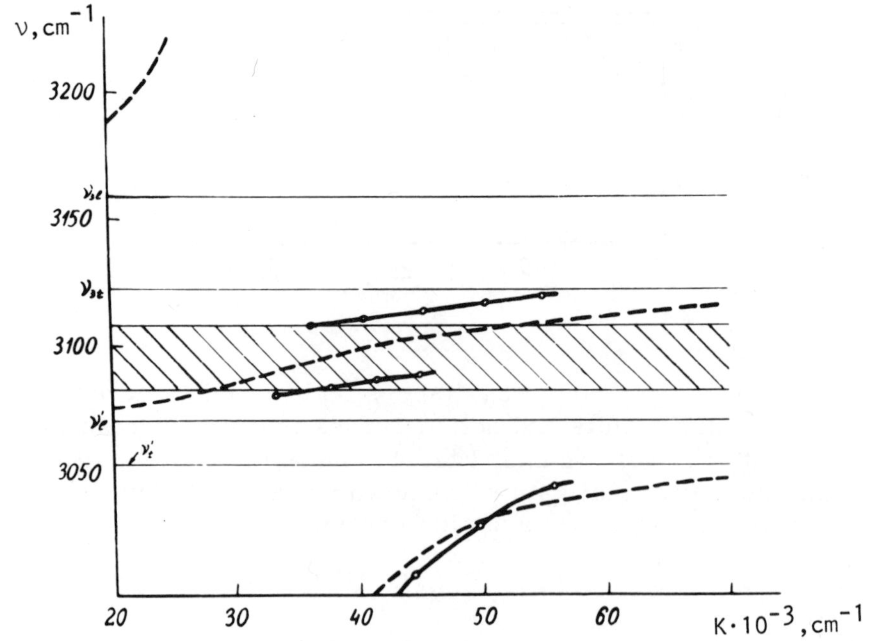
ν, cm⁻¹
3200
3150
3100
3050
20
30
40
50
60
K·10⁻³, cm⁻¹

Fig. 4c

law obtained for the polaritons. Also Fig. 4 gives (dotted lines) the dispersion law, calculated by means of the true formula for cubic crystals:

$$\nu = \frac{K}{2\pi\sqrt{\epsilon(\nu)}} \; ; \; \epsilon(\nu) = \epsilon_\infty \prod_j \frac{\nu_{j\ell}^2 - \nu^2}{\nu_{jt}^2 - \nu^2} \tag{3}$$

In the last expression we take into account three fundamental polar vibrations (F_2-type) and the polar bound states (ν_t', ν_ℓ'); the frequencies in (3) are: ν_{jt} = 188, 1400, 3052, 3112 cm^{-1}, $\nu_{j\ell}$ = 275, 1418, 3070, 3159 cm^{-1}. Thus, in (3) we have not taken into account two-particle vibrational states.

As can be seen from Fig. 4, there is a distinct gap near the polar bound state (ν_t', ν_ℓ') for dotted and solid lines which confirms theoretical predictions [3, 4]. Calculations according to (3) are in good agreement with the experimental results everywhere except in two-particle state regions. In the last case there is a number of gaps (for $\nu_4(F_2) + \nu_L(F_2)$, $2\nu_4(F_2)$ and $\nu_2(E) + \nu_4(F_2)$) in the polariton curves. At the same time there is no distinct gap near the $\nu_2(E) + \nu_L(F_2)$ band.

So we have observed in NH_4Cl crystals a gap near the polar bound state in the polariton Raman spectrum and in the polariton dispersion curve in accordance with theory [3, 4]. Besides, we have found that gaps might also occur for two-particle bands, but such an effect did not take place for all the two-particle bands. It should be noted that the nature of gaps for polar bound states and for two-particle bands is quite different. In the first case we have a strong reflection of infrared light in the region of the TO-LO gap; the second case corresponds to intense damping processes for photons with energies in the two-particle band.

We are grateful to Dr. V. M. Agranovich for discussing the results of this work and for fruitful remarks, and also to J. I. Galinkovskii and A. I. Pisanskii for growing perfect NH_4Cl monocrystals.

REFERENCES

1. J. Ruvalds, A. Zawadowski, Phys. Rev. B2, 1172 (1970).

2. J. Ruvalds, A. Zawadowski, Light Scat. Sp. in Sol., p. 29, Paris, 1972.
3. V. M. Agranovich, I. I. Lalov, PTT (Sov. Phys.), 13, 1032 (1971).
4. V. M. Agranovich, I. I. Lalov, JETP (Sov. Phys.) 61, 656 (1971).
5. J. F. Scott, Phys. Rev. Lett., 21, 907 (1968).
6. J. F. Scott, Phys. Rev. Lett., 24, 1107 (1970).
7. M. H. Cohan and J. Ruvalds, Phys. Rev. Letters, 23, 1378 (1969).
8. G. G. Matin, V. S. Gorelik, L. A. Culevskii, J. N. Polivanov, M. M. Sushchinskii, JETP (Sov. Phys.), 68, 1757 (1975).
9. G. G. Matin, V. S. Gorelik, M. M. Sushchinskii, PTT (Sov. Phys.) 16, 2956 (1974).
10. G. G. Matin, V. S. Gorelik, M. M. Sushchinskii, A. A. Halezov, Grat. soob. po phys. (Sov. Phys.) 10, 8 (1974).

LIGHT ABSORPTION AND SCATTERING BY NONLINEAR LOCAL AND QUASI-LOCAL OSCILLATION

M. I. Dykman

Institute of Semiconductors
Ukrainian Academy of Sciences
Kiev, USSR

and

M. A. Krivograz

Institute of Metal Physics
Ukrainian Academy of Sciences
Kiev, USSR

The influence of nonlinearity of particular oscillators interacting with a medium (e.g., local or quasi-local vibrations) on their time correlation functions and spectral distributions is investigated. The modulational broadening which is not connected with the finite lifetime of the oscillator and the fine structure of the spectral distribution are considered.

The calculation of the line shape of light absorption or scattering by a particular oscillator κ interacting with a medium (e.g., a local or quasi-local vibration near a defect in a crystal) to a good approximation is reduced to the calculation of the spectral representation $Q_\kappa(\omega)$ of the time correlation function

$$Q_\kappa(t) = \langle q_\kappa(t) q_\kappa(0) \rangle \tag{1}$$

of the oscillator κ under consideration with normal coordinate q_κ. The problem of determination of $Q_\kappa(t)$ arises also in the theory of lasers [1-3] and plasma. It is connected with the general statistical-mechanical problem of investigating the process of approaching the equilibrium state in a subsystem.

The relaxation of a linear oscillator was considered in detail in [4-6]. In this case $Q_\kappa(\omega)$ has the well-known Lorentzian shape with the half-width equal to the reciprocal lifetime of the oscillation κ. However real local oscillations are to some degree nonlinear. It is essential to note that even small nonlinearity produces qualitative changes in the character of these oscillations and can significantly affect the shape of their spectral distribution.

In terms of classical theory these effects are due to the dependence of the nonlinear oscillator frequency on the amplitude and their continuous and random change in time (as a consequence of interaction with a medium). The time dependence of the frequency leads to a specific broadening of $Q_\kappa(\omega)$. This broadening is of a modulational nature and is not caused directly by the relaxation of the local oscillation, that is, by frictional forces.

In terms of the quantum theory, the nonlinearity leads to nonequal spacing of the energy levels of the oscillator. This is evident from Fig. 1, where ω_κ is the harmonic frequency, and V_κ is a parameter of nonlinearity. The transitions between adjacent levels have different frequencies. If $|V_\kappa| \gg I_\kappa$, where I_κ is the characteristic width of the energy level due to relaxation, then the lines corresponding to different transitions do not overlap and the spectral distribution has fine structure. If $|V_\kappa| < I_\kappa$, there appears a single distribution of essentially non-Lorentzian shape.

The calculation of $Q_\kappa(\omega)$ is rather complicated because a nonlinear oscillator has an infinite number of almost equidistant energy and the corresponding quantum states interfere. That is why the use of the Green's function method provides the opportunity to obtain $Q_\kappa(\omega)$ only in the extreme cases $|V_\kappa| \gg I_\kappa$ and $|V_\kappa| \ll I_\kappa$ [7, 8]. The development of new asymptotic classical [9] and quantum [10, 11] methods allowed us to determine $Q_\kappa(t)$ at arbitrary V_κ / I_κ (but $|V_\kappa|, I_\kappa \ll \omega_\kappa$). It was found, somewhat unexpectedly, that $Q_\kappa(t)$ had a simple form (an elementary function). At the same time a corresponding expression for $Q_\kappa(\omega)$ was rich enough to lead to various spectra of essentially non-Lorentzian shape, including those with fine structure.

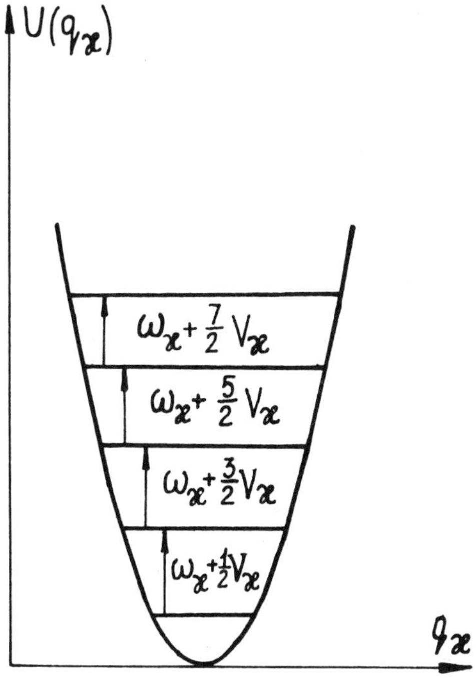

Fig. 1. Schematic of the nonlinear oscillator energy levels.

In this paper we give a short review of our results on the theory of nonlinear oscillators interacting with a medium obtained in [9-14].

The Hamiltonian of a single nonlinear oscillator, interacting with a medium, is of the form

$$H = \tfrac{1}{2}(p_\kappa^2 + \omega_\kappa^2 q_\kappa^2) + \tfrac{1}{4}\gamma_\kappa q_\kappa^4 + \tfrac{1}{2}\sum_k (p_k^2 + \omega_k^2 q_k^2) + \sum_k \epsilon_{\kappa k} q_\kappa q_k$$

$$+ \sum_{kk'} \epsilon_{\kappa k k'} q_\kappa q_k q_{k'}$$

Here q_κ and p_κ, q_k and p_k are the normal coordinates and momenta of the particular oscillation κ and of the medium vibration k; the frequencies ω_k belong to the continuous spectrum; γ_κ is the nonlinearity parameter ($V_\kappa = .75\ \hbar\gamma_\kappa / \omega_\kappa^2$); $\epsilon_{\kappa k}$ and $\epsilon_{\kappa k'}$ determine the interaction with the medium linear in the coordinate of the singled-out oscillator.

In classical considerations one can exclude continuous spectrum oscillations and thus obtain a nonlinear integro-differential equation for $q_\kappa(t)$ [see Eq. (5) in [9]]. This equation is essentially stochastic since the initial amplitudes and phases of the continuous spectrum oscillations are random quantities. Therefore, the application of the usual asymptotic methods of nonlinear oscillation theory is not enough for its solution. One must also make a rather complex, non-trivial averaging. A special stochastic asymptotic method [9] allows one to reduce the expression for $Q_\kappa(t)$ at large t to the explicit form:

$$Q_\kappa(t) = 2\mathrm{Re}[\exp(-i\tilde{\omega})_\kappa |t| \overline{Q}_\kappa(|t|)];$$

$$\overline{Q}_\kappa(t) = \frac{kT}{2\omega_\kappa^2} \exp(I_\kappa t)\left[\frac{I_\kappa(1+2i\alpha)}{a} \mathrm{sh}(at) + \mathrm{ch}(at)\right]^{-2} \tag{3}$$

$$t \gg \omega_\kappa^{-1}; \quad a^2 = I_\kappa(1+4i\alpha); \quad \alpha = \frac{3kT}{8\omega_\kappa^3} \frac{\gamma_\kappa}{I_\kappa}.$$

Here

$$I_\kappa = \frac{\pi}{4} \sum_k \epsilon_{\kappa k}^2 \delta(\omega_\kappa - \omega_k) + I_{\kappa\ell}$$

(essentially $I_{\kappa\ell} \sim T \sum_{kk'} \epsilon_{\kappa kk'}^2 \delta(\omega_\kappa - \omega_k \pm \omega_{k'})$ for local vibrations) defines the broadening due to interaction with the medium and characterizes the inverse lifetime of the oscillatory states; the parameter $\alpha \sim \gamma T / I_\kappa$ describes the ratio between the nonlinearity and the relaxation broadening.

The light absorption or scattering line shape $Q_\kappa(\omega)$ is determined, according to (3), by a single parameter α and may be easily obtained by numerical Fourier-transformation at arbitrary α (Fig. 2). At $\alpha = 0$, $Q_\kappa(\omega)$ is a Lorentzian curve with the half width $2 I_\kappa$. Such a distribution is typical for a linear oscillator, interacting linearly with a medium, when relaxation processes are simply reduced to an effective friction proportional to the velocity. However, at $\alpha \neq 0$ the picture becomes more complex

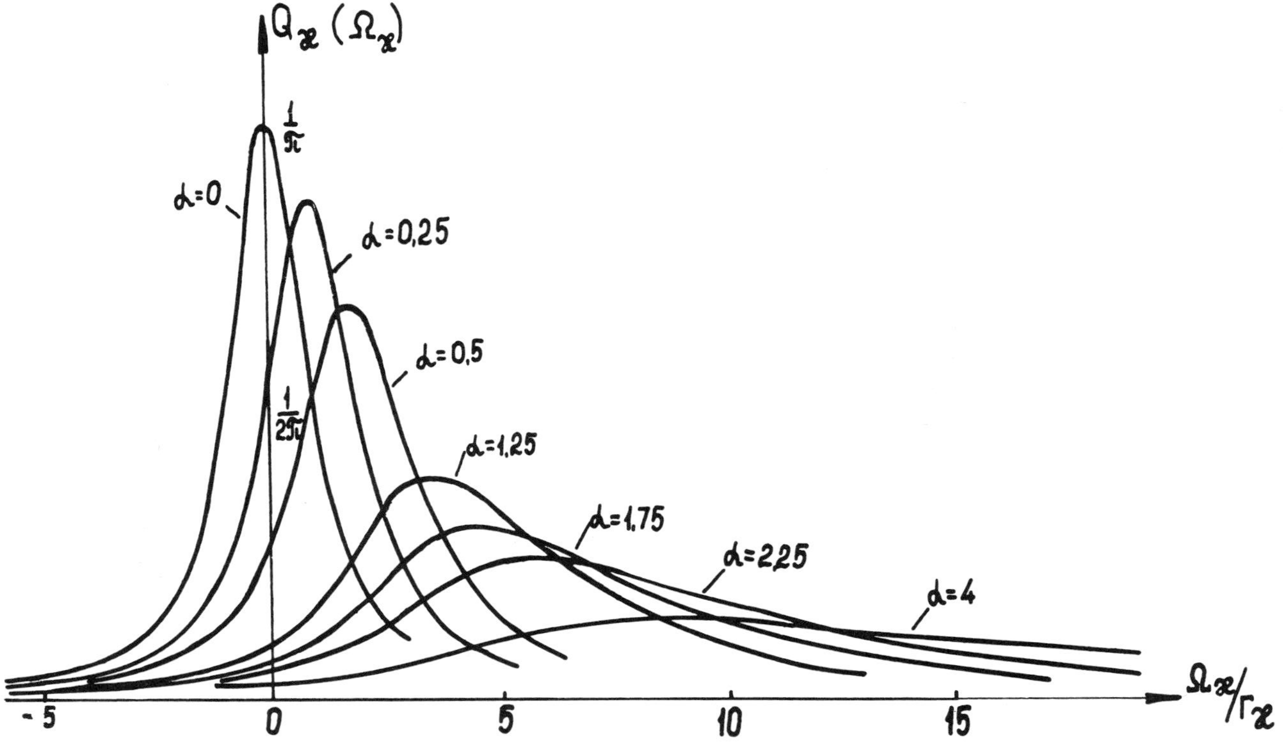

Fig. 2. Spectral distributions $Q_\kappa(\omega)$ for different values of the dimensionless parameter α in the classical limit; $\Omega_\kappa = \omega - \omega_\kappa$, and the intensity is normalized to unity.

and the distribution becomes non-Lorentzian and asymmetric. As α grows, the asymmetry increases, the distribution width becomes larger and its maximum is shifted. At $|\alpha| \gg 1$ the width is determined only by the nonlinearity parameter γ_κ and, therefore, has a purely modulational nature. The detailed analysis of the temperature dependence of the form and width of the spectral distribution for quasi-local and local vibrations is presented in [9].

If the condition $kT \gg \hbar\omega_\kappa$ does not hold, then quantum considerations are needed. It is necessary also for the investigation of the spectral fine structure. The latter is connected with the discreteness of the energy levels, that is, with a pure quantum effect.

In order to calculate the time correlation function of impurity oscillations in quantum theory, a special asymptotic method was developed [10].

This method enabled the following expression for $Q_\kappa(t)$ to be obtained (cf. Eq. (4) in [10]):

$$Q_\kappa(t) = \exp\left[-i(\omega_\kappa - \tfrac{1}{2}V_\kappa)t + \frac{\hbar\omega_\kappa}{kT}\right](\bar{n}_\kappa + 1)^{-1}$$

$$\times \sum_{m=1}^{\infty} mF(m,t)\exp\left(-\frac{\hbar\omega_\kappa m}{kT} - iV_\kappa mt\right), \qquad t \gg \omega_\kappa^{-1},$$

where $\bar{n}_\kappa$ is the Planck occupation number.

The Fourier transform of $F(m, m-1;t) = F(mt)\exp\left(-\frac{\hbar\omega_\kappa m}{kT}\right)$ is connected with a light absorption line, corresponding to the transition between the (m-1)-th and m-th oscillator levels. According to (4), $Q_\kappa(\omega)$ may be interpreted as the sum of partial spectra. As all energy levels are almost equidistant, the lines [F(m, t)] with different m interfere and for F(m, m-1;t) in [10, 11] an equation was obtained (cf. Eq. (6) in [11]) which is an analog of the quantum kinetic equation

$$\frac{\partial F(n,m;t)}{\partial t} = \sum_{j=-1}^{1} D_j(n,m)\exp\left[-iV_\kappa j(n-m)t\right]F(n+j, m+j;t). \qquad (5)$$

D_j depend on the damping I_κ and frequency shift of the singled-out oscillator due to interaction with the vibrations of the continuous spectrum. It is possible to find a generating function for F and to show that it satisfies an equation in partial derivatives. It is essential that the exact solution of this equation may be obtained for our problem. It was found that in quantum theory the time correlation function may be formally described by the same formula as in classical theory. The only difference is that in the quantum case

$$\alpha = \frac{3\gamma_\kappa \hbar}{16\omega_\kappa^2 I_\kappa}(2\bar{n}_\kappa+1); \; a^2 = I_\kappa^2 [1+4i\alpha - \frac{4\alpha^2}{(2\bar{n}_\kappa+1)^2}]. \qquad (6)$$

Detailed analysis of $Q_\kappa(\omega)$ as a function of temperature and the parameters of the interaction is given in [10] on the basis of (3), (6). We should note that the most interesting quantum effect, namely the fine structure of $Q_\kappa(\omega)$, appears at $|V_\kappa| >> I_\kappa(2\bar{n}_\kappa+1)$ and exists within a limited temperature range. The fine structure may be due also to nonlinear interaction between the singled-out oscillator considered and another one [10].

The classical [9] and quantum [10] asymptotic methods allow also for consideration of the shape of $Q_\kappa(\omega)$ near combined frequencies of the singled-out oscillations [12, 13]. The effects of nonlinearity there are even more pronounced.

Some of the above-mentioned effects caused by nonlinearity have been already observed in experiments on infrared absorption by local and quasi-local oscillations. For example, the increase in temperature leads to strong asymmetry in the absorption lines in the systems NaCl:Cu^+ [15, 16] and KI:Ag^+ [17] (quasi-local oscillation) which is probably of a modulational nature. The width of the lines also grows strongly (by an order of magnitude for the former system at $kT \sim \hbar\omega_\kappa$). Fine structure was found in the spectrum of quasilocal oscillation in MnF_2:Eu^{2+} [18]. Its character and temperature dependence agree with theoretical results [10]. In [19] modulational broadening of the high-frequency local vibration of the impurity complex SO_4^{2-} and Ca^{2+} in a KCl crystal due to interaction with low-lying quasi-local oscillation was observed.

Even strong modulational broadening does not lead to a change of the infrared absorption peak intensity with temperature

[14]. With the use of the unambiguous procedure of separation of the peak in [14] an expression (Eq. (8) in [14]) for the peak intensity change containing only relaxation parameters was obtained. The strong temperature broadening of the absorption line under a slight change of its intensity, observed in [16], is typical for modulational broadening and may be used to distinguish it.

New effects arise in the case of nonlinear friction when the energy of interaction between a particular singled-out oscillation and a continuous spectrum of oscillations contains not only the above terms

$$q_\kappa(\sum_k \epsilon_{\kappa k} q_k + \sum_{kk'} \epsilon_{\kappa kk'} q_k q_{k'})$$

but also the terms

$$q_\kappa^2 \sum_k \epsilon_{\kappa kk'} q_k$$

proportional to q_κ^2, corresponding to decay with participation of two quanta of the oscillator κ [11]. Here, in terms of classical theory, the friction coefficient depends on amplitude, i.e., it changes with time even for a linear singled-out oscillator. Therefore, $Q_\kappa(t)$ decays nonexponentially and its spectral representation is non-Lorentzian even for a linear oscillator.

In [11] the asymmetric complex $Q_\kappa(\omega)$ is investigated for an arbitrary ratio of the parameters of linear and nonlinear friction and nonequal spacing, on the basis of generalizations of Eq. (4) and (5). In the absence of nonequal spacing nonlinear friction leads to narrowing of $Q_\kappa(\omega)$ as compared to a Lorentzian curve with the same wings and integral intensity.

REFERENCES

1. L. Mandel, E. Wolf, Rev. Mod. Phys. 37, 231 (1965).
2. M. O. Scully, W. E. Lamb, Jr., Phys. Rev. 159, 208 (1967).
3. M. Lax, Fluctuation and Coherence Phenomena in Classical and Quantum Physics, NY, 1968.
4. N. N. Bogoliubov, O nekotorikh statisticheskikh metodakh v matematicheskoi fizike. Izd. Akad. Nauk Ukr. SSR, Kiev, 1945.

5. J. Schwinger, J. Math. Phys. 2, 407 (1961).
6. B. Ya. Zeldovich, A. M. Perelomov, and V. S. Popov, Zh. eksper. teor. fiz. 55, 589 (1968); 57, 196 (1969).
7. M. A. Ivanov, L. B. Kvashnina, M. A. Krivoglaz, Fiz. tverd. tela 7, 2047 (1965); Soviet Phys.-Solid State 7, 1652 (1966).
8. M. A. Krivoglaz and I. P. Pinkevich, Ukr. fiz. zh. 15, 2039 (1970).
9. M. I. Dykman, M. A. Krivoglaz, Phys. Stat. Sol. (b), 48, 497 (1971).
10. M. I. Dykman, M. A. Krivoglaz, Zh. eksper. teor. fiz. 64, 993 (1973).
11. M. I. Dykman, M. A. Krivoglaz, Phys. Stat. Sol. (b), 68, 111 (1975).
12. M. I. Dykman, M. A. Krivoglaz, Ukr. fiz. zh. 17, 1971 (1972).
13. M. I. Dykman, Fiz. Tverd. Tela, 15, 1075 (1973).
14. M. I. Dykman, M. A. Krivoglaz, Ukr. fiz. zh. 19, 125 (1974).
15. R. Weber, P. Nette, Phys. Lett. 20, 493 (1966).
16. R. Weber, F. Siebert, Zs. Phys. 213, 273 (1968).
17. S. Takeno, A. J. Sievers, Phys. Rev. Lett. 15, 1020 (1965).
18. R. W. Alexander, Jr., A. E. Hughes, A. J. Sievers, Phys. Rev. B1, 1563 (1970).
19. M. P. Lisitsa, G. G. Tarasov, L. I. Bereginskij, Fiz. tverd. tela, 16, 1302 (1974).

INTERVALLEY PHONON RAMAN SCATTERING IN MANY-VALLEY SEMICONDUCTORS

P. J. Lin-Chung and K. L. Ngai

Naval Research Laboratory

Washington, D.C. 20375 USA

Theoretical calculations of the Raman cross-sections for resonant intervalley phonon scattering near the indirect gap and the higher direct gap energies in many-valley semiconductors are reported, and the possibility of observing these effects experimentally is discussed. Band structure calculations are employed to compute the intervalley deformation potentials and the cross-sections of light scattering processes for typical semiconductors.

INTRODUCTION

Recently, a considerable number of papers[1-4] have appeared in which the first and second order resonant Raman scattering in many-valley semiconductors has been measured over a wide energy range covering the direct band gap, the indirect gap,as well as higher direct gaps. Several features of the Raman spectrum obtained from those experiments seem to correspond to scattering from intervalley-phonons. We report here theoretical considerations of several possibly observable resonant intervalley phonon scattering phenomena. The Raman amplitudes and their dependences on the incident light frequency near the higher direct gap and the indirect gap energies, respectively, for each possible process are derived in the third section. In order to estimate the Raman cross-sections for typical semiconductors, we have employed band structure calculations to compute the intervalley deformation potentials which are given in the next section.

INTERVALLEY DEFORMATION POTENTIALS

Before we proceed to determine the resonance intervalley phonon scattering amplitude, it is essential to know the intervalley deformation potential constants for the many-valley semiconductors. In the following, we shall describe briefly the procedure used to compute the intervalley deformation potentials in Si and Ge.

The intervalley deformation potential, d, associated with a scattering process involving a phonon with wave vector $\underset{\sim}{Q}$ and polarization λ is defined by

$$d \equiv |\langle \Psi_{ks}, n, Q, \lambda \,|\, H_{ep} \,|\, \Psi_{kt}, n+1, Q, \lambda \rangle |(a_o / |\vec{A}|) \tag{1}$$

where a_o is the lattice constant, and Ψ_{ks}, Ψ_{kt} are the electron wave functions belonging to different valleys located at k_s and k_t respectively; n and n+1 denote the phonon state with n and n+1 phonons respectively, and $|\vec{A}|$ is the displacement amplitude of an atom in the crystal, when a phonon with momentum Q is present, (hereafter, the phonon polarization index λ will be omitted for convenience). In a homopolar crystal with a center of inversion, $|\vec{A}|$ is independent of the equilibrium position of the atom in a unit cell.

To first order in the expansion of the potential in terms of the ion displacement, H_{ep} in eq. (1) can be expressed as

$$H_{ep} = - \sum_{\alpha, q} i(\vec{q}\cdot\vec{A}_\alpha) \exp[-i\vec{q}\cdot(\vec{r}^{\,o}_\alpha - \vec{r})]\, V_{q\alpha} [a(Q)\,\delta_{\vec{Q}-\vec{q},\vec{K}} + a^+(Q)\,\delta_{-\vec{Q}-\vec{q},\vec{K}}] \tag{2}$$

where $\vec{r}^{\,o}_\alpha$ is the equilibrium position vector of the αth atom in a unit cell, $V_{q\alpha}$ is the atomic potential form factor of the αth atom, K are reciprocal lattice vectors, and a(Q) and $a^+(Q)$ are the annihilation and creation operators of phonon Q, respectively.

If the symmetry operation P_R brings $\vec{k}_s$ into $\vec{k}_t$, we have

$$\vec{k}_t = \beta \vec{k}_s \tag{3}$$

where β is the rotational part of the symmetry element $R = (\beta, \vec{t})$ of a non-symorphic group; $\vec{t}$ is the translation operation. The wave functions ψ_{kt} and ψ_{ks} are then related by

$$\Psi_{kt} = (P_\beta \Psi_{ks}) \exp[-i(\beta \vec{k}_s)\cdot \vec{t}] \tag{4}$$

Because of the momentum conservation condition, $\vec{k}_t - \vec{k}_s = \vec{Q} + \vec{K}$, a phonon of momentum Q participates in the intervalley scattering process.

To a very good approximation, the matrix elements $\langle \psi_{ks} | H_{ep} | \psi_{kt} \rangle$ can be determined by replacing the true wave functions, ψ_{ks}, ψ_{kt} and the true atomic form factor, $V_{q\alpha}$ with pseudo-wave functions ψ^P_{ks}, ψ^P_{kt} and pseudopotential form factor $V^P_{q\alpha}$. We may write

$$\Psi^P_{ks} = (N\Omega)^{-\frac{1}{2}} \sum_j A_j \exp[i(\vec{k}_j + \vec{k})\cdot \vec{r}] \tag{5}$$

$$\begin{aligned} \Psi^P_{kt} &= (P_\beta \Psi_{ks}) \exp[-i(\beta \vec{k}_s)\cdot \vec{t}] \\ &\simeq (N\Omega)^{-\frac{1}{2}} \sum_n A_n \exp[i\beta(\vec{k}_n + \vec{k})\cdot \vec{r}] \exp[-i\beta(\vec{k}_n + \vec{k})\cdot t] \end{aligned} \tag{6}$$

Thus, the general expression for the intervalley deformation potential becomes

$$\begin{aligned} d = -i \sum_{j,n} a_o A_j^* A_n \exp[-i\beta(\vec{k} + \vec{k}_n)\cdot \vec{t}] \sum_\alpha (\vec{Q} + \vec{K})\cdot (\vec{A}_\alpha / |\vec{A}|) \\ \times V^P_{\vec{Q}+\vec{K},\alpha} \exp[-i(\vec{K}+\vec{Q})\cdot r^o_\alpha] \delta(\vec{K}, \vec{k}_j + \vec{k} - \beta(\vec{k}_n + \vec{k}) - \vec{Q}) \end{aligned} \tag{7}$$

Selection rules for electron intervalley scattering have been determined previously for Si and Ge.[6] Equation (7) can then be evaluated by employing band structure calculation results for A_j, A_n and $V^P_{\vec{Q}+\vec{K},\alpha}$ [7] and phonon dispersion calculation results for $\vec{A}_\alpha$.[8] Our results obtained in the manner described above are listed in Table 1. As shown in Table 1, electron scattering between the L-valleys in Si and Ge (i.e. $L_1 \rightarrow L_{1t}$) are rather weak compared with hole-scattering ($L'_3 \rightarrow L'_{3t}$). On the other hand, electron scattering between the Δ-valleys in Si are of two types; g-processes and f-processes represent scattering between valleys with opposite wave vectors and between adjacent valleys, respectively. The deformation potentials for both g and f processes are very large.

Table 1. Intervalley Deformation Potentials Obtained by Pseudopotential Band Structure Calculations

	Phonons	d (eV)
Ge		
$L_1 \rightarrow L_{1t}$	LO and LA (X_1)	0.77
$L'_3 \rightarrow L'_{3t}$	LO and LA (X_1)	14.79
Si		
$\Delta_1 \rightarrow \Delta_1$	g: LO(Δ'_2)	23.36
	f: TO(Σ_1)	17.38
	f: LA(Σ_1)	30.98
$L_1 \rightarrow L_{1t}$	LO and LA(X_1)	

INTERVALLEY-PHONON RESONANCE SCATTERING

We consider here several possibilities for observing first order and second order intervalley phonon scattering in many-valley semiconductors. We will discuss separately the case for which the incident photon energy $\hbar\omega_i$ is near an indirect band gap E_g and that for which it is near a higher direct gap E_h.

(I) $\hbar\omega_i$ near an indirect band gap

In a typical indirect gap semiconductor, e.g., Si, the forbidden band gap is an indirect gap which occurs between Γ'_{25} and one of the Δ_1 valleys. Other indirect gaps, $L_3'-X_1$, $L_3'-\Gamma_{15}$ etc., have higher energies. Ge has a similar band structure except that the smallest indirect gap is between Γ_{25}' and L_1. When the incident photon energy $\hbar\omega_i$ is near any of the indirect band gaps, three prominent intervalley phonon scattering processes are in resonance. They are indicated by cases (B), (C) and (E) of the second column of Fig. 1. A phonon-assisted indirect gap transition of an electron (hole) takes the electron (hole) to any of the valleys associated with the indirect gap which is in resonance with the incident photon. The electron (hole) may then emit a single phonon Γ, as in case (B), or it may be scattered by an intervalley photon to any of the other equivalent valleys as in case (C) in Fig. 1. Subsequently, the electron (hole) de-excited in a phonon assisted process with the emission of the scattered light. The process in essence corresponds to Raman scattering from an intervalley phonon. The Stokes and anti-Stokes components in case (C) will occur at frequencies ($\pm\Omega_Q$) different from those of the first order zone center phonons and may be observable. In both cases (B) and (C), a virtual phonon is involved. The phonon which assists the indirect gap transition of an electron (hole) is emitted and then re-absorbed, whereas in case (E), two such phonons are emitted (or are absorbed) which results in an overtone two-phonon Raman shift.

The measured intensity of the scattered light is proportional to the square of the Raman tensor $R_{\rho\sigma}$

$$I \propto \Big[\sum_{\rho,\sigma} \vec{\epsilon}_i^{\,\rho} R_{\rho\sigma} \vec{\epsilon}_s^{\,\sigma}\Big]^2 \tag{8}$$

where $\vec{\epsilon}_i$ and $\vec{\epsilon}_s$ are unit vectors in the direction of the polarization of the incident and scattered light respectively.

The general form of the Raman tensor in a given process is

$$R_{\rho\sigma} = \sum_{a,b,\ldots} \frac{\langle f|\vec{\epsilon}_i^{\,\rho}\cdot\vec{p}|e\rangle\langle e|H_{ep}|d\rangle ---- \langle b|H_{ep}|a\rangle\langle a|\vec{\epsilon}_s^{\,\sigma}\cdot\vec{p}|i\rangle}{(E_a-\hbar\omega_i)(E_b-\hbar\omega_i) ----- (E_e-\hbar\omega_i)} \tag{9}$$

where E_a, E_b - - - are the energies of the intermediate states a, b, - - - in a given process.

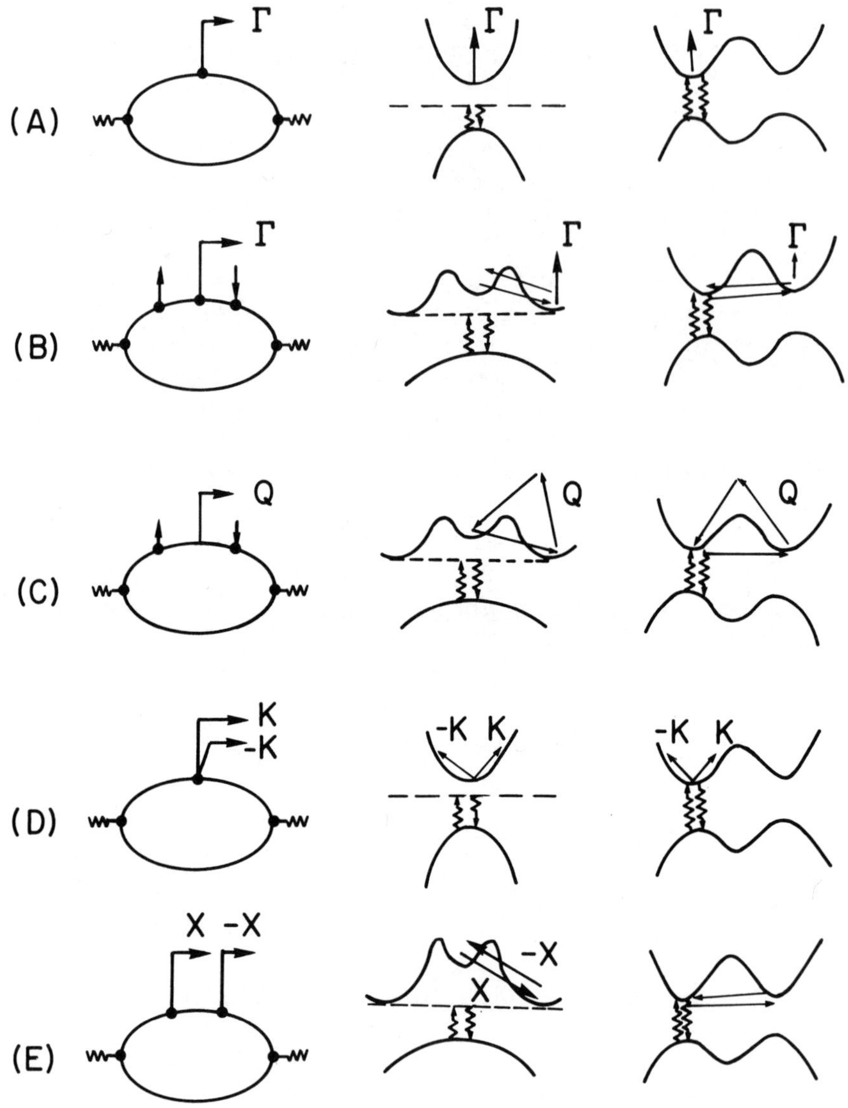

Fig. 1. One and two phonon scattering processes (1st column); $\hbar\omega_i$ near indirect gap (2nd column); $\hbar\omega_i$ near higher direct gap (3rd column).

Case (B)

From eq. (9), the Raman amplitude for case (B) near the indirect gap Eg has the following form.

$$R(\omega_i) = N_Q (H_{eR})^2 (H^{ep}_{cc'}) (H^{ep}_{c'c'}) (H^{ep-}_{c/c}) V^2 (2\pi^3 \hbar^6)^{-1}$$

$$\times (m_c^* m_v^*)^{3/2} E_m F_1 / (\hbar \Omega_o \overline{E}^2) \tag{10}$$

where N_Q is the number of valleys associated with the indirect transitions, H_{eR} is the electron-photon interaction matrix element, $H^{ep}_{cc'}$ is the electron-one-phonon interaction matrix element between the states c (i.e., Γ_{15} in Si) and c' (Δ_1 in Si). (A minus sign placed next to the superscript ep in H^{ep} indicates a phonon annihilation, rather than creation), V is the volume of the crystal, and m_c^*, m_v^* are the effective masses at the conduction band minimum and valence band maximum, respectively. $\hbar\Omega_o (\hbar\Omega_Q)$ is the energy of the zone center phonon (phonon Q). The rest of the quantities are defined as follows.

$$F_1 = I_5(X_2) - I_5(X_3)$$

$$E_m = \frac{\hbar^2}{9} \left(\frac{2\pi}{a_o}\right)^2 \left(\frac{1}{2m_c^*} + \frac{1}{2m_v^*}\right)$$

$$X_2 = (E_g + \hbar\Omega_Q - \hbar\omega_i) / E_m$$

$$X_3 = (E_g + \hbar\Omega_Q + \hbar\Omega_o - \hbar\omega_i) / E_m \tag{11}$$

$$I_5(X) \equiv \tfrac{1}{4} - X/2 + X^2 I_1(X)$$

$$I_1(X) \equiv \begin{cases} \tfrac{1}{2} (\ln|1+X| - \ln|X|), & X > 0 \\ \tfrac{1}{2} (\ln|1+X| - \ln|X|) - i\pi/2, & X < 0 \\ \infty & X = 0 \end{cases}$$

$$\overline{E} = E_G + E_m/2 \qquad E_G = \text{direct gap energy}$$

Equation (10) is for the electron scattering process. A similar expression describes the hole scattering process except that the intermediate state energies and matrix elements must be changed. In fact, the electron scattering for case (B) in Si is forbidden by a selection rule. Only hole scattering is allowed. In the latter process, the intermediate states involve an electron at the conduction band minimum, Δ_1, and a hole at Γ'_{25}. A single Γ-phonon is emitted by the hole during the process. Using the value of $H^{ep}_{cc'}$, deduced from indirect optical absorption[9] we have calculated the ratio of the Raman amplitudes of this process and the non-resonant zone-center phonon first order process [case (A) in Fig. 1] at $\hbar\omega_i = E_g$. The ratio R(B)/R(A) is 7.4×10^{-4}. The observation of resonant cancellation in Si has been reported by Ralston et al.[3] A decrease of a factor of five in the scattering cross-section of a single Γ phonon as $\hbar\omega_i$ approaches E_g was attributed to the cancellation of case (A) with case (B) by these authors. Our calculated value, however, is too small to produce such cancellation in Si, and therefore contradicts this interpretation. Although near the $\Gamma'_{25} - \Delta_1$ gap of Si, the process in case (B) gives an insignificant contribution, it becomes important near the $L'_3 - X_1$ gap because of the highly anisotropic energy surface at L'_3, and there it gives rise to a stronger resonance. Renucci et al[4] have observed a shoulder in the first order Raman spectrum near 2.9 eV. This is most probably associated with the resonance of case (B) near the $L'_3 - X_1$ gap.

Case (C)

The Raman amplitude for case (C), at frequency shift $\Omega_{Q'}$, and at $\hbar\omega_i \sim E_g$ is

$$R(\omega_i) = N_Q(N_Q - 1)(H_{eR})^2(H^{ep}_{cc'})(H^{ep}_{c'c''})(H^{ep^-}_{c''c})V^2(2\pi^3\hbar^6)^{-1}$$

$$\times (m_c^* m_v^*)^{3/2} E_m F_1/(\hbar\Omega_{Q'}, \overline{E}^2) \qquad (12)$$

where

$$X_3 = (E_g + \hbar\Omega_Q + \hbar\Omega_{Q'} - \hbar\omega_i)\Big/ E_m ,$$

and the rest of the quantities have the same definitions as in Eq. (11).

Using the deformation potentials in Table 1 and Eq. (12), we have determined the ratio R(C)/R(A) for the Raman amplitudes in case (C) and case (A). The ratios are 9.02×10^{-4}, 2.69×10^{-3} and 4.79×10^{-3} for LO(Δ_2') TO (Σ_1) and LA (Σ_1) phonons, respectively. These values are again too small to be detected experimentally. However, in a similar manner to the way that case (B) becomes important near the $L_3' - X_1$ gap we expect the appearance of an L intervalley phonon scattering [case (C)] to become important near 2.9 eV.

Case (E)

Perhaps the most often observed intervalley Raman scattering is the (E) process. The Raman amplitude derived from Eq. (9) for this process near $\hbar\omega_i \sim E_g$ is

$$R(\omega_i) = N_Q (H_{eR})^2 (H^{ep}_{cc'})(H^{ep}_{c'c}) V (2\pi^2)^{-1} k_m^3 F_2 / (E_m \overline{E}^2) \tag{13}$$

where

$$k_m = \frac{1}{3}\left(\frac{2\pi}{a_o}\right)$$

$$E_m = \hbar^2 k_m^2 \left(\frac{1}{2m_c^*} + \frac{1}{2m_v^*}\right) \tag{4}$$

$$F_2 = I_2(X) = \begin{cases} 1 - (\sqrt{-X}/2)\,\ell n\,|(1-\sqrt{-X})/(1+\sqrt{-X})| + i\pi\sqrt{-X}/2, & X < 0 \\ 1 - \sqrt{X}\,\tan^{-1}(1/\sqrt{X}), & X > 0 \end{cases}$$

$$X = (E_g + \hbar\Omega_Q - \hbar\omega_i)/E_m$$

In Si, the enhancement of both 2TO (0.85X) and 2TA (0.85X) phonons has been observed[1] near the $\Gamma_{25}' - \Delta_1$ gap. In addition, near the $L_3' - X_1$ gap in Ge[2] selective enhancement of 2TO (L) phonons has also been reported. We believe from inspection of the data of Ref. 4 that the same enhancement occurs near the $L_3' - \Delta_1$, $L_3' - \Gamma_{15}$ gaps in Si. These enhancements all arise from the (E) process. Again, we use the result in Table 1 and Eq. (13). The ratio of the scattering amplitudes for Si of the process (E) to the non-resonant two-phonon process (D) in Fig. 1, is R(E)/R(D) = 23.9 for 2TO (0.85X) phonons at the $\Gamma_{25}' - \Delta_1$ gap. This theoretical result further confirms the importance of intervalley Raman scattering in many-valley semiconductors.

(II) $\hbar\omega_i$ near a higher direct gap

Another possible resonant intervalley-phonon Raman scattering is that at a photon frequency near a higher direct gap [e.g., E_1 - $(E_1+\Delta)$ or E_2 gap in Si and Ge, etc.]. As in the third column of Fig. 1, the processes of cases (B) and (C) near higher direct gaps involve the emission (absorption) of one intervalley phonon followed by the emission of a zone-center phonon (B) or another intervalley phonon (C); the processes are completed by the absorption (emission) of an intervalley phonon which brings the electron (hole) back to its original valley. Cases (B) and (C) are three-phonon processes, but in effect they result in a single-phonon Raman-shift. The mechanism which contributes to the two-intervalley phonon process in case (E) of the third column in Fig. 1 is the analogue of the iterative two near-zone-center phonon process previously discussed by Renucci et al.[2] The electron (hole) in the intermediate/state is scattered to any of the several equivalent valleys and back to the original valley. Since in these cases, the energies of all the intermediate states are close to the higher direct gap E_h, the Raman amplitude for the processes (B), (C) and (E) in the third column of Fig. 1 will have two more poles than the corresponding cases in the second column near resonance. Thus even more dramatic resonance enhancement for higher direct gaps is expected.

At present, only a few of the higher direct gaps in semiconductors are accessible to resonance Raman work owing to the lack of lasers available in this energy region. To our knowledge, only the E_1 and $E_1+\Delta$ gaps at $\Lambda(\nu,\nu,\nu)$ in Si and Ge have been investigated. We shall restrict our discussion to these gaps.

Case (B)

When the higher direct gap is at a point along a symmetry direction (e.g., E_1 and $E_1+\Delta$ gap in Ge), the electron (hole) may have a cylindrical-like energy surface. The reduced effective mass of the electron and hole then is highly anisotropic.

$$[\mu_{\parallel} >> \mu_{\perp} \text{ where } \frac{1}{\mu} = (\frac{1}{m_e^*} + \frac{1}{m_h^*})].$$

To obtain the Raman amplitude for this process, we perform the summation over intermediate states in Eq. (9) using cylindrical coordinates. We first sum over the electron momentum $k_{\parallel}$, along the symmetry direction in the intermediate state, and then integrate over the $k_{\perp}$. We obtain

$$\begin{aligned} R(\omega_i) = {} & N_Q(N_Q - 1)(H_{eR})^2 (H^{ep}_{cc'})(H^{ep}_{c'c'})(H^{ep-}_{c'c}) V^2 k^2_{\parallel m} (\pi \hbar^6 \Omega_0^2 \alpha)^{-1} \\ & (4m^*_{e\perp} m^*_{h\perp}) \{\ln (X_4/X_1)[I_1(X_2) - I_1(X_3)] + I(X_2, X_1) \\ & - I(X_3, X_1) - I(X_2, X_4) + I(X_3, X_4)\} \end{aligned} \tag{15}$$

where

$$E_m = \frac{1}{9}\left(\frac{2\pi}{a_o}\right)^2 \left(\frac{\hbar^2}{2m^*_{e\perp}} + \frac{\hbar^2}{2m^*_{h\perp}}\right), \quad k_{\parallel m} = \frac{1}{3}\left(\frac{2\pi}{a_o}\right)$$

$$\alpha = \left(1 + \frac{m^*_{e\perp}}{m^*_{h\perp}}\right)$$

$$X_1 = (E_h - \hbar\omega_i)/E_m \tag{16}$$

$$X_2 = (E_h + \hbar\Omega_Q - \hbar\omega_i)/E_m$$

$$X_3 = (E_h + \hbar\Omega_Q + \hbar\Omega_o - \hbar\omega_i)/E_m$$

$$X_4 = (E_h + \hbar\Omega_o - \hbar\omega_i)/E_m$$

$$I(X_i, X_j) = \tfrac{1}{2}\int_o^1 d\beta \ln(X_j + \alpha\beta)/(X_i + \beta)$$

The above equation is also valid for hole scattering. However, in that case $m^*_{e\perp}$ and $m^*_{h\perp}$ in Eq. (15) and Eq. (16) should be interchanged.

It is difficult to separate the contributions from case (B) and case (A) experimentally, since at the higher direct gap, both cases are resonant at the same frequency shift.

Case (C)

The Raman amplitude for case (C) can be obtained by multiplying Eq. (15) by another factor $(N_Q - 1)$ and changing the quantities $(H^{ep}_{c'c'})$ and Ω_o in that equation to $H^{ep}_{c'c''}$ and Ω_Q, respectively. It is possible to separate case (C) from case (A) because of the different frequency shifts. The frequency shifts in case (C) correspond to the energies of any branch of the phonons at the $\Delta(0, 2\nu, 0)$ point.

Case (E)

Although the intervalley deformation potential between the Λ valleys in Ge is smaller than the intravalley deformation potential at Λ (by approximately a factor of five), the factor $N_Q(N_Q - 1)$ enters into the Raman amplitude of the intervalley scattering and makes process (E) as important as the intravalley scattering observed in Ref. 2. The Raman amplitude in case (E) near the higher direct gap E_h, when the electron (hole) has cylindrical-like energy surface, is given by the following expression.

$$R(\omega_i) = N_Q(N_Q - 1)(H_{eR})^2 (H^{ep}_{cc'})(H^{ep}_{c'c}) V (2\pi^3 E^3_m)^{-1} k^2_{rm} k_{zm} F_3 \tag{17}$$

where

$$\begin{aligned} k_{zm} &= k_{rm} = \frac{1}{3}\left(\frac{2\pi}{a_o}\right) \\ E_m &= \hbar^2 k^2_{rm} \left(\frac{1}{2m^*_{e\perp}} + \frac{1}{2m^*_{h\perp}}\right) \\ X_1 &= (E_h - \hbar\omega_i)/E_m \\ X_2 &= (E_h + \hbar\Omega_Q - \hbar\omega_i)/E_m \\ X_3 &= (E_h + 2\hbar\Omega_Q - \hbar\omega_i)/E_m \\ F_3 &= \sum_{i=1}^{3} [I_1(X_i) / \prod_{j\neq i} (X_j - X_i)] \end{aligned} \tag{18}$$

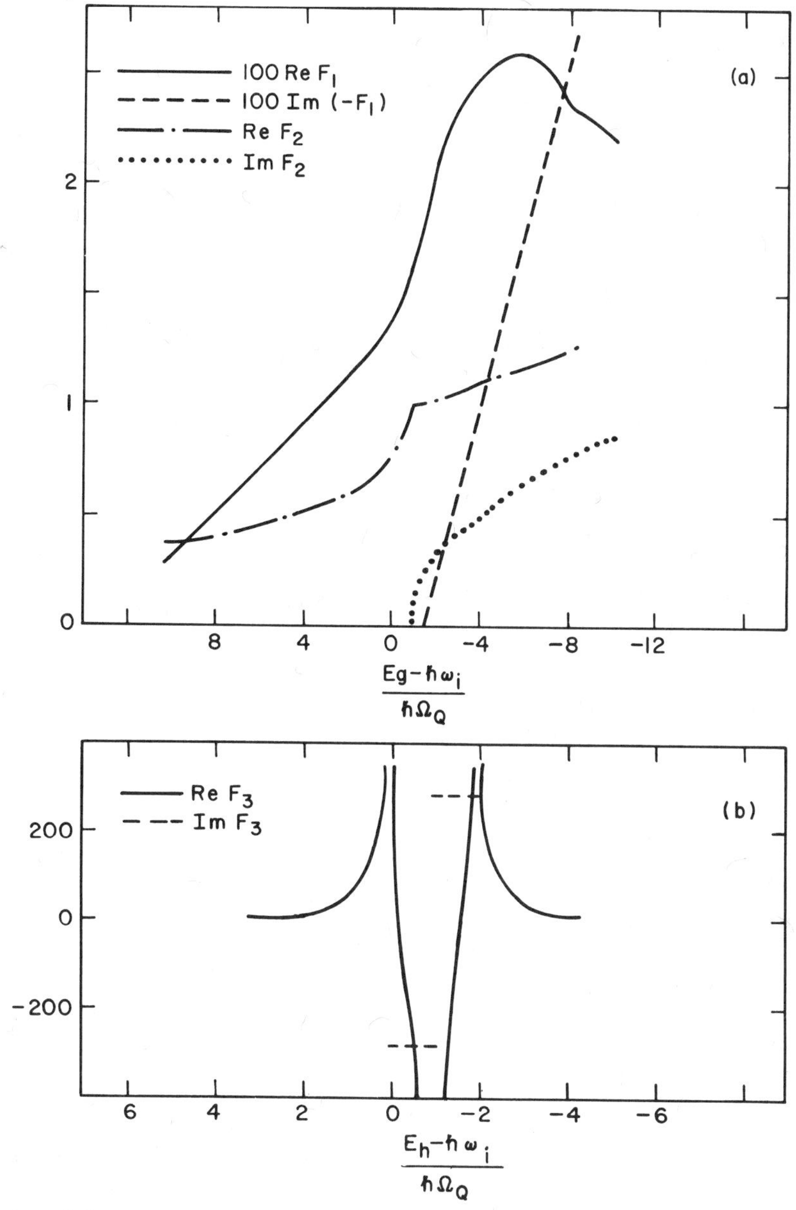

Fig. 2. (a) The quantities F_1 and F_2 which appear in Eqs. (10, (12) and (13) are plotted as functions of $(E_g - \hbar\omega_i)/\hbar\Omega_Q$. Here full curve, dashed curve, dot-dashed curve, and dotted curve represent the quantities 100 $\mathrm{Re}F_1$, 100 Im $(-F_1)$, $\mathrm{Re}F_2$ and $\mathrm{Im}F_2$, respectively. (b) F_3 in Eq. (17) is plotted as a function of $(E_h - \hbar\omega_i)/\hbar\Omega_Q$. Here full curve and dashed curve represent the quantities $\mathrm{Re}F_3$ and $\mathrm{Im}F_3$, respectively.

In Ge and Si only hole-scattering is important in the intervalley (E) process near E_1. The two phonons involved in this process are the 2LO(0, 2ν, 0) and 2LA(0, 2ν, 0) phonons. We note that some enhancement near the 2LO and 2LA frequency shifts have been observed previously.[2] In order to compare the dispersion relation of the Raman amplitude for this case [Eq. (17) with Eqs. 10, (14), and (13), we illustrate in Fig. 2 the two quantities F_1 and F_2 as functions of $(E_g - \hbar\omega_i)/\hbar\Omega_Q$ and F_3 as a function of $(E_h - \hbar\omega_i)/\hbar\Omega_Q$. It is apparent that, near E_h, case (E) (Fig. 2b) has sharper resonance than those cases in Fig. 2a. On the other hand, near E_g, the resonant enhancement extends considerably beyond $\hbar\omega_i = E_g$.

CONCLUSION

We have investigated the possibilities of observing first order and second order intervalley phonon scattering in many-valley semiconductors. The intervalley deformation potentials for Si and Ge were determined by employing pseudopotential band structure calculations. Using these results, we find that the process (E) of Fig. 1 has the largest Raman cross-section for Si and Ge. The dispersion relation derived here for the Raman amplitude $R(\omega_i)$ of each process, holds for all many-valley semiconductors. It would be of great interest to observe the change of Raman cross-section of the intervalley phonons over a wider range of incident photon energies near the indirect band gaps and higher direct gaps in order to confirm the relative importance of these enhancements in various semiconductors.

REFERENCES

1. P. B. Klein, H. Masui, J. J. Song and R. K. Chang, Solid State Commun. 14, 1163 (1974).
2. M. A. Renucci, J. B. Renucci, R. Zeyher and M. Cardona, Phys. Rev. B10, 4309 (1974).
3. J. M. Ralston, R. L. Wadsack and R. K. Chang, Phys. Rev. Lett. 25, 814 (1970).
4. J. B. Renucci, R. N. Tyte and M. Cardona, Phys. Rev. B11, 3885 (1975).
5. L. J. Sham, Proc. Phys. Soc. London 78, 895 (1961).
6. M. Lax and J. J. Hopfield, Phys. Rev. 124, 115 (1961); H. W. Streitwolf, Phys. Stat. Sol. 37, K47 (1970).
7. M. L. Cohen and T. K. Bergstresser, Phys. Rev. 141, 789 (1966).

8. N. K. Pope, Proc. Intern. Conf. on Lattice Dynamics, Pergamon Press, 1963, p.147.
9. P. B. Klein, thesis, Yale University 1975 (unpublished).

Section III

Scattering from Electrons and Electron-Hole Drops

THE INVESTIGATION OF EXCITON CONDENSATION IN GERMANIUM BY THE LIGHT SCATTERING METHOD

V. S. Bagaev, H. V. Zamkovets, N. N. Sybeldin
and V. A. Tsvetkov

P. N. Lebedev Physical Institute
Academy of Sciences
Moscow, USSR

At low temperatures and rather high concentrations of excitons in semiconductors electron-hole drops (EHD) are formed [1, 2] as a result of the interaction between excitons. In EHD the electrons and holes are collectivized and bound by the forces of mutual interaction.

The electron and hole concentrations in EHD are substantially larger than their average in the volume of the sample. Owing to this fact, the EHD index of refraction differs slightly from that of the crystal and, consequently, the drops of condensed phase will scatter electromagnetic radiation. The light scattering by EHD was discovered experimentally by Pokrovsky and Svistunova [3].

Using the Rayleigh-Jeans scattering theory [4] one can show that the intensity of light scattered by drops with radius R into the angle θ is [5]:

$$I_s \sim n_0^2 N R^6 Y^2(R\,\theta) \qquad (1)$$

where n_0 is the concentration of the electron-hole pairs in EHD, N is the concentration of drops in the crystal and $Y(R\,\theta)$ is a standard function. From this formula one can see that the angular distribution of the scattered light intensity is determined only by the size of scattering particles. The energy of the light that is absorbed by the carriers, bound in EHD, is

$$W \sim n_0 N R^3 \qquad (2)$$

As the coefficients of proportionality in (1) and (2) are sufficiently well known, having determined the radius of the drops owing to the angular distribution of the scattered light and absorption, one can find the concentration of the EHD in the crystal and the density of the electron-hole liquid.

In this work the dependence of drop size and concentration in germanium on the temperature, intensity of excitation and rise time of excitation pulses were investigated in order to obtain information on the kinetics of EHD nucleation.

EXPERIMENTAL PROCEDURE

The scheme of the experiment is presented in Fig. 1 [6].

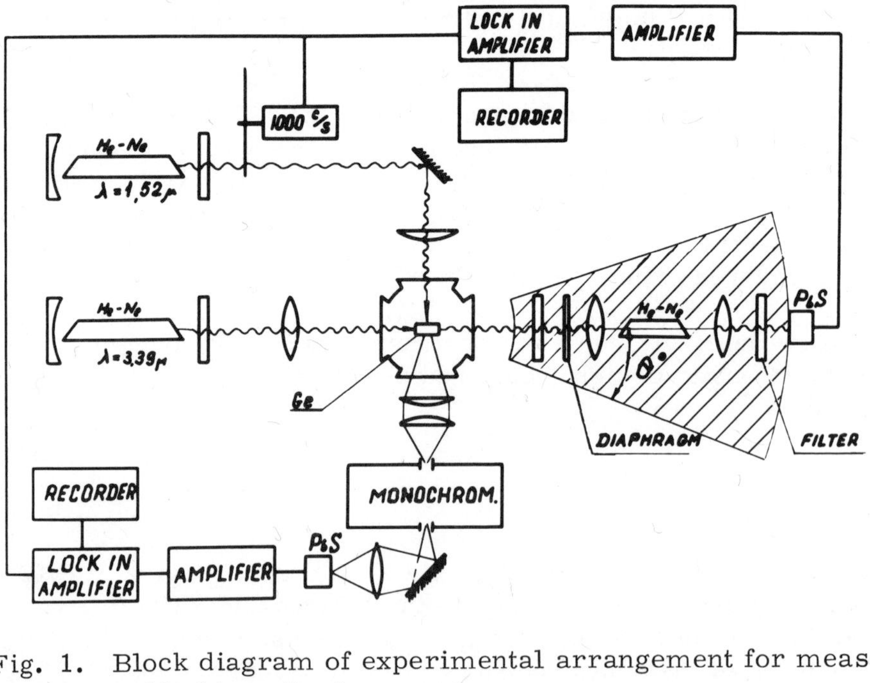

Fig. 1. Block diagram of experimental arrangement for measurements of light scattering.

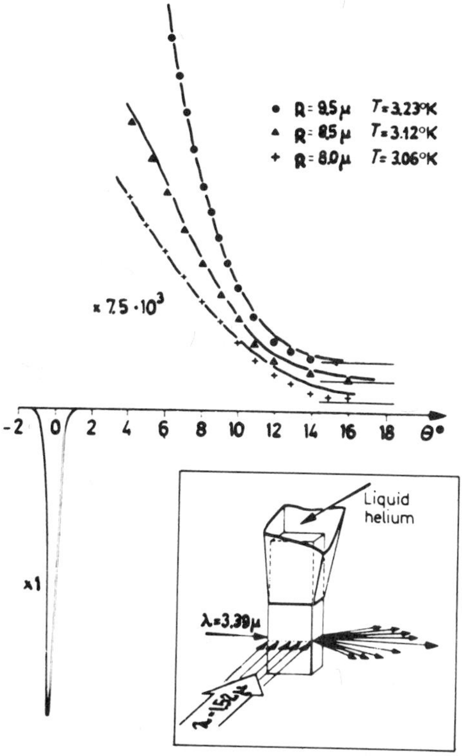

Fig. 2. Angular distribution of scattered light intensity at three different temperatures; the lower part shows the absorption signal at T = 3.12 K. Insert: a sketch of the sample illumination.

The source of excitation was a He-Ne laser of power ~ 10 mW at wavelength 1.52 μ. Since the absorption coefficient of such radiation in germanium is comparatively small at low temperatures, the excitation would be practically uniform along the sample. The exciting radiation was modulated at a frequency of 100 sec^{-1} and focussed on the wide side of a crystal, by means of a cylindrical lens, into a narrow line parallel to a laser beam with a wavelength of 3.39 μ (the insert in Fig. 2).

The scattering of He-Ne laser radiation at the wavelength 3.39 μ was investigated. The scattered light leaving the crystal was amplified by an optical quantum amplifier on a He-Ne mixture and registered by a cooled PbS photodetector. The PbS detector together with the quantum amplifier were moving in the horizontal plane along the arc of a circle, in the center of which there was a sample under investigation. At the same time the function $Y^2(R\theta)$ was recorded on the tape in relative units as a function of scattering angle. In Fig. 2 recordings of the angular dependence of the scattered light intensity are shown for three different temperatures. The calculated values of the function $Y^2(R\theta)$ are plotted as points on the recordings. The parameter R is chosen in order to fit the calculated and experimental data. In the left bottom corner in Fig. 2 the record of an absorption signal is shown, with the shape of this signal reproducing the intensity distribution in the laser beam at the wavelength 3.39 μ.

While measuring the surface tension and the dependence of EHD size and concentration on the excitation level, the beams of both lasers passed through the front side of the sample. In this case the excitation beam was focussed into a spot $\sim$ 200 μ in diameter. At such focussing a high level of excitation was achieved. The spectrum of radiative recombination was recorded simultaneously with scattered and absorbed signals. In the case when it was necessary to have a short rise-time of the excitation pulse, the laser beam was focussed sharply at the wing of the chopper.

RESULTS AND DISCUSSION

Figure 3 shows the temperature dependence of the light intensity scattered by EHD at an angle of 8^0 (in vacuum) to the incident beam, together with those of the absorption signal and the intensities of radiative recombination of EHD and excitons. One can see that as temperature decreases to $\sim$ 3.5 K (the threshold temperature for a given level of excitation used in this experiment) the signal of scattering and that of radiative recombination of drops appear simultaneously with the increase of absorption, while the intensity of the exciton line begins to decrease. Thus as the temperature decreases the volume of the liquid phase (that is the total number of particles bound in the drops) increases due to the lessening of the exciton gas density. The increase in absorption is conditioned by the longer lifetime of the carriers in the drop ($\sim$ 40 μ sec) as compared to that of free excitons ($\leq$ 8 μ sec).

The intensity of scattered light first increases with the lowering of temperature, and then begins to decrease owing to the decrease of the EHD radius. The simultaneous appearance of

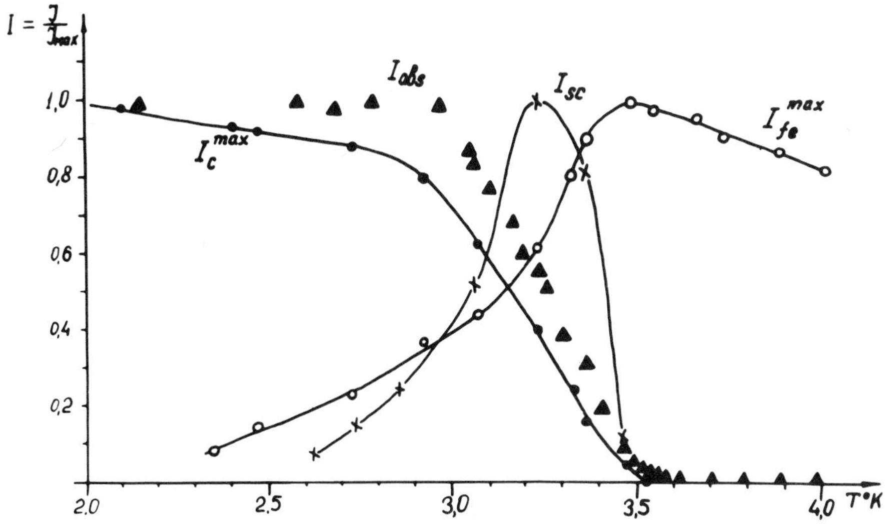

Fig. 3. Signals of scattered light, absorption and radiation vs. temperature.

drop luminescence and scattered light means that the same particles emit and scatter the light.

The estimate of carrier concentration in the liquid phase using formulae (1) and (2) gives the value of $n_0 \simeq 1.9 \times 10^{17}$ cm^{-3}. This value is in good agreement with the results of other experiments [7-11] and theoretical calculations [12, 13].

The dependence of the drop radius on the temperature is shown in Fig. 4. The drop radius decreases with temperature. The upper curve relates to a rise-time of the exciting pulses of $t_0 \simeq 100\,\mu$ sec, and the lower one to $t_0 \leq 3\,\mu$ sec. The durations of the pulses were equal in both cases. One can see that with the increase in the rate of the light intensity growth the drop radius decreases.

The dependence of drop concentration on the temperature, measured for excitation pulses with short and long rise, are shown in Fig. 5. The drop concentration increases with a decrease of temperature, but in the case of steep pulses this

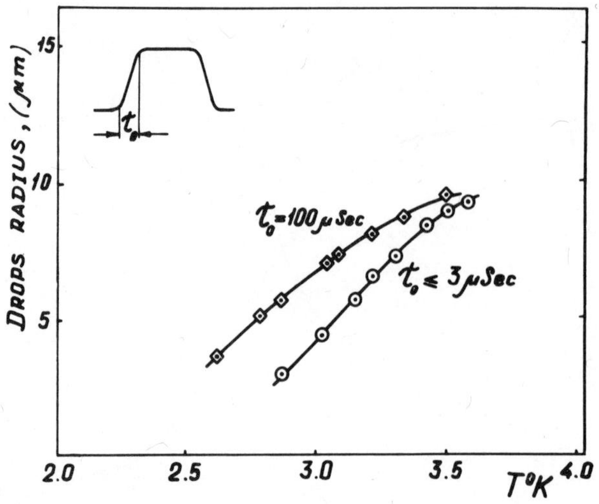

Fig. 4. EHD radius for pulse excitation with long ($t_0 \simeq 100\,\mu$ sec) and short ($t_0 \leq 3\,\mu$ sec) rise-times vs. temperature.

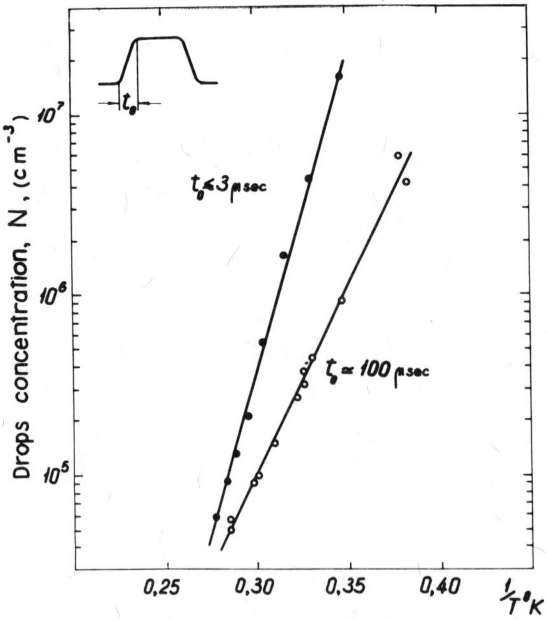

Fig. 5. Drop concentration for pulse excitation with long and short rise-times vs. temperature.

dependence becomes more pronounced. It should be emphasized that in spite of a large duration of the pulse ($\sim 500\mu$ sec) as compared to the rise-time, the drops "remember" the rise-time of the pulse; this means that the nucleation and growth of the drops takes place mainly at the switching on of the excitation.

One can explain the growth of the drop radius with the increase of temperature by the decrease of drop concentration. The dependence of the ratio of the generation rate to the volume of the liquid phase on the drop radius is shown in Fig. 6 at three different levels of excitation. In the range of small R (low temperatures), when the recombination takes place mainly in the liquid phase, the value g/NR^3 almost does not depend on R. This means that the radius R is proportional to $(g/N)^{1/3}$, that is the radius is determined by drop concentration. At large R (high temperatures) the drop radius hardly depends on drop concentration.

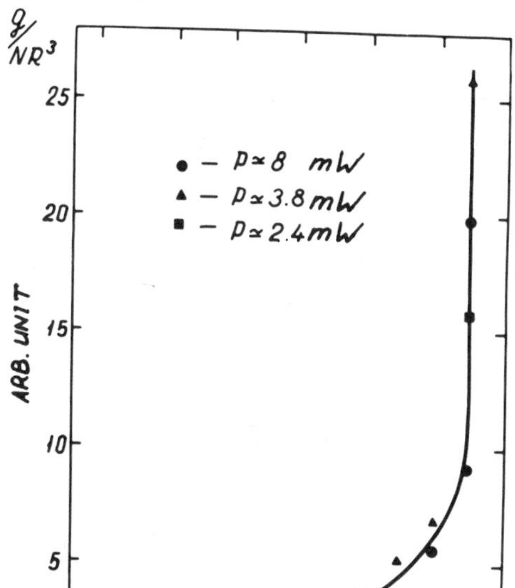

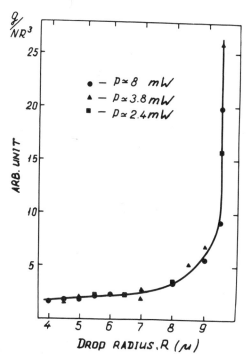

Fig. 6. The ratio of the rate of generation to the volume of the liquid phase (g/NR^3) vs. EHD radius.

The drop concentration turned out to be strongly dependent on the level of excitation: While the level of excitation decreases, the drop concentration decreases as well (Fig. 7). With the increase of temperature this dependence gets stronger. The slope of the curves increase, when steep pulses of excitation are used (Fig. 7b).

Thus, we have observed a strong dependence of the drop concentration on the temperature, intensity and method of excitation. This means that the kinetics of exciton condensation is determined mainly by the process of nucleation, and not by the growth of drops as it was suggested earlier [14-16].

The kinetics of condensation, taking into account the diffusion of excitons towards the surface of EHD and surface tension of an electron-hole liquid, was analyzed by Keldysh [17]. In the steady-state one can obtain the expression for the mean exciton concentration far away from the EHD:

$$n = n_{0T} e^{2\sigma/n_0 R kT} + \frac{n_0 R}{3 v_T \tau_0}\left(1 + \frac{R v_T}{D}\right) \qquad (3)$$

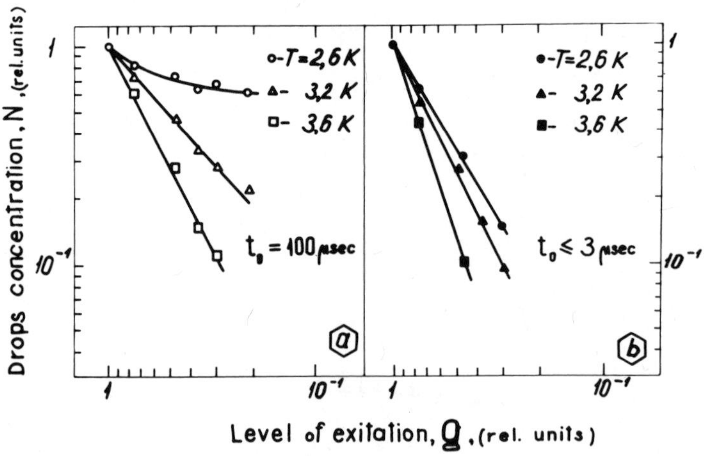

Fig. 7. EHD concentration vs. the level of excitation for three temperatures: a) Pulse excitation with long rise-time, b) Pulse excitation with short rise-time.

where n_{0T} is the thermodynamic equilibrium concentration of excitons at a flat surface between phases; σ is the coefficient of surface tension of an electron-hole liquid, τ_0 is the lifetime of carriers in the drop, D is the diffusion constant of excitons and

$$v_T = \left(\frac{kT}{2\pi m^*}\right)^{\frac{1}{2}}$$

is their thermal velocity. One can see from this expression, that for large radii of EHD, the contribution of diffusion can be considerable; at small R and $\sigma = 0$ this expression coincides with the usual one [2, 8, 14-16]. This expression (3) allows one to determine the radius of EHD in the steady-state, if the temperature and density of the exciton gas are known. This function for three different temperatures is shown graphically in Fig. 8. At $R < R_{min}$ (R_{min} is the position at the minimum of the curve) the solution turns out to be unstable and determines the radius of the "critical embryo" [18] at definite exciton density. The solution at $R > R_{min}$ gives the steady-state radius of EHD, which is in stable equilibrium with the exciton gas of density n.

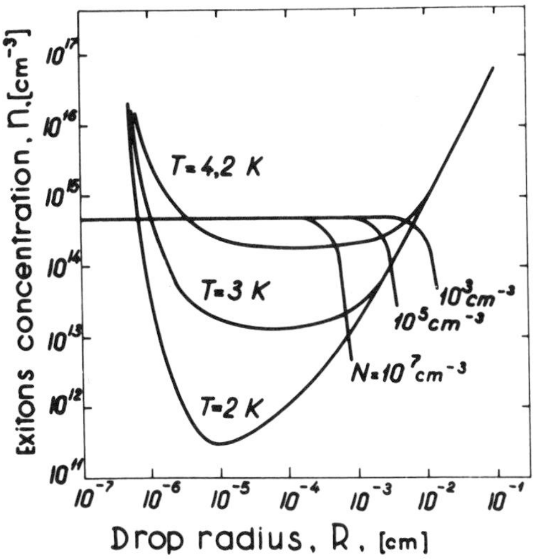

Fig. 8. The graphic solution of eq. (3) for three different temperatures and of (4) for three different concentrations of EHD in the sample.

For specific instances R can be found from eq. (3) together with the balance equation for the number of particles

$$g = \frac{n}{\tau} + \frac{4}{3}\pi R^3 \frac{n_0 N}{\tau_0} \tag{4}$$

here g is the rate of generation, and τ is the lifetime of excitons.

The joint solution of eqs. (3) and (4) gives the stationary radius of EHD at the given concentration of the drops. The solutions of eq. (4) at different concentrations of EHD are also shown in Fig. 8. The stationary radius of EHD is determined by the point of intersection of these curves with the stable branch of the curve which is described by eq. (3).

It is seen from Fig. 8 that the steady-state concentration of excitons can considerably exceed their smallest concentration, required for the formation of drops. Consequently, it is quite natural to suggest the concentration of EHD to increase in the course of time. This has to lead to the decrease of exciton concentration and, consequently, to the decrease of EHD sizes, so that in the limit, the drops must have the minimal stable radius R_{min}, and their concentration will be determined by the value of this radius. Such a situation was analyzed by Silver [19]. However, numerical estimates show that the EHD radii, observed in experiments, are considerably larger than R_{min}. Besides, in such a quasi-equilibrium state the drops do not have "to remember" the rise-time of the excitation pulse, which is inconsistent with the results of our experiments, that is, under experimental conditions the value of N does not reach the maximum value that follows from eqs. (3) and (4).

For the adequate description of experimental results one has to assume the following picture of exciton condensation: upon switching on the excitation pulse the exciton concentration n increases with the increase of the level of excitation (Fig. 9). The rate of nucleation of EHD, dN/dt, is slow till the exciton concentration only slightly exceeds n_{0T}. With the increase of n the value of dN/dt and, consequently, the drop concentration N increases. As the EHD concentration increases, the drops begin "to swallow" the larger number of excitons. Therefore, the concentration of excitons together with the rate of nucleation goes through the maximum and begins to decrease. As the exciton gas density reaches some value n_{st}, when dN/dt is insignificant, the

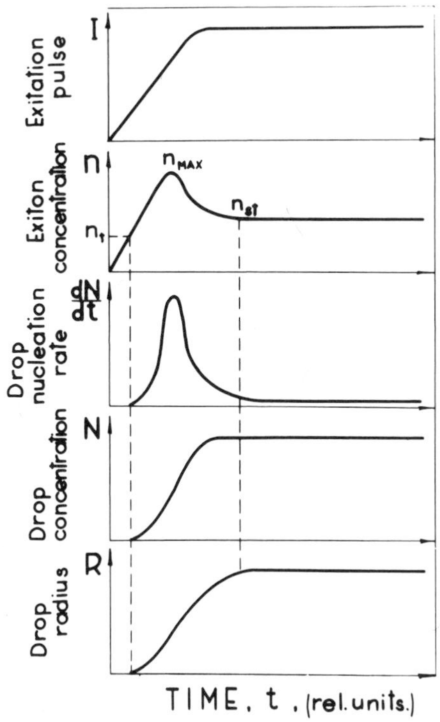

Fig. 9. The qualitative picture of the kinetics of exciton condensation relevant for the obtained experimental results.

nucleation process develops extremely slowly, and those drops which have already been formed, grow to some steady-state size R, which is determined by n_{st}.

The theoretical expressions describing this process were obtained by Keldysh [17] on the basis of the Becker-Döring theory [20]. These expressions are in qualitative and reasonable quantitative agreement with our results and describe properly the observed dependences of drop concentration on the temperature, level of excitation and rise time of the excitation light pulse. At low temperatures and high supersaturation of exciton vapor the expression for N [17] reduces to the usual form of Becker-Döring.

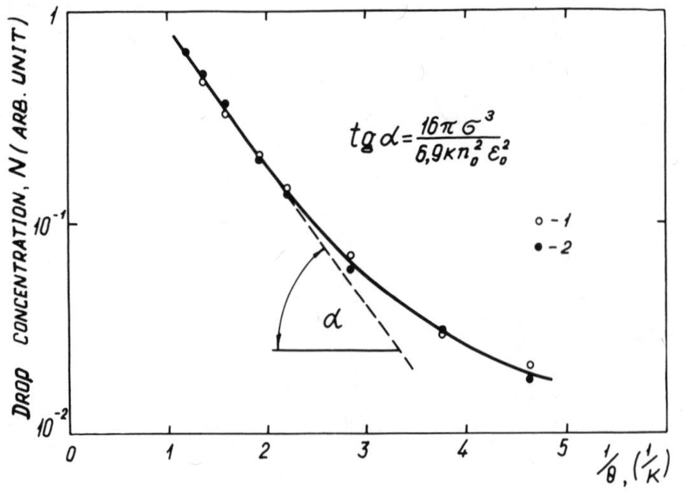

Fig. 10. EHD concentration vs. the value

$$\frac{1}{\theta} = \frac{1}{T}\left(\frac{T_0 - T}{T_0} + \frac{3}{2}\frac{kT}{\epsilon_0}\ln\frac{T_0}{T}\right)^2$$

From the slope of the linear region of this dependence, one can get the value $\sigma = 1.8 \times 10^{-4}$ dn/cm.

But reconstructing the experimental data according to this formula, one cannot estimate the value of σ. In Fig. 10 the dependence of the drop concentration on the value of

$$\frac{1}{\theta} = \frac{1}{T}\left(\frac{T_0 - T}{T_0} + \frac{3}{2}\frac{kT}{\epsilon_0}\ln\frac{T_0}{T}\right)^2$$

is shown, where ϵ_0 is the binding energy for particles in the liquid phase with respect to the exciton level, and T_0 is the threshold temperature of condensation at the given density of exciton gas and in the interface plane between the phases. The light circles are related to the drop concentration obtained from the scattering measurements, and the dark circles to that obtained from the ratio of the volume of the liquid phase (measured from the radiative recombination data) to the value R^3. As seen from Fig. 10, at $1/\theta < 2 (T < 3K)$ the experimental points fit

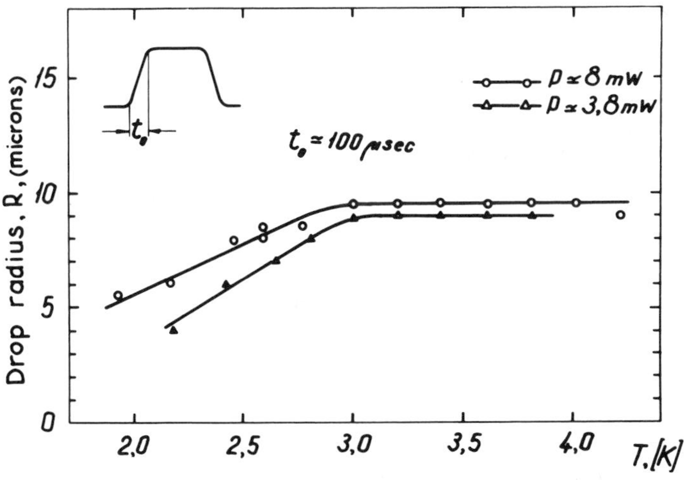

Fig. 11. EHD radius vs. the temperature for pulse excitation with long rise-time.

the straight line (dotted line) fairly well. The slope of the line gives the coefficient of surface tension for the electron-hole liquid $\sigma = 1.8 \times 10^{-4}$ dn/cm.*

In conclusion it should be noted that the estimates of drop sizes with the help of eqs. (3) and (4) and using the drop concentration obtained experimentally, agree quite satisfactorily with experimental data at low temperatures. But at high temperatures eqs. (3) and (4) predict a stronger dependence of the EHD radius on the level of excitation than the experimental data. The dependence of the drop radius on the temperature is shown in Fig. 11 for two different levels of excitation; these data were obtained for excitation pulses with long rise-time. At high temperatures the drop radius is almost independent of the level of excitation and remains constant with change of temperature. The evident tendency of the drop radius to have a constant value (~10 μ) and the slight dependence of the radius on the level of excitation at high temperatures are not well understood.

---

*We are indebted to Dr. J. Hensel for the discussion of this experiment.

For the explanation of this behavior of a drop radius two mechanisms were considered, that can lead to a decrease of diffusion coefficient D and, consequently, to that of the growth rate of drops. They are: exciton-exciton collisions and exciton-drag effect by the phonon "wind". The latter arises inside the drops due to Auger-recombination. Both these mechanisms lead to an exponential dependence of exciton concentration on the drop radius and allow one to explain qualitatively the behavior of the EHD radius in the range of high temperatures. But quantitative estimates do not agree with experiment.

One should note that the assumption of the "phonon wind" can explain the results of a number of experiments, where the increase of the volume occupied by the EHD with the increase of level of excitation was observed [21-24].

We are deeply grateful to L. V. Keldysh for constant attention and numerous discussions.

REFERENCES

1. L. V. Keldysh, The Proc. of the IX Intern. Conf. on Sem. Phys., Moscow, 1969.
2. L. V. Keldysh, "Excitony v poluprovodnikach", Nauka, Moscow, 1971, p. 5.
3. Ja. E. Pokrovsky, K. I. Svistunova, JETP Letters 13, 297 (1971).
4. H. C. van de Hulst, Light Scattering by Small Particles, New York, 1957.
5. N. N. Sybeldin, V. S. Bagaev, N. A. Penin, V. A. Tavetkov, FTT, 15, 177 (1973).
6. V. S. Bagaev, N. V. Zamkovets, N. A. Penin, V. A. Tsvetkov, PTE, N2, 242 (1974).
7. V. S. Vavilov, V. A. Zajats, V. N. Mursin, JETP Letters, 10, 304 (1969).
8. J. E. Pokrovsky, K. I. Svistunova, FTP, 4, 491 (1970).
9. V. S. Bagaev, T. I. Galkina, N. A. Penin, V. B. Stopachinsky, M. N. Churaeva, JETP Letters, 16, 120 (1972).
10. G. A. Thomas, T. G. Phillips, T. M. Rice, J. C. Hensel, Phys. Rev. Lett., 31, 386 (1973).
11. T.K. Lo, Solid St. Comm. 15, 1231 (1974).
12. W. P. Brinkman, T. M. Rice, Phys. Rev. B7, 1508 (1973).
13. M. Combescot, P. Nozieres, J. Phys. C. Solid St. Phys. 5, 2369 (1972).

14. Ya. Pokrovskii, Phys. Stat. Sol. 11a, 385 (1972).
15. V. S. Vavilov, V. A. Zayats, V. N. Murzin, Proc. X. Int. Conf. Phys. Semicond. 1970, p. 509.
16. J. C. Hensel, T. G. Phillips, T. M. Rice, Phys. Rev. Lett. 30, 227 (1973).
17. V. S. Bagaev, N. V. Zamkovets, L. V. Keldysh, N. N. Sybeldin, V. A. Tsvetkov, JETP (in press).
18. L. D. Landau, E. M. Lifschitz, Statisticheskaya physika, Nauka, Moscow, 1964, p. 544.
19. R. N. Silver, Phys. Rev. B 11, 1569 (1-75).
20. Ja. I. Frenkel, Kineticheskaya teoria jidkostei, AN, Moscow, 1945.
21. R. W. Martin, Phys. St. Sol. (bI, 61, 223 (1974).
22. Ya. E. Pokrovsky, K. I. Svistunova, Proc. XII Int. Conf. Phys. Semicond., 1974, p. 71.
23. B. J. Feldman, Phys. Rev. Lett., 33, 359 (1974).
24. M. Voos, K. L. Shaklee, T. M. Worlock, Phys. Rev. Lett., 33, 1161 (1974).

EXPERIMENTS ON LIGHT SCATTERING FROM ELECTRON HOLE DROPS IN GERMANIUM

J. M. Worlock

Bell Telephone Laboratories

Holmdel, New Jersey 07733 USA

It is a real pleasure and a distinct privilege to be here in Moscow to talk about electron-hole drops (EHD). For not only was this condensation phenomenon conceived here, by Keldysh,[1] but it was also discovered here, by Pokrovskii and Svistunova,[2] as well as by Asnin and Rogachev in Leningrad.[3]

Light scattering was first used by Pokrovskii and Svistunova[4] to provide strong evidence that the condensate was in the form of tiny droplets; and Sybel'din, Bagaev, and their coworkers[5] have studied many properties of the droplets using the light-scattering technique. The latest of these experiments have been described this afternoon by Dr. Sybel'din.[6]

About three years ago, we began some experiments on scattering of light from electron-hole drops in germanium. Happily, the Soviet and American experiments tend to be complementary. We inject electron hole pairs solely by optical absorption at the surface, whereas the Lebedev group can excite deep in the bulk of the crystal; temperature is a variable in the Soviet experiments, while we have done more with time-dependent phenomena.

I will describe three of our experiments, all of which are done in a geometry shown in Fig. 1. A rectangular Ge crystal of high purity, immersed in superfluid He is illuminated on a Syton polished surface to create electron-hole pairs. A few milliwatts

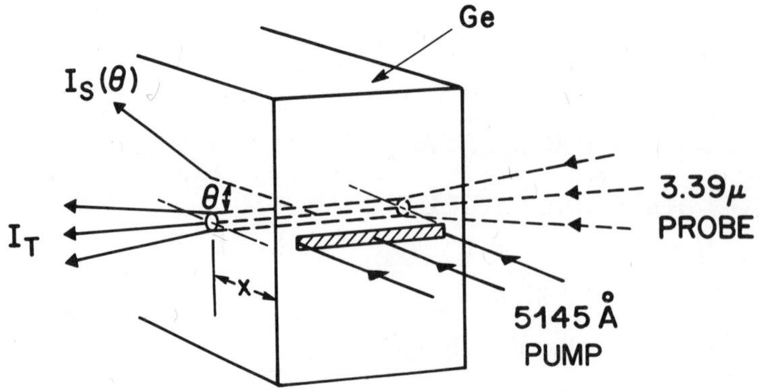

Fig. 1. Standard experimental geometry for investigating scattering and absorption of light of EHD.

of 3.39 μ light from a Ne Ne laser is focussed into the crystal to probe the droplet properties, and we measure directly the transmission I_T and the scattering $I_S(\Theta)$. The important variables at our disposal are: 1. Pump geometry: a) point focus, b) stripe focus and c) unfocussed; 2. Temporal properties of pump: a) steady state, chopper frequency < 1 kH_z, b) pulsed, the time resolution is ~ 1 μsec; 3) Pump power; 4. Probe position: a) depth and b) height measured from pump region, the spatial resolution is 50-70 μm.

In our first experiment,[7] we discovered an unanticipated Fabry Perot interference effect which allowed us to determine not only the characteristic droplet sizes but also the absolute concentration and optical indices of refraction of the liquid state. Previously,[4,5] scattering and absorption measurements had to be analyzed using an index of refraction with a theoretical real part and an imaginary part taken from measurements on doped p-type germanium.

Because our crystals were cut slightly wedge-shaped, the sample thickness varied as we swept the depth x. The Fabry

Perot modulation of the transmitted beam is shown as the upper curve in Fig. 2. The change in transmission ΔI_T or apparent absorption in the presence of EHD is shown in the lower curve, and it undergoes violent oscillations, even becoming negative. Negative absorption would imply gain which we would love to have, but do not. The oscillations in ΔI_T are caused by the real part of the macroscopic index of refraction of the droplet cloud, while the bias toward positive absorption is given by the imaginary part.

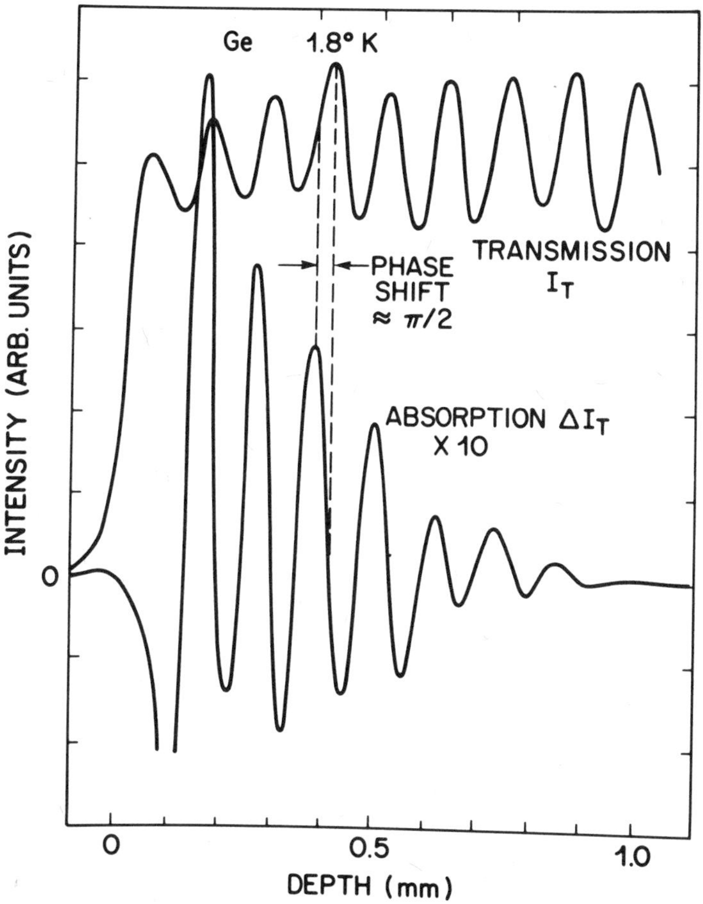

Fig. 2. Transmission I_T, and modulated transmission ΔI_T, as function of depth, showing strong Fabry-Perot effects.

The model connecting these measurements with the microscopic properties of the droplets involves assuming that the droplet cloud contains only monodispense spheres of condensate, and that the exciton and free electron-hole-pair gas is optically ineffective. Our analysis gave a negative real index change in the liquid which was $1x10^{-3}$, fully consistent with the expected optical response of an electron-hole plasma of density $2x10^{17}$ cm^{-3}, and with the measured EHD plasma frequency of 130 cm^{-1} obtained by Vavilov, Zayats and Murzin.[8] Our imaginary index change was $1x10^{-4}$, about a factor of three smaller than that calculated from the p-Ge measurement of Kaiser, Collins and Fan.[9]

We found that droplet radii were 3-4 μm, roughly independent of pump power and depth. Droplet densities varied nearly exponentially with depth, with 1/e length of 300 μm at an input power of 100 mwatts of 5145 Å light. This length increases with laser power, while the surface EHD density appears to remain constant, at about 10^8 EHD/cm^3.

In the second experiment,[10] the pump geometry is changed to a point focus, and the cloud properties are remarkably changed. Figure 3 shows how both absorption and scattering depend on distance from the pumped surface. The solid line shows that these data are consistent with a hemispherical cloud of uniform density out to a very sharp boundary at about 1 mm radius. Droplet sizes in this experiment were 2 μm, and the density invoked about 10^9 cm^{-3}. Attempts to explain this result[11] have only been partially successful. Recently, we have repeated this experiment on another crystal,[12] scanning not only in depth but in height, and this two dimensional scan allows us a much more reliable determination of the droplet density map. The new results show a cloud which is neither hard-edged nor hemispherical. There is obviously much work to be done before we understand the shapes of steady state EHD clouds.

In the final experiment,[13] we attempt to study, in a time resolved experiment, the behavior of a cloud which results from pulsed excitation. Here the excitation source is a 300 nsec pulse from a Nd:YAG laser, and the absorption and scattering are studied as a function of time using a boxcar circuit with gate time $\sim$ 1 μsec.

Figure 4 shows the time-development of the signal at various depths in the crystal. The signal here is scattered intensity.

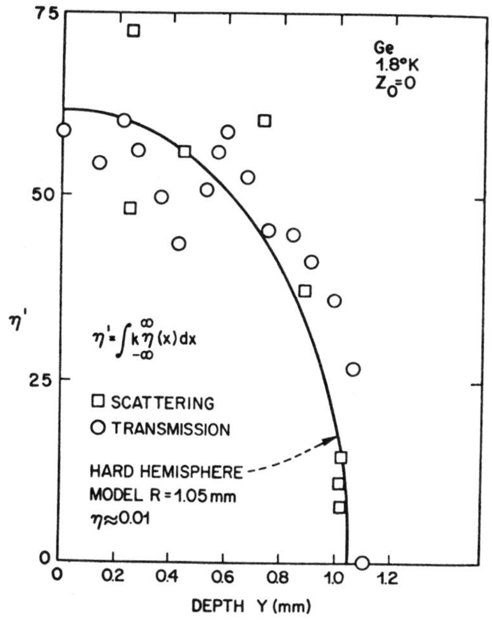

Fig. 3. Variation of absorption and scattering signals with depth: point focus excitation.

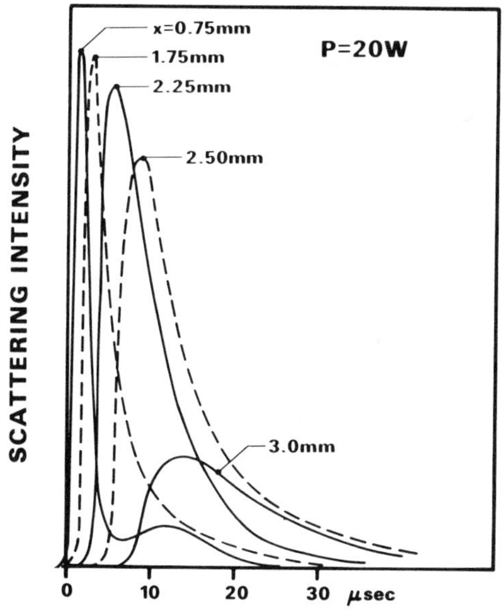

Fig. 4. Time development of scattered intensity, with depth in the crystal as a parameter.

The correction for variation of droplet size is rather small, so we take this signal to represent droplet density. As one might expect, the deeper one takes the probe beam, the later is the arrival of EHD. What is not entirely expected is that the drops appear and then disappear very rapidly. This phenomenon is shown more graphically in Fig. 5, which is simply a transcription of Fig. 4, so as to give snapshots of droplet density. It appears that a cloud of drops moves into the crystal with rather high speed and comes to rest at about 2.5 mm, leaving a hollow place behind.

One measure of the velocity of penetration is given in Fig. 6, where we have shown the depth of the crystal versus the times of arrival of the leading edge of the signal in Fig. 4. These curves taken at face value suggest droplet clouds moving at very high initial velocities (indeed faster than sound), subsequently slowing, and finally decaying.

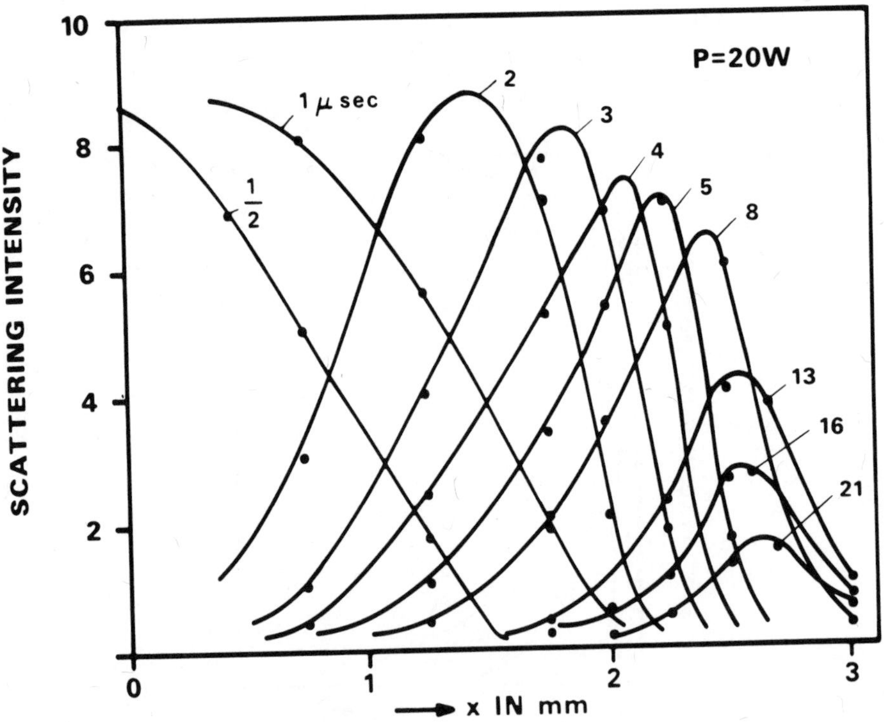

Fig. 5. Spatial dependence of scattered intensity, at different times following exciting pulse.

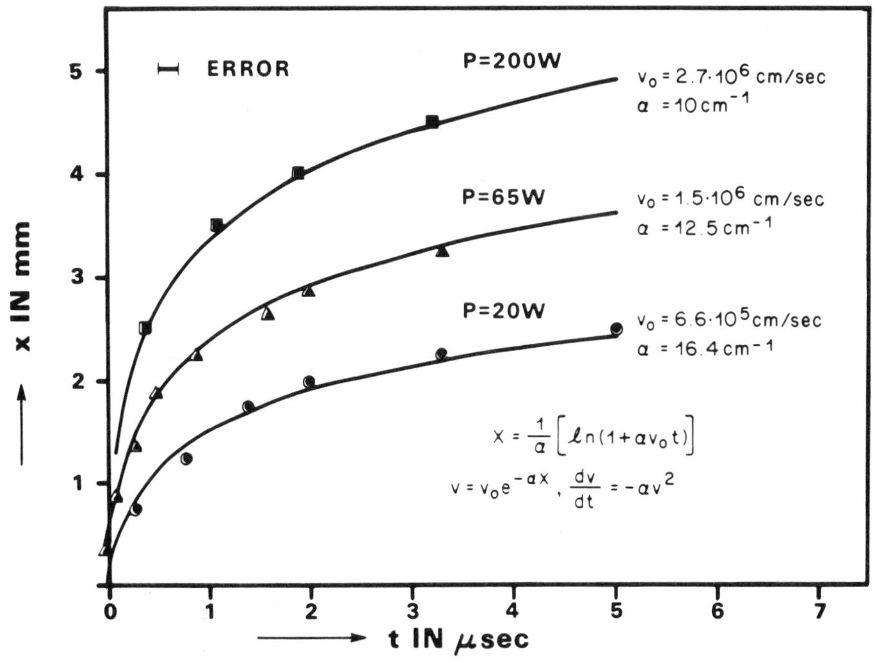

Fig. 6. Arrival time versus depth in crystal, for three different pump power values.

Obviously one would like to have a way of measuring actual droplet velocities, in order to determine the validity of such ideas. We feel that the best way of doing that is by doppler spectroscopy, i.e., measuring the spectrum of doppler shifted light scattered from moving drops. As a result most of our efforts are currently going toward this type of experiment.

Meanwhile the experiments I have described here today are largely unexplained, and in fact there exists very little theoretical work oriented toward understanding the spatial distributions of EHD either in steady state or pulsed experiments. Perhaps the best gift to a convocation composed primarily of theoretical physicists is a package of unsolved problems.

I would like to acknowledge those who have been my coworkers at Bell Laboratories, K. L. Shaklee (now at University of Campinas, Brazil), M. Voos (from Ecole Normale Superieure, Paris), J. C. V. Mattos (on leave from University of Campinas, Brazil), T. C. Damen, and J. P. Gordon.

REFERENCES

1. L. V. Keldysh in Proc. Ninth International Conf. on Physics of Semiconductors, Moscow, 1968, M. S. Ryvkin, Editor (Nauka, Leningrad, 1968) p.1303.
2. Y. E. Pokrovskii and K. I. Svistunova JETP Letters 9, 261 (1969) [Pisma ZhETF 9, 435 (1969)].
3. V. M. Asnin and A. A. Rogachev JETP Letters 9, 248 (1969) [Pisma ZhETF 9, 415 (1969)].
4. Y. E. Pokrovskii and K. I. Svistunova JETP Letters 13, 212 (1971) [Pisma ZhETF 13, 297 (1971)].
5. N. N. Sybel'din, V. S. Bagaev, V. A. Tsvetkov and N. A. Penin, Soviet Physics - Solid State 15, 121 (1973).
6. N. N. Sybel'din, V. S. Bagaev et al. Proc. First USA, USSR Seminar-Symposium on Theory of Light Scattering in Condensed Matter; preceding paper.
7. J. M. Worlock, K. L. Shaklee, T. C. Damen and J. P. Gordon, Phys. Rev. Letters 33, 771 (1974).
8. V. S. Vavilov, V. A. Zayats, and V. N. Murzin, JETP Letters 10, 192 (1969) [Pisma ZhETF 10, 304 (1969)] and in Proc. Tenth Int'l. Conf. on Physics of Semiconductors, Cambridge 1970 S. P. Keller, J. C. Hensel and F. Stern ed. (USAEC, 1970).
9. W. Kaiser, R. T. Collins and H. Y. Fan, Phys. Rev. 91, 1380 (1953).
10. M. Voos, K. L. Shaklee, and J. M. Worlock, Phys. Rev. Letters 33, 1161 (1974).
11. M. Combescot, Phys. Rev. B12, 1591 (1975).
12. J. C. V. Mattos, K. L. Shaklee, M. Voos, T. C. Damen and J. M. Worlock, to be published.
13. T. C. Damen and J. M. Worlock, Proc. Third Int'l Conf. on Light Scattering on Solids, Campinas, Brazil 1975, to be published.

COMMENT

(by L. M. Falicov)

I would like to take this opportunity to show you a photograph made at Berkeley[1] of an electron-hole drop in germanium (Fig. 1). This shows a single large electron-hole drop which is stably formed in suitably stressed discs of pure germanium. The

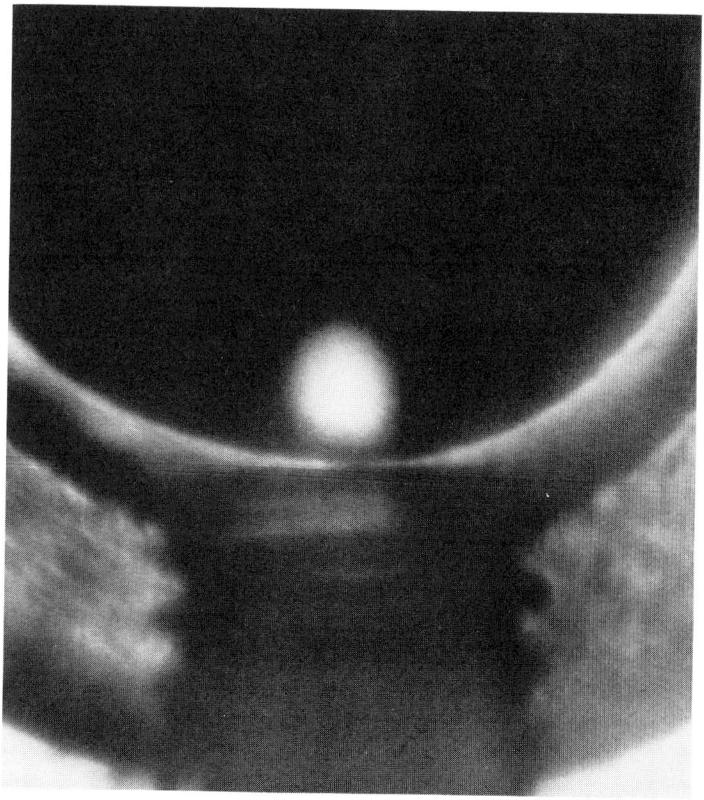

Fig. 1

photo shows the image of a drop obtained by focusing its luminescence onto the surface of an infrared-sensitive vidicon image tube. The drop, of approximately 0.6 mm diameter, is located at a point of minimum drop energy (maximum stress point) in the crystal. The lifetime of the drop has been measured to be 490 μsec from the decay of the total luminescence intensity. The long lifetime is a consequence of the reduced density of electron-hole pairs in the drop under stress.

REFERENCE

1. J. P. Wolfe, W. L. Hansen, E. E. Haller, R. S. Markiewicz, C. Kittel and C. D. Jeffries, Phys. Rev. Lett. 34, 1292 (1975).

INELASTIC X-RAY RAMAN SCATTERING

P. M. Platzman and P. Eisenberger

Bell Laboratories

Murray Hill, New Jersey 07974 USA

Historically x-ray scattering is one of the oldest and most useful tools available to scientists for investigating the properties of solid state systems.[1] However, it is also true that historically work utilizing x-rays has focused almost exclusively on elastic or Bragg scattering. Recently there has been a renewal of interest in the phenomenon of inelastic x-ray scattering.[2] This rebirth originates because of the fact that new and more powerful sources are becoming available. With the recent advent and utilization of synchrotron sources to the list of conventional sources we now have available highly intense and in some cases tunable source of x-rays.[3]

The type and variety of phenomenon and systems which have been intensively investigated in the past few years is extremely large.[2,3] In this talk I would like to focus on certain aspects of the inelastic scattering of x-rays from simple metals. In a typical inelastic scattering process (see Fig. 1) the incoming monochromatic x-ray, frequency ω_1, polarization ϵ_1, is scattered through a fixed angle φ and the spectrum of the scattered radiation frequency ω_2, polarization ϵ_2, is analyzed. Each different frequency shift (to lower frequency) is due to the excitation of the electronic system from its ground state. Thus momentum (k) and energy (ω) are transferred in the scattering process and information is contained in both the angle and frequency dependence of the scattering event.[4]

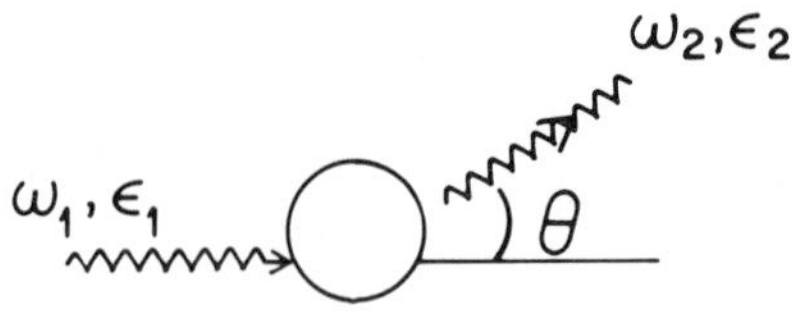

$$\omega = \omega_1 - \omega_2$$

$$k = k_1 - k_2$$

Fig. 1. General schematic of an inelastic scattering experiment. ω and k are the energy and momentum transfer.

In order to qualitatively understand the phenomenon in these systems it is important to keep in mind their electronic excitation spectrum. For simple metals the electronic excitations spectrum is shown schematically in Fig. 2. The low lying core states labelled as K and L are filled and the conduction band shown schematically as the parabola, is partially filled to the Fermi-energy. For most simple metals there is a rather nice separation of energies in the problem. The K shell is bound by thousands of electron volts while the electrons in the L shell are bound by hundreds of electron volts. The Fermi-energy is typically of the order of five electron volts. X-rays which scatter from such a system may leave it in varying states of excitation which are rather well separated in energy.

The excitation of a conduction electron leads to so-called plasmon and Compton scattering.[4] Much work has been done in this area recently but we will not touch on it here. The excitation of an electron from the L shell to an unoccupied state above the Fermi-surface leads to an inelastic scattering event which we chose to call X-Ray Raman Scattering.[5] We will be directly concerned with understanding the nature of this Raman-like process.

In an actual situation both processes, Compton and Raman occur simultaneously. A typical schematic spectrum is sketched in Fig. 3. Due to the large separation in energies for the two processes (as is the case in Fig. 3) there are regions of momentum and frequency where the two spectra are clearly dis-

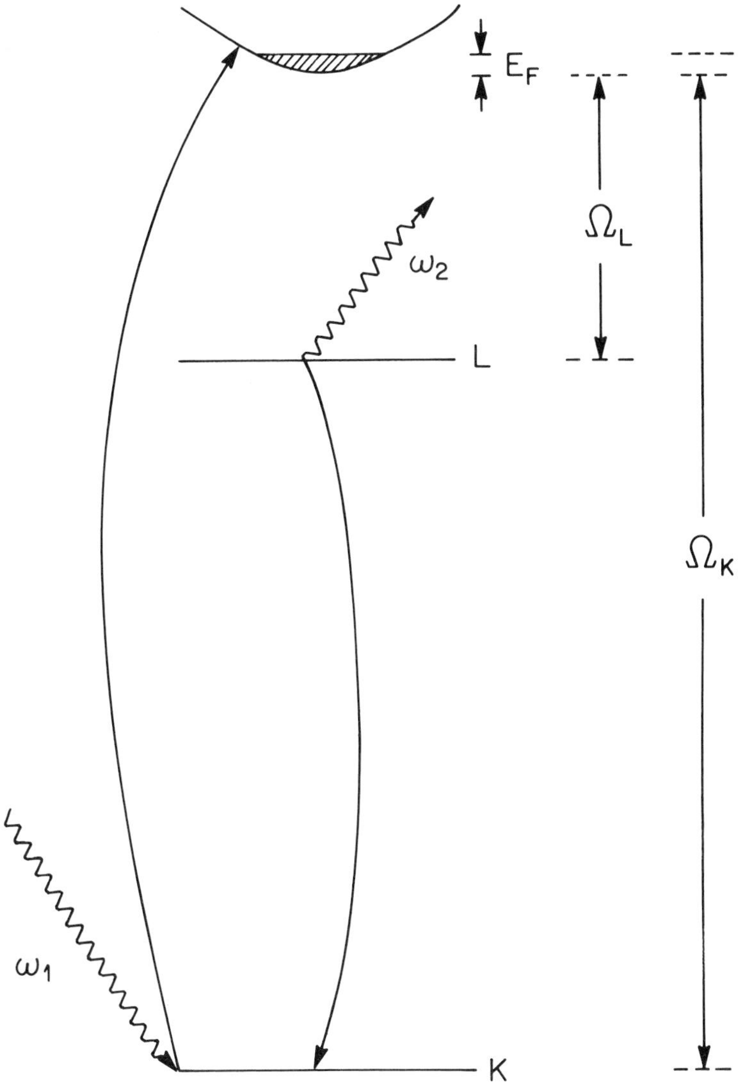

Fig. 2. Relevant energy levels for a simple metal.

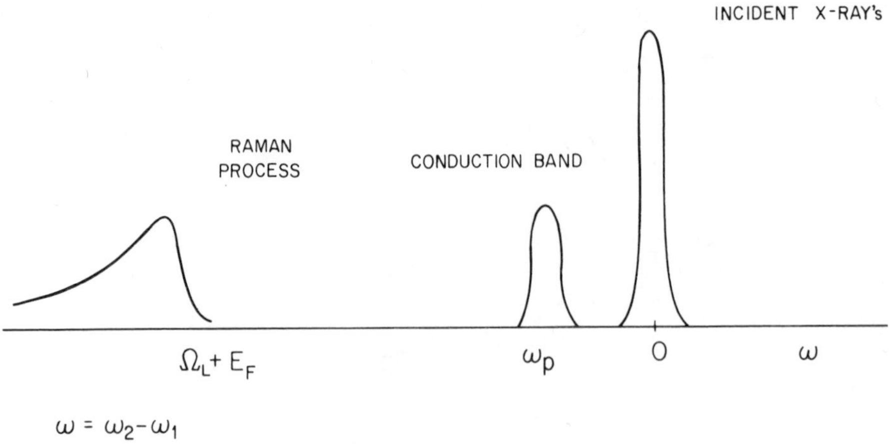

Fig. 3. Idealized spectrum for x-ray scattering from simple metals.

tinguishable. Thus we can in fact focus on the Raman process which appears as an edge beginning at the energy required to excite an L electron to the Fermi-surface and extending indefinitely to lower photon energy.

The complete theory of the x-ray effect has not been written down. In this paper, however, we will discuss certain general aspects of the theory which will give us a rather good understanding of the magnitude of the effects which are involved, the kinds of physics which are contained in such experiments and a hint of the kind of theoretical calculations which must be performed in order to get an accurate description of such experiments.

X-rays to a very good approximation, only couple to the electrons. Insofar as the x-rays and the electrons are nonrelativistic the coupling of the electrons to the photons is described by,

$$H_c = \int d\tau \left(\frac{e}{c} \vec{j} \cdot \vec{A} + \frac{e^2}{2mc^2} \rho A^2 \right) \tag{1}$$

where $\vec{A}$ is the vector potential of the x-ray field and ρ and j are the number and current densities of the electrons. There are clearly two distinct terms in Eq. (1), [A^2 and $\vec{j} \cdot \vec{A}$]. Since the scattering is second order, i.e., involves two photons, we must use A^2 to lowest order and $\vec{j} \cdot \vec{A}$ to second order.

The basic microscopic processes which are involved are shown in diagramatic form in Fig. 4. The wiggly lines represent photons, a double solid line indicates a tightly bound core state and the single line is a representation for a weakly bound conduction electron. Such an electron is weakly bound only in the sense that its wave function is modified by the presence of the periodic ionic potentials in the solid. In Fig. 4a the A^2 term annihilates the incoming photon, creates the final photon and simultaneously produces a hole in the L shell and an excited conduction electron. The $\vec{j} \cdot \vec{A}$ term (Fig. 4b) contributes to a quite distinct second order process. The first photon is absorbed creating a hole in the K shell along with the final conduction electron. An electron from the L shell falls into the K shell producing a L hole and creating the final photon. The matrix element for this simple lowest order process is

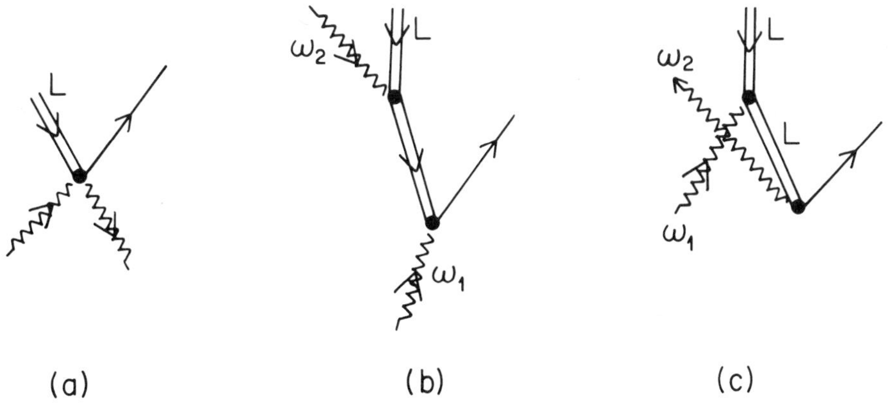

Fig. 4. Lowest order perturbation diagrams for x-ray Raman scattering.

$$M_{fi} = \frac{\langle \underline{P} \,|\, \vec{p}\cdot\vec{\epsilon}_2 \,|S\rangle\langle k\,|\vec{p}\cdot\vec{\epsilon}_1\,|S\rangle}{m(\omega_1 - \epsilon_k - \Omega_K + i\Gamma_K)} \tag{2}$$

The states $\langle \underline{P}$ and $\langle S|$ are the bound hydrogenic-like L and K shell states and the energies Ω_K and ϵ_k are defined in Fig. 2. The quantity Γ_K is a phenomenological lifetime for the K hole. This matrix element is (Eq. (2)) resonant when the energy of the incoming photon ω_1 is above threshold. The matrix element for the diagram shown in Fig. 4c is non-resonant and we will neglect it.

The simple resonant matrix element is corrected by a host of "many body effects". All the particles and holes created may in fact interact via coulomb interactions with all the other electrons and holes in the system. In addition the conduction electron will interact with the ion cores and will become Bloch electrons. We will discuss some of these high order effects in the next section.

However, the details of the matrix element are unimportant for an estimate of the relative size of A^2 and p.A terms. If we call the contribution of the A^2 term 1 then the second term (Eq. (2)) is of order

$$M^{p.A} = \left(\frac{E_F}{E_B}\right)^{\frac{1}{2}}\left[\frac{E_B}{\omega_1 - \epsilon_k - \Omega_K + i\Gamma_K}\right] \tag{3}$$

Here E_F is the Fermi-energy and E_B is the binding energy of the L state. The quantity E_F/E_B is small, however, the resonance as one approaches the energy required to promote an electron from the K shell to the Fermi-surface can more than compensate for this factor. In fact the resonant term can, in many cases, with the use of tunable radiation, be made to dominate the cross section. In order to get some idea of the physics contained in these processes we would like to consider the resonant and nonresonant cases separately.

NONRESONANT CASE

In the nonresonant case[2] the cross section comes solely from the A^2 term in Eq. (1) and the rate for nonresonant scattering can be written, in the nonrelativistic case as

$$\frac{dR(\omega_1, \omega_2)}{d\omega_2} = \left(\frac{\omega_2}{\omega_1}\right)\left(\frac{e^2}{mc^2}\right)^2 \frac{c}{V} \sum_f |\langle f| \sum_{x_i} e^{i\vec{k}\cdot\vec{x}_i} |i\rangle|^2$$

$$\times\ \delta(E_F - E_i - \omega) \equiv \left(\frac{\omega_2}{\omega_1}\right)\left(\frac{e^2}{mc^2}\right) \frac{c}{V}\ S(k, \omega) \tag{4}$$

where $\omega = \omega_1 - \omega_2$ and $\langle f|$ and $|i\rangle$ are the exact many body states of the system. The coordinates x_i are the coordinates of all the electrons in the system. In principle Eq. (4) contains all of the many body corrections to the lowest order process shown in Fig. 4a. Since the momentum transfer $k \cong 2k \times \sin\theta/2$ vanishes for forward scattering it is usually true that by looking close enough to the forward direction we can make $\vec{k}\cdot\vec{x} << 1$. In this case Eq. (4) is directly proportional to the optical absorption coefficient i.e., the leading term in the expansion of $(e^{i\vec{k}\cdot\vec{x}})$ which exists is the dipole term $\vec{k}\cdot\vec{x}$.

While Eq. (4) looks simple it obviously contains an enormous variety of physics and many complications. However, if we for the moment focus on those transitions i.e., final states involving an L hole and a conduction electron (See Fig. 4a) we can access the importance of these kinds of experiments.[6] The forward scattering amplitude or dipole approximation has been studied both experimentally[7] and theoretically[8] in great detail. Roughly speaking the transition may be represented schematically as shown in Eq. (5);

$$\left.\frac{dR(\omega_1, \omega_2)}{d\omega_2}\right|_{k\to o} = A(\omega) \tag{5}$$

where

$$A(\omega) = A_{\ell-1}(\omega) + A_{\ell+1}(\omega)$$

The index ℓ on the amplitude refers to the angular momentum of the initial state. For the case of excitation from an L state $\ell = 1$. Equation (5) is simply a consequence of angular momentum conservation.

The basic process is shown in Fig. 4a. Many body effects, or more precisely coulomb interactions among the electrons leads to a very interesting frequency dependence of $A(\omega)$ near threshold. The influence of coulomb interactions may be roughly divided into three categories; (see Fig. 5)

1) The suddenly produced hole may create multiple electron hole pairs in the conduction band, Fig. 5a .
2) The outgoing electron can interact with the hole which has been left behind, Fig. 5b.

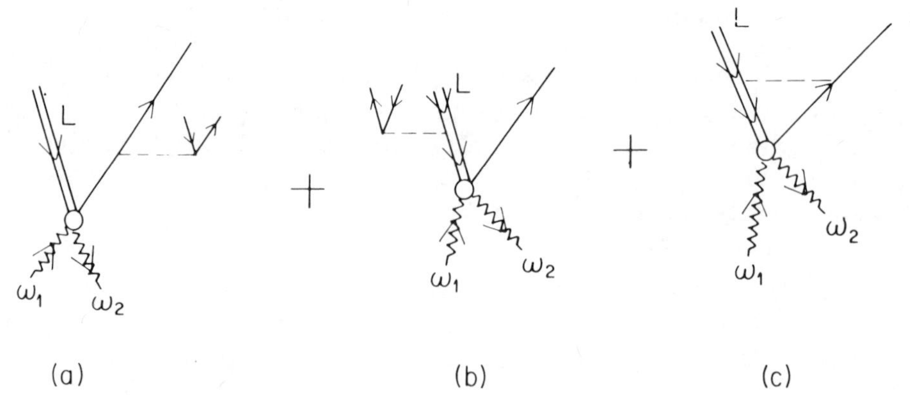

Fig. 5. Higher order coulomb corrections to the x-ray Raman process.

3) The outgoing electron can scatter off the remaining conduction electrons and create additional electron hole pairs, Fig. 5c.

The first two effects have been analyzed in great detail.[8] The third effect has not been considered. If one considers the first two effects one finds that near threshold, i.e., $\omega \cong E_F + \Omega_L \equiv \omega_T$ the amplitude is expected to have a singular behavior which may be written as

$$A_{\ell-1} = F_{\ell-1}(\omega)\left[\frac{\xi_0}{\omega-\omega_T}\right]^{\alpha_{\ell-1}} \tag{6}$$

where

$$\alpha_\ell = 2\delta_\ell - g \tag{7}$$

and

$$g = 2\sum_{\ell'} (2\ell' + 1)(\delta_{\ell'}/\omega)^2 \tag{8}$$

The quantities δ_ℓ are the phase shifts for an electron scattering at the Fermi surface from the hole. The function $F_\ell(\omega)$ is some smoothly varying function. The cross section clearly has, as can be seen from Eq. (6), an interesting dependence on the sign of the quantity α_ℓ. Depending on the size and sign of the phase shifts the cross section may have a pronounced peak behavior or dip behavior near threshold. In Fig. 6, we show a plot of the optical absorption spectra near the L edge in magnesium.[7] The solid curve is the experimental results while the dotted curve is a fit using Eq. (6). The fit employs phase shifts which are consistent with estimates made from simple models of the electron hole potential. The theoretical formula Eq. (6) has also been smeared with an appropriate lifetime broadening effect.

While the agreement between theory and experiment looks reasonable there is still a good deal of controversy as to the real origin of this peaking effect. The temperature dependence of these curves, consistency of this data with other data obtained near K absorption levels, the importance of band structure effects have all entered into the arguments concerning the many body question.[9] We will not go into these in detail;

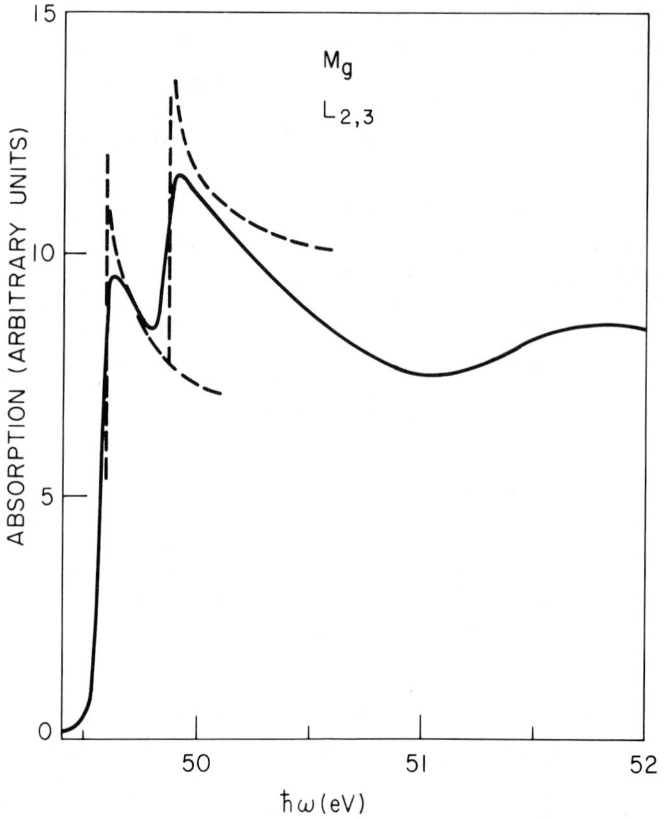

Fig. 6. The absorption edge of magnesium. The solid curve is the experiment of Kunz et al, Ref. 7. The dashed curve is a phenomenological fit using the theory of Nozieres et al, Ref. 8.

however, we would like to point out that the Raman scattering offers an additional experimental handle on the nature of these singularities.

The essential idea involved in the analysis of the Raman process is simply that the angular dependence of the spectrum, i.e., different spectra for each momentum transfer is unambiguously determined by expressions of the form given in Eq. (5).[6] More concisely stated the operator $\exp(i\vec{k}.\vec{x})$ in Eq. (4) mixes in higher spherical harmonics into the amplitudes purely on the basis of angular momentum considerations. We may quite generally write

$$S(\underset{\sim}{k}, \omega) = \sum_{\ell} R_{\ell}(k)\, A_{\ell}(\omega) \tag{9}$$

The quantities $R_{\ell}(k)$ are determined quite accurately by angular momentum considerations; i.e.

$$\begin{aligned} R_{\ell}(\vec{k}) = \sum_{m} \Big| (4\pi)^2 \int r^2 dr [j_{\ell}(k_F r)\, j_{\ell}(kr) \varphi^*(r) \\ \times Y_{\ell m}(\hat{k}) - (4\pi)^{3/2}\, \delta_{m,o}\, \delta_{\ell,o} |\varphi(r)|^2 \tilde{\varphi}(k_F)\, j_o(kr)] \Big|^2 \end{aligned} \tag{10}$$

where

$$\tilde{\varphi}(k) = \int d^3 r e^{ik\cdot r} \varphi(r) \tag{11}$$

is the fourier transform of the atomic wave function $\varphi(r)$ characterizing the occupied deep level.

As one varies the momentum transfer k the quantities R_{ℓ} vary in magnitude in a well defined way. Thus for the L edge in Mg for example, we start in the forward direction with S and D wave amplitudes. The dipole operator promotes an electron from its P-like L state to a mixture of S and D waves. By changing the angular momentum given to the final electron state we can begin to mix in P and F waves. This means that once the complete set of scattering phase shifts are calculated then the angular dependence of the cross section is completely determined by Eq. (9). Utilizing estimates of the phase shifts we have found that a material such as Mg (see Fig. 6) should exhibit a rather striking change in the Raman spectra as a function of momentum transfer. In fact such a behavior has very recently been observed.[10] Qualitatively these results seem to be consistent with such a many body picture, however a detailed analysis of this very preliminary data seems to indicate that there is an internal difficulty. It arises because there seems to be no set of phase shifts which can accurately represent the data. Our object here is not to get bogged down in the details of this analysis but simply to point out the additional physics contained in the resonant Raman process. It is considerable.

RESONANT CASE

In Resonant Raman scattering the theoretical situation is much more rudimentary. Likewise the experimental results are just now becoming available. The first observation of this effect was reported in Ref. 5. One subsequent unpublished experiment utilizing synchrotron radiation has appeared in preprint form.[11]

Since both the theoretical and experimental situations are in such preliminary phases we will look at the process strictly from the point of view of one electron theory and avoid any real discussion of the many body effects which were so important in the case of the nonresonant cross section near threshold. In addition we, for the purposes of this paper, will only consider the case where the resonance i.e. the amplitude shown in Fig. 4b completely dominates the process. This, of course, occurs only close enough to resonance (see Eq. (3)). Of course, there will be interesting effects associated with the interference between these two kinds of amplitudes.

When the above conditions are satisfied and we treat the problem in the one electron approximation the rate for resonant Raman scattering is given by

$$\frac{dR(\omega_1, \omega_2)}{d\omega_2} = \left(\frac{\omega_2}{\omega_1}\right)\left(\frac{e^2}{mc^2}\right)^2 c \int \frac{d^3k}{(2\pi)^3}(1-n_{\underset{\sim}{k}})\,|M_{fi}|^2\,\delta\left(\omega_1-\omega_2-(\epsilon_{\underset{\sim}{k}}+\Omega_K)\right) \tag{12}$$

The matrix element M_{fi} is defined in Eq. (2). Each separate photon event involves a pure dipole transition. This is the case because the wave length of the x-rays is long compared to the size of the spatial extent of the K shell wave function. This behavior is to be contrasted with the nonresonant case where the appropriate length which must be considered is the spatial extent of the L electrons wave function. In Eq. (12) the quantity n_k is the occupation number, typically the Fermi function for the conduction band so that the integral over k extends over unoccupied states. The phenomenological lifetime Γ_K imitates the effect of radiative and nonradiative decays of the K hole. There are two regions to consider:

1) When the real part of the energy donominator in Eq. (13) vanishes (incident energy above threshold) the second order process factors into an absorption followed by emission, i.e. (K_α characteristic lines)

$$R(\omega_1, \omega_2) = (\Gamma_K)^{-1} W^{Abs}(\omega_1) W^{Emis}(\omega_2) \tag{14}$$

with $\omega_2 = \Omega_K - \Omega_L$. Equation (14) simply states that the excited state of the system builds up at a rate W^{Abs} for a time $(\Gamma_K)^{-1}$ where-upon it fluoresces at a fixed frequency $\Omega_K - \Omega_L$. Above threshold Eq. (14) quite accurately describes the fluorescence spectrum.

2) Below threshold we may rewrite Eq. (1) as

$$\frac{dR(\omega_1, \omega_2)}{d\omega_2} = W^{Abs}(\omega_1 - \omega_2 + \Omega_K - \Omega_L) \frac{\omega_2}{(\Omega_K - \Omega_L - \omega_2)^2 + \Gamma_K^2} . \tag{15}$$

Since the absorption rate depends solely on the K absorption process it varies on an energy scale determined by the K shell binding energy. In this experiment this means that it is slowly varying and may be evaluated at threshold, i.e. $W^{Abs}(0)$. The frequency dependence which is present comes from the energy denominator in Eq. (15). A typical idealized spectrum in this case is shown in Fig. 3.

The peak position and intensity of the line obtained in a synchrotron Raman experiment performed on Cu are shown in Figs. 7 and 8. The absorption edge in Cu is roughly speaking at the break in the curve in Fig. 7. The dotted straight line is a plot of the shift expected if the experimental resolution function were infinitely narrow. The nonlinear solid curve is the theoretical curve obtained using Eq. (15) but putting in finite resolution effects. The fit is excellent. An integral over Eq. (15) yields the solid line in Fig. 8. The quantity ΔE is the energy from threshold.

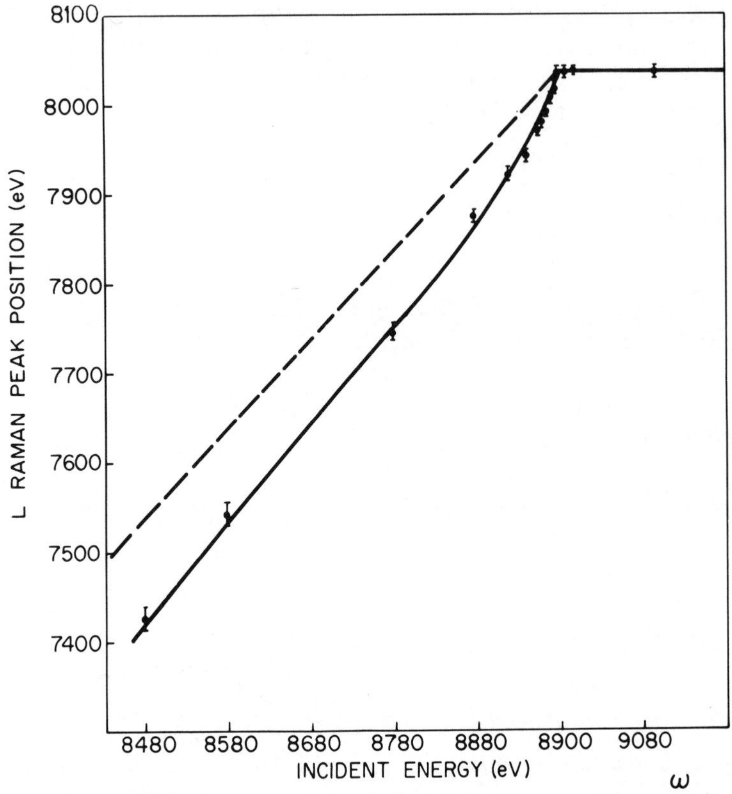

Fig. 7. The dependence of the peak position of the L electron to continuum transition as the incident frequency ω_1 is tuned in the region of the K absorption edge. The solid line is the result of theoretical calculations using Eq. (15) smeared with an experimental Gaussian resolution function to take account of natural line width and spectrometer effects.

Within the limitations on resolution inherent in this study we see that the agreement between theory and experiment for the L resonant Raman transition is very good. However, the importance of this study lies not so much in this agreement as it does in revealing that high resolution studies will be possible with synchrotron radiation. Such studies could provide additional information about absorption edge phenomena and extended fine structure. In addition, we should point out that these techniques need not be limited only to transitions involving the L electrons but can also be used to study other more weakly bound electrons with the conduction electrons being of special interest.

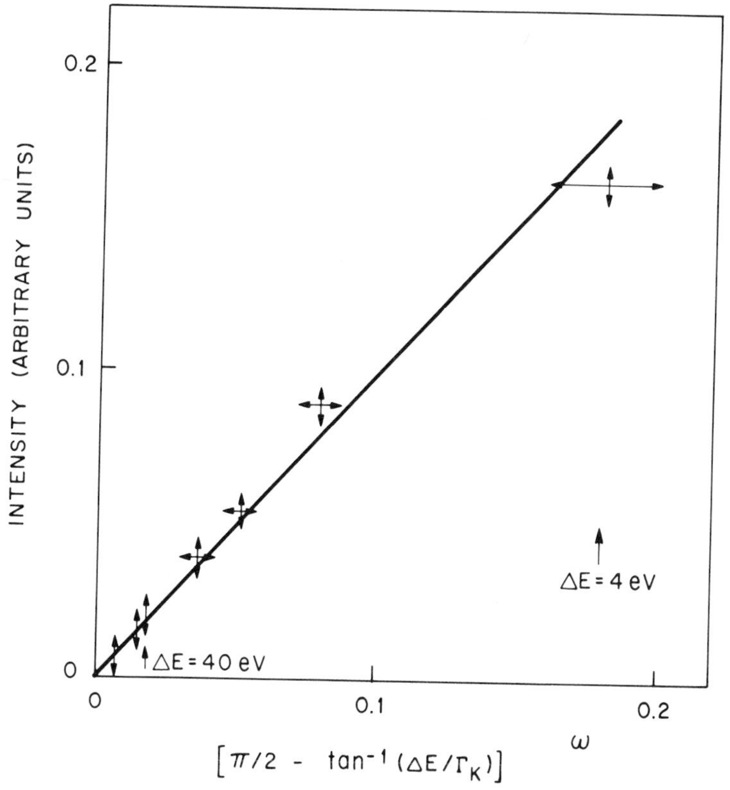

Fig. 8. The integrated intensity of the L electron to continuum transition as the incident frequency ω_1 is tuned in the region of the K absorption edge. The solid line is the result obtained by integrating Eq. 15 over ω_2.

REFERENCES

1. R. W. Jones, The Optical Principles of the Diffraction of X-Rays. (Cornell U.P. New York 1965).
2. P. M. Platzman and N. Tzoar, Phys. Rev. 139, 410 (1965). W. Phillips and R. J. Weiss, Phys. Ref. 171, 790 (1968); M. Cooper and J. A. Leake, Phil. Mag. 15, 1201 (1967); R. J. Weiss and W. Phillips, Phys. Rev. 176, 900 (1968); P. Eisenberger and P. M. Platzman, Phys. Rev.
3. P. Eisenberger and W. C. Marra, Phys. Rev. Lett. 27, 1413 (1971); P. M. Platzman and P. Eisenberger, Phys. Rev. Lett. 33, 152 (1974); P. Eisenberger and W. A. Reed, Phys. Rev. 9B, 3237 (1974).

4. P. M. Platzman, Lectures at NATO Advanced Study Institute on Elementary Excitations in Solids, Molecules and Atoms, Ed.; J. Devreese, A. B. Kunz and T. C. Collins, Plenum Press, New York (1974).
5. C. J. Sparks, Jr., Phys. Rev. Lett. 33, 262 (1974); Y. B. Bannett and I. Freund, Phys. Rev. Lett. 34, 372 (1975).
6. S. Doniach, P. M. Platzman and J. Yue, Phys. Rev. 4B, 3345 (1971).
7. C. Kunz, R. Haensel, G. Keitel, P. Schreiber and B. Sonntag, "Electronic Density of States"; L. H. Bennett ed. (Nat. Bur. Stand. Spec. publication No. 323 U.S. Govt. Print. Office Washington, D.C.).
8. P. Nozieres and C. deDominics, Phys. Rev. 178, 1097 (1969).
9. J. Dow, B. F. Sonntag, Phys. Rev. Lett. 31, 1461 (1973).
10. P. Gibbons and S. Schnatterly (private communications).
11. P. Eisenberger, P. M. Platzman and H. Winick, Phys. Rev. (to be published).

RAMAN SCATTERING BY LOCAL PLASMONS AND ELECTRON-HOLE DROPS

Yu. E. Lozovik, V. N. Nishanov and V. I. Yudson

Institute of Spectroscopy
Academy of Sciences of the USSR
Podolskii r-n, P.O. Academgorodok, Moscow, USSR

INTRODUCTION

In the last few years a number of works appeared dealing with electron-hole drops in semiconductors, small metallic and semiconductor particles in a host lattice and gas cavities (voids) generated in metals under irradiation. In this report the localized plasma oscillations (LPO) in the above-mentioned systems and Raman scattering of light or x-rays by excitation of localized plasmons will be considered. The main attention will be given to some general questions of the nonuniform electron gas theory and to the investigation of LPO damping.

LINEAR RESPONSE IN FINITE AND NONUNIFORM SYSTEMS

Collective characteristics of a nonuniform* system may be investigated by means of the dielectric function $\hat{\epsilon}(\Omega, \vec{x}, \vec{x}')$ describing the "density-density" linear response. Normal oscillations (collective excitations) of the system can be obtained with the use of the $\hat{\epsilon}$ operator from the equation

$$\int \hat{\epsilon}(\Omega, \vec{x}, \vec{x}')\, n^{eff}\, d\vec{x}' = 0 \qquad (1)$$

*The general questions of inhomogenious electron gas theory, e.g., are investigated in [2, 3].

where Ω is frequency, $n^{eff}(\vec{x})$ is density of the effective charge in the system (for a detailed analysis of linear responses in uniform systems see [1]).

With the aid of the matrix $\epsilon(\Omega, \vec{x}, \vec{x}') \equiv \langle \vec{x} | \epsilon(\Omega) | \vec{x}' \rangle$ one can express also the other characteristics of a nonuniform system such as the correlation energy* E_c:

$$E_c = -\frac{i}{4\pi} \int_{-\infty}^{\infty} d\Omega \, \mathrm{Sp}[\ln \hat{\epsilon}(\Omega) + \hat{\epsilon}(\Omega) - 1] \, ,$$

the photoabsorbtion cross-section:

$$\sigma(\Omega) = \frac{\omega}{c} \mathrm{Im} \int d\vec{x} \, d\vec{x}' \, \langle \vec{x} | \hat{\epsilon}^{-1}(\Omega) | \vec{x}' \rangle$$

and the Raman light scattering cross-section (per unit solid angle):

$$\sigma'(\Omega, \vec{q}) = - r_0^2 (\vec{e}_i \cdot \vec{e}_2) \frac{\omega_2}{\omega_1} \frac{q^2}{4\pi} \int d\vec{x} \, d\vec{x}' \, e^{i\vec{q}(\vec{x}-\vec{x}')} \mathrm{Im} \langle \vec{x} | \hat{\epsilon}^{-1}(\Omega) | \vec{x}' \rangle \tag{2}$$

In Eq. (2), $\vec{e}_1$, $\vec{e}_2$ are polarization vectors of incident and scattering photons; ω_1, ω_2 are their frequencies; $\vec{q}$ is the transmitted momentum; and r_0 is the Compton radius of an electron.

The dielectric function $\hat{\epsilon}(\Omega, \vec{x}, \vec{x}')$ is connected with the polarization operator of the electron system $\hat{\Pi}$ by the relation $\hat{\epsilon} = 1 - \hat{\Pi}\hat{V}$ (where $\hat{V}$ is the operator of the Coulomb interaction) or in the coordinate representation:

$$\epsilon(\Omega, \vec{x}, \vec{x}') = \delta(\vec{x} - \vec{x}') - \int d\vec{x}'' \, \Pi(\Omega, \vec{x}, \vec{x}'') \frac{1}{|\vec{x}' - \vec{x}''|} \tag{3}$$

For the investigation of LPO it is more convenient to use instead of Eq. (1) another equation obtained from Eq. (1) by passing from the "oscillating" charge density $n(\vec{x})$ to the corresponding potential $\Phi(\vec{x})$:

---

*Below we put $e = \hbar = m = 1$ (unless otherwise specified).

$$-\frac{1}{4\pi}\Delta\Phi(\vec{x}) = n(\vec{x}) \tag{4}$$

Putting Eqs. (3) and (4) in Eq. (1), integrating by parts and using the equality

$$\Delta_{x_2}\frac{1}{|\vec{x}_1-\vec{x}_2|} = -4\pi\delta(\vec{x}_1-\vec{x}_2),$$

we obtain the equation for the localized oscillations:

$$\Delta\Phi(\vec{x}) - 4\pi e^2\int\Pi(\Omega,\vec{x},\vec{x}')\Phi(\vec{x}')d\vec{x}' = 0 \tag{5}$$

In deducing Eq. (5) we may omit the surface integrals (appearing from integrating by parts) if the integrated functions vanish at infinity; so Eq. (5) holds only for the local modes.

Note in conclusion that in a nonuniform system (in contrast to a uniform one) there exists a number of nonequal dielectric operators. In particular, the "potential-potential" linear response $V^{ext} = \hat{\epsilon}'(\Omega)V^{eff}$ is characterized by the operator $\hat{\epsilon}(\Omega) = 1 - \hat{V}\hat{\Pi}$, connected with the operator $\hat{\epsilon} = 1 - \hat{\Pi}\hat{V}$ (characterizing the "density-density" linear response) by a unitary transformation. Indeed, we have $\hat{\epsilon}(\Omega)n^{eff} = (1 - \hat{\Pi}\hat{V})n^{eff} = n^{ext}$. Introduce the operator $\hat{V}^{-1} = -1/4\pi\,\Delta$ inverse to the Coulomb operator ($\hat{V}^{-1}\hat{V} = 1$). Then $n^{ext} = \hat{V}^{-1}\hat{V}^{ext}$; $n^{eff} = \hat{V}^{-1}\hat{V}^{eff}$, so that: $(1 - \hat{\Pi}\hat{V})\hat{V}^{-1}\hat{V}^{eff} = \hat{V}^{-1}\hat{V}^{ext}$. Acting with the operator $\hat{V}$ from the left on the last equation we obtain: $(1 - \hat{V}\hat{\Pi})V^{eff} = \hat{V}^{ext}$ so, that indeed $\hat{\epsilon}'(\Omega) = 1 - \hat{V}\hat{\Pi} = \hat{V}\hat{\epsilon}(\Omega)\hat{V}^{-1}$; (different questions concerning linear responses in nonuniform systems will be analyzed elsewhere [4]).

EQUATION FOR LPO FREQUENCY AND DAMPING

One of the most important aspects of the LPO theory is the analysis of the LPO damping, because only the smallness of the damping Γ in comparison with the LPO frequency ω allows one to consider LPO as a well defined elementary excitation; at the same time an additional channel (in comparison with a uniform medium) of damping is opened for LPO and so the LPO level width may become significant.

As is known, in a uniform dense electron gas when the random phase approximation RPA holds, the decay of a plasmon into an "electron-hole" pair is strictly forbidden (at temperature $T = 0$) by energy and momentum conservation if the plasmon momentum is smaller than some critical k_c; $k_c \sim k_{TF}$, where k_{TF} is the Thomas-Fermi screening momentum.

The situation changes in the nonuniform systems under consideration because the decay into one pair is always now possible. It is associated with the fact that due to the nonuniformity the momentum becomes a bad quantum number and is not conserved for the decay of LPO. In other words, a loss or gain of a momentum of the order of $\hbar/R$ during scattering in nonuniform media is always possible (R is the nonuniformity length). The absence of momentum conservation allows a plasmon decay into one pair if $\hbar/R$ is not much smaller than $k_c \sim k_{TF} = \hbar/r_{TF}$.

The polarization operator of a system is, in general, complex and its imaginary part corresponds to the different channels of the excitation damping. Therefore, the eigenvalue of Eq. (5) is complex: $\Omega = \omega + i\Gamma$ where ω is the frequency, Γ is the damping. When $|\mathrm{Im}\,\hat{\Pi}| << |\mathrm{Re}\,\hat{\Pi}|$, the damping Γ is small ($\Gamma << \omega$) and the problem of finding ω and Γ may be solved in two steps. First ω is obtained from the equation:

$$\Delta\Phi_o(\vec{x}) - 4\pi e^2 \int \mathrm{Re}\,\hat{\Pi}(\omega, \vec{x}, \vec{x}')\Phi_o(\vec{x}')d\vec{x}' = 0 \tag{6}$$

then the damping is given by the relation, deduced by expansion of Eq. (6) for small $\mathrm{Im}\,\Pi$ and Γ (taking into account Eq. (6)).

$$\Gamma = -\frac{\int d\vec{x}d\vec{x}'\Phi_o(x)\,\mathrm{Im}\,\Pi(\omega, \vec{x}, \vec{x}')\Phi_o(\vec{x}')}{\int d\vec{x}d\vec{x}'\Phi_o(\vec{x})(\frac{\partial}{\partial\omega}\mathrm{Re}\,\Pi(\omega, \vec{x}, \vec{x}'))\Phi_o(\vec{x}')} \tag{7}$$

We now use the RPA. In the RPA the polarization operator of the system is taken in the lowest order in the Coulomb interaction and characterized by a simple "particle-hole" bubble. We have:

$$\mathrm{Re}\,\Pi(\vec{x}, \vec{x}', \omega) = P\sum_{\nu,\mu} \frac{\theta_\nu - \theta_\mu}{\epsilon_\mu - \epsilon_\nu + \omega} \chi_\nu(\vec{x})\chi^*_\mu(\vec{x})\chi^*_\nu(\vec{x}')\chi_\mu(\vec{x}') \tag{8}$$

$$\operatorname{Im}\Pi(\vec{x},\vec{x}',\omega) = \pi \sum_{\nu,\mu} (\theta_\nu - \theta_\mu)\delta(\epsilon_\mu - \epsilon_\nu + \omega)\chi_\nu(\vec{x})\chi^*_\mu(\vec{x})\chi^*_\nu(\vec{x}')\chi_\mu(\vec{x}) \tag{9}$$

where $\chi_\nu(\vec{x})$ and ϵ_ν are the wave function and energy level of the electron in the Hartree-Fock approximation; $\theta_\nu = \theta(\epsilon_\nu - \epsilon_F)$ is the Fermi-distribution function at T = 0.

The equation for the LPO frequency contains $\operatorname{Re}\Pi(\omega,\vec{x},\vec{x}')$. The latter is given by the infinite double sum Eq. (8) of quantities which should be found from Hartree-Fock equations. It is clear that even the problem of the analytical calculation of $\operatorname{Re}\Pi$ cannot be solved for an arbitrary nonuniform system. But for finite systems we can use the quasi-electrostatic approximation (QEA) to find the LPO spectrum [5, 6]. The dispersion relation for LPO in this case is the following

$$\epsilon_2(\omega) = -\frac{L}{L+1}\epsilon_1(\omega) \tag{10}$$

Putting in Eq. (10) $\epsilon_1(\omega) = \epsilon_\infty(1 - \omega_0^2/\omega^2)$, where $\omega_0^2 = 4\pi ne^2/\epsilon_\infty m$, one finds [5, 6] the LPO frequencies in a metallic particle (in vacuum) or in an electron-hole drop (in the latter case m* is the reduced mass of electron and hole):

$$\omega = \omega_0\sqrt{\frac{L}{2L+1}} \tag{11}$$

and in a void ($\epsilon_1 = 1$, $\epsilon_2(\omega) = 1 - \omega_p^2/\omega_2$)

$$\omega = \omega_0\sqrt{\frac{L+1}{2L+1}} \tag{12}$$

Analogously, for the LPO near a boundary of a conducting particle placed into another conductor we have:

$$\omega = \sqrt{\frac{\omega_{01}^2 L + \omega_{02}^2(L+1)}{2L+1}} \tag{13}$$

It is easy to find the polarization operator $\Pi(\omega)$ in the quasi-electrostatic approximation (with regard to $\epsilon(\omega) = 1 - \omega_p^2/\omega^2$, $\omega_p^2 = 4\pi ne^2/m$).

$$\Pi(\omega) = -\frac{n}{\omega^2}\Delta - \frac{n}{\omega^2}\delta(r-R)\nabla \tag{14}$$

Note that the QEA, completely neglecting the spatial dispersion effects in an electron gas, corresponds formally to the case $r_{TF} \to 0$, i.e., it holds for the problem considered when $r_{TF}/R \ll 1$.

Equation (14) can be derived in the RPA by expanding $\Pi(\omega, \vec{x}, \vec{x}')$ [7-9] into series in powers of $(\epsilon_\mu - \epsilon_\nu)/\omega$. Using simple sum rules we easily find:

$$\mathrm{Re}\,\hat{\Pi}(\omega, \vec{x}, \vec{x}') = \lim_{\substack{\vec{x}_1 \to \vec{x}' \\ \vec{x}_2 \to \vec{x}}} 2 \sum_{k=1}^{\infty} \frac{1}{\omega^{k+1}} \sum_{\ell} C_k^{\ell} (-1)^{\ell} \hat{H}^{\ell}_{\vec{x}_1} \rho(\vec{x}_1, \vec{x}) \hat{H}_{\vec{x}_2} \delta(\vec{x}_2 - \vec{x}') \tag{15}$$

where $\hat{H}$ is the one-particle Hamiltonian in the Hartree-Fock approximation, and $\rho(\vec{x}, \vec{x}') = \sum_\nu \theta_\nu \chi_\nu^*(\vec{x}) \chi_\nu(\vec{x}')$ one-particle density matrix. In the lowest approximation in ω^{-2} we obtain from Eq. (15):

$$\mathrm{Re}\,\hat{\Pi}(\omega) = -\frac{n(\vec{x})}{\omega^2}\Delta - \frac{\nabla n}{\omega^2}\cdot\nabla \tag{16}$$

where $n(\vec{x})$ is electron density distribution. Then Eq. (5) for the LPO has the form:

$$\Delta \Phi_o(\vec{x}) - \frac{4\pi}{\omega^2}\nabla(n(\vec{x})\nabla\Phi_o(\vec{x})) = 0 \tag{17}$$

Note that in this approximation spatial dispersion is absent and therefore Eq. (16) corresponds to QEA. Indeed when $n(\vec{x}) = n_0\theta(r-R)$, Eq. (16) coincides with Eq. (14). Obviously Eq. (16) may be represented in the form $\nabla(\vec{E}\epsilon(\omega)) = 0$ (where $\vec{E} = -\nabla\Phi$, and $\epsilon(\omega) = 1 - \omega_p^2/\omega^2$), which illustrates the sense of the QEA.

In the next approximation ($\sim 1/\omega^4$) the correction to the polarization operator already takes into account the spatial dispersion.

The resulting equation for the LPO has the form:

$$\Delta \Phi_o(\vec{x}) - \frac{1}{\omega^2} \nabla (n(\vec{x}) \nabla \Phi_o(\vec{x})) - \frac{12\pi v_F^2(\vec{x}) n(\vec{x})}{5 m \omega^4} \Delta^2 \Phi_o(\vec{x}) = 0 \qquad (18)$$

where $v_F(\vec{x})$ is the Fermi-velocity, $v_F(\vec{x}) = \frac{\hbar}{m}[3\pi^2 n(\vec{x})]^{1/3}$.

Deducing Eq. (18) from Eq. (15) we neglected in Eq. (15) terms small in the interaction strength parameter r_s, and used the convergence condition of the series Eq. (15), namely $\ell_1 >> r_{TF}$, where ℓ_1 is the characteristic length for the function $\Phi_o^1(\vec{x})$ (see also [9]). Equation (18) differs from the LPO equation, deduced in [7, 8]. However if one neglects the gradients of $n(\vec{x})$ in Eq. (18) (i. e., it is assumed that $R >> \ell_1$, where R is the characteristic length for $n(\vec{x})$), then Eq. (18) can be written in the form $\nabla(\hat{L}\Phi_o(\vec{x})) = 0$ so we obtain the equation for the mode L = 0:

$$(1 - \frac{4\pi}{\omega^2} n(\vec{x})) \nabla \Phi_o(\vec{x}) - \frac{12\pi v_F^2(\vec{x}) n(\vec{x})}{5 m \omega^4} \nabla^3 \Phi = 0 \qquad (19)$$

Equation (19) goes into the LPO equation deduced in [7, 8] if one assumes $n(\vec{x})$ = const. in the last term. Note, however, that the assumption $\ell_1 >> R$ ($\ell_1 >> r_{TF}$) necessary for passing to Eq. (19), does not hold for the plasma oscillation considered in [7, 8] localized near a charged impurity (details in [9]); but the results [7, 8] are qualitatively correct.

CALCULATION OF THE LPO DAMPING

In order to find the width of the LPO levels it is necessary to calculate the LPO decay probability determined by the expression (7). We derive this quantity by calculating corresponding sums with the use of quasiclassical Hartree-Fock functions $\chi_\nu(\vec{r})$:

$$\chi_\nu(\vec{r}) = B_{\ell m} \sqrt{\frac{2}{\pi} \frac{\partial \epsilon_{\ell n}}{\partial n}} (2\epsilon_{\ell n} - \frac{\ell^2}{r^2})^{\frac{1}{4}} \cos \Phi_{n\ell}(r) p_\ell^m(\cos \theta) e^{im\varphi} \qquad (20)$$

where

$$\Phi_{n\ell}(r) = \int_{r_{n\ell}}^{r} (2\epsilon_{n\ell} - \frac{\ell^2}{r^2})^{\frac{1}{2}} dr - \frac{\pi}{4}$$

the normalizing constant is

$$B_{\ell m} = \frac{1}{\sqrt{2\pi}} \sqrt{\frac{2\ell+1}{2} \frac{(\ell-m)!}{(\ell+m)!}}$$

$r_{n\ell}$ are turning points, $\epsilon_{\ell n}$ quasiclassical energies. The use of $X_\nu(\vec{r})$ is justified here when:

$$k_R R >> 1 \tag{21}$$

We omit here rather long calculations (7, 9). The resulting formulas for the damping are, however, quite simple.

For a void:

$$L = 1, \quad \Gamma = 16e^2(\epsilon_F^2 - \omega_o^2 \sqrt{2/3})/3 R\omega_o^2 \tag{22}$$

For the LPO in a metallic particle (in vacuum) and in electron-hole drops:

$$L = 1, \quad \Gamma = 8e^2\epsilon_F^{3/2}(\epsilon_F + \omega_o\sqrt{1/3})^{\frac{1}{2}}/3\omega_o^2 R \tag{23}$$

where $\omega_o^2 = 4\pi ne^2/\epsilon_\infty m$, m is the electron mass (or reduced mass for an electron-hole drop). Calculations of Γ for $1 \ll L \ll k_F R$ show that for sufficiently large L the LPO broadening becomes comparable with the distance between the LPO levels.

For small R it is easily seen that the main contribution to the LPO damping is made by the LPO decay into an electron-hole pair (Landau damping) and not by the one-particle damping γ connected with the scattering by impurities, etc. ($\gamma \sim 10^{12}$ sec^{-1}). Indeed, the relation

$$\frac{\Gamma}{\gamma} \sim \frac{\omega_p}{\gamma} \frac{r_{TF}}{R},$$

so that $\Gamma > \gamma$, when $R < 10^3$ Å. Then estimation according to the above formula shows that the relative damping quantity

$$\frac{\Gamma}{\omega} \sim \frac{r_{TF}}{R}$$

is not small, for example, in the case of a void with radius $R \sim 20$ Å we emphasize that the value of Γ deduced in the RPA corresponds only to the channel of plasmon decay into one particle-hole pair; for the intermediate values of the density, e.g., in the case of an electron-hole drop and electron gas in metals, strictly speaking, it is necessary, to take into account the decay into a number of pairs.

One should expect, however, the calculated values of Γ to be correct in the order of magnitude even in these cases. For the electron gas in a semiconductor the density condition is held even for not extremely large electron concentrations (see e.g., [10]) and therefore the obtained values Γ are sufficiently reasonable.

RAMAN SCATTERING BY LPO

Raman scattering of x-rays and light in a uniform medium with excitation of bulk plasma oscillation was calculated microscopically (see [10, 11] and literature given there) and phenomenologically [12]. We calculate here Raman scattering with an excitation of LPO. For this purpose we may use expression (2) for a differential cross-section of Raman scattering. Intending to restrict ourselves to the quasielectrostatic approximation, it is convenient to use $\sigma(\omega)$ expressed in terms of a "density-density" correlation function [10]:

$$\frac{\partial^2\sigma}{\partial\Omega\,\partial\omega} = (\vec{e}_1\vec{e}_2)^2 r_o^2 \frac{\omega_2}{\omega_1} \int_{-\infty}^{\infty} \frac{dt}{2\pi} e^{i\omega t} \int \langle n(\vec{r}', t)n(\vec{r}', 0)\rangle e^{-i\vec{q}(\vec{r}-\vec{r}')} d\vec{r}d\vec{r}'$$

The above calculated x-dependence of the density n is determined to within the accuracy of normalization factors, as the density n which enters the uniform equation. Consequently, we must normalize it, so that the corrèsponding classical LPO energy [13]:

$$E = \frac{\partial}{\partial \omega}(\epsilon(\omega)\cdot\omega)\,|\nabla\varphi_\omega|^2$$

should be the equantum energy

$$E = \hbar\omega$$

where

$$\Delta\varphi_\omega = 4\pi e n_\omega$$

For qR << 1 (where $\vec{q}$ is the photon momentum transferred to the lattice) we obtain:

$$\frac{\partial^2\sigma}{\partial\omega\partial\Omega} = r_o^2(\vec{e}_1\vec{e}_2)^2\,\frac{\omega_2}{\omega_1}\,\frac{\pi\hbar\omega_p}{4e^2q}\,\frac{(qR)^{2L+1}\,C_L}{2^{2L+1}[\Gamma(L+\frac{3}{2})]^2}\,\delta(\omega-\omega_L)$$

$$C_L = \begin{cases} (L+1)^{\frac{1}{2}}(2L+1)^{3/2} & \text{a bubble in metal} \\ L^{\frac{1}{2}}(2L+1)^3/[L(\epsilon_o+1)+\epsilon_o]^{3/2} & \text{a particle} \\ \dfrac{L^{\frac{1}{2}}}{\epsilon_\infty^{3/2}}(2L+1)^{3/2} & \text{electron-hole drop} \end{cases}$$

REFERENCES

1. D. Pines, P. Nozières, The Theory of Quantum Liquids, W. A. Benjiamin, Inc., New York, 1966.
2. D. A. Kirzhnitz, Yu. E. Lozovik, G. V. Shpatakovskaya, Uspekhi Fys. Nauk 117, 3 (1975).
3. D. A. Kirzhnitz, Field-Theoretical Methods in Many-Body Systems, Oxford, Pergamon Press, 1967.
4. Yu. E. Lozovik (to be published).

5. C. Kittel, Quantum Theory of Solids, New York, 1963.
6. A. A. Lucas, Phys. Rev. B7, 3527 (1973).
7. E. A. Sziklas, Phys. Rev. 138, A 1070 (1965).
8. L. J. Sham, Localized Excitation in Solids, Ed. R. Wallis, Plenum Press, New York, 1968.
9. Yu. E. Lozovik, V. N. Nishanov (to be published).
10. P. M. Platzman, P. A. Wolf, Waves and Interactions in Solid State Plasmas, Academic Press, New York - London, 1973.
11. I. I. Sobel'man, E. L. Feinberg, ZhEFT 34, 494 (1958).
12. V. M. Agranovich, V. L. Ginzburg, ZhETF 40, 913 (1961).
13. L. D. Landau, E. M. Lifshitz, Electrodynamics of Continious Media, New York, Pergamon Press, 1960.

RAMAN SCATTERING IN METALS AND HEAVILY DOPED SEMICONDUCTORS

I. P. Ipatova and A. V. Subashiev

A. F. Ioffe Physico-Technical Institute
Leningrad, USSR

INTRODUCTION

Raman scattering by lattice vibrations occurs through the electrons. In metals and heavily doped semiconductors the electron-phonon interaction defines not only the intensity of the scattering, but also the shape of the spectra.

First experiments on metals showed the line width to be larger than in the case of dielectrics.[1] Subsequent measurements on heavily doped semiconductors revealed considerable concentration dependence of the Raman frequency.[2,3] Therefore, the study of the Raman effect in these materials enables us to understand both the formation of the phonon spectrum by electrons and the electron-phonon interaction itself.

To describe the electron-phonon interaction we shall make use of a Hamiltonian of the Fröhlich type:

$$H_{int} = \sum_{mn;\vec{p}\vec{q};j} \langle \vec{p}n | V_j e^{i\vec{q}\vec{r}} | \vec{p}+\vec{q};m \rangle A^+_{\vec{p}n} A_{\vec{p}+\vec{q};m} (B^+_{-\vec{q}j} + B_{\vec{q}j}) \tag{1}$$

where

$$\langle\vec{p};n|V_j(\vec{r})e^{i\vec{q}\vec{r}}|\vec{p}+\vec{q};n\rangle$$

is a matrix element of the electron-phonon interaction, calculated using Bloch wave functions,

$$V_j(\vec{r}) = \sum_s V_s^{\alpha}(\vec{r})\xi_s^{\alpha}(\vec{q}j) \qquad (2)$$

where s labels the atom in the unit cell, $\xi_s^{\alpha}(\vec{q}, j)$ is the polarization vector of the lattice vibrations of the branch j with momentum $\vec{q}$; $B^{+}_{\vec{q}j}$ and $B_{\vec{q}j}$ are phonon creation and annihilation operators; $A^{+}_{\vec{p}n}$ and $A_{\vec{p}n}$ are electron creation and annihilation operators for electrons of the band n with momentum $\vec{p}$.

Due to the momentum conservation law only the long-wave length optical phonons contribute to the Raman scattering. In the long-wavelength limit the optical vibrations reduce to a uniform displacement of the sublattices, which changes the electron wave functions. Therefore, in general, the interaction with optical vibrations should not vanish when $\vec{q} \to 0$.

The matrix element is defined by both the contribution of the short-range forces and long-range forces. In the case of short-range forces the diagonal matrix elements $\langle\vec{p}n|V_j(\vec{r})|\vec{p}n\rangle$ depend essentially on the electron quasi-momentum. This dependence is determined by the symmetry of the long-wave optical phonon.[4]

In a crystal with a center of inversion, for lattice vibrations with $\vec{q} = 0$, corresponding to even nonidentical representations of the group of the wave vector, the quantity $V_j(\vec{p})$ should satisfy the conditions

$$\sum_{\vec{p}} \langle\vec{p}n|V_j(\vec{r})|\vec{p}n\rangle_{sh-r}\, \delta(\epsilon - \epsilon_{\vec{p}}) = 0 \qquad (3)$$

$$\sum_{\vec{p}} \langle\vec{p}n|V_j(\vec{r})|\vec{p}n\rangle_{sh-r}\, \frac{\partial\epsilon_{\vec{p}}}{\partial\vec{p}}\, \delta(\epsilon - \epsilon_{\vec{p}}) = 0 \qquad (4)$$

The condition (3) implies the absence of a change of electron density in a uniform displacement of the sublattices. The condition (4) corresponds to the absence of an electron current in the uniform displacement of the sublattices. It has been shown in (4) that Eqs. (3) and (4) guarantee no screening for short-range electron-phonon interaction effects.

For long-range forces the diagonal matrix element is also non-zero in the long-wavelength limit. In homopolar crystals of diamond symmetry one has[5]

$$\lim_{\vec{q} \to 0} \langle \vec{p}n | V_j(\vec{r}) | \vec{p} + \vec{q}, n \rangle_{\ell\text{-}r} = W_j^{\alpha\beta}(n) \frac{q^\alpha q^\beta}{q^2} . \qquad (5)$$

Through this interaction lattice vibrations create the electron density vibrations and all the effects are screened.

In n - Ge, for example, the long-wavelength phonon of short-range nature could shift valleys in such a way that it provides the electron density and density of the current to be zero when averaged over the Brillouin zone (see Fig. 1a). The effect of long-range forces is shown in Fig. 1b. Vibration of electron

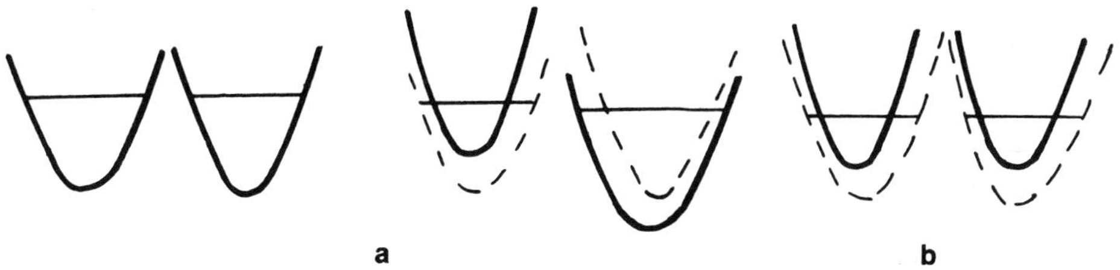
a b

Fig. 1

density and density of current accompanies the valley shift and all the effects of the electron-phonon interaction are screened.

EFFECTS OF ELECTRON-PHONON INTERACTION IN THE CASE OF NONDEGENERATE ELECTRON BANDS

We consider first the case of nondegenerate electron bands, when the electron band separation near Fermi level is much larger than the phonon frequency. The electronic contribution to phonon renormalization is defined by the interband electronic transitions.

The renormalization of the long-wavelength optical phonon due to short-range interaction (1) has been studied in Ref. 4. It was shown that the dimensionless parameter ζ resembles the one in Migdal's theory[6] of the acoustic phonon renormalization in metals

$$\zeta = \frac{1}{2\pi^2}\left(\frac{D}{\hbar\omega_o}\right)^2 \frac{m}{M}\,\frac{p_oA}{\hbar} \tag{6}$$

where D is the deformation potential constant, ω_o is the frequency of the long-wavelength optical phonon. In metals ζ is of the order of unity. In semiconductors $p_o << \hbar/A$ and $\zeta << 1$. Therefore, perturbation theory can be used. In the lowest approximation of perturbation theory, the phonon damping and phonon frequency shift is determined by the imaginary and real parts of the polarization operator π_j,[4]

$$\pi_j(q) = \frac{4}{\hbar\omega_{qj}} \sum_{\vec{p}} \frac{V_j^2(\vec{p})(n_{\vec{p}+\vec{q}} - n_{\vec{p}})}{\hbar\omega - \epsilon_{\vec{p}+\vec{q}} - \epsilon_{\vec{p}} + isqn\omega} \,. \tag{6'}$$

It follows from (6') that for small q' there should be a region of strong dispersion of the optical branch. This results from the fact that for $q > \omega_o/V_F$ the phase velocity of the phonon is less than V_F and the electron distribution easily follows slow vibrations of the ions. There is also Landau-type damping of the phonon due to the electron moving in phase with the wave.

However, there is nonadiabatic region $\omega_o/q > V_F$ where electrons are slow and the lattice is quick. There is no renormalization and no Landau-damping. The qualitative behavior of the phonon spectrum is shown in Fig. 2.

There is the anomaly of the phonon spectrum due to the threshold of Landau-damping at $q = \omega_o/V_F$. The singularity is of the type $\zeta \ln(\omega - qV_F)$. For arbitrary direction q, the singularity is screened as the result of the sharp anisotropy of the electron distribution over the Fermi surface produced by the phonon. Therefore, only the singularity in the derivative

$$\left(\frac{\partial\omega}{\partial q}\right)_{\hbar\omega = qV_{max}}$$

remains. If q is along a symmetry direction there is no screening, similar to the case of zero sound in metals.

The region of the nonadiabatic renormalization of the optical vibration spectrum corresponds to small momenta

$$q \lesssim \frac{\omega_o}{V_F} << q_{max} .$$

It is precisely this region of momenta that is manifested in light scattering experiments. Investigations of this kind have become possible recently in connection with the use of laser light sources.

Light scattering in metals and heavily doped semiconductors occurs in the thin layer defined by the depth of penetration of the light into the crystal. In a metal this is the skin-depth and in a semiconductor it is the length determined by the magnitude of the absorption coefficient. Violation of the momentum conservation law for the wave vector component perpendicular to the surface of the sample leads in this case to the result that all phonons with $q \lesssim \hbar/\delta$ make a contribution to the light scattering cross-section.[7]

In the case of metals usually $\delta \approx 10^{-6}$ cm in the optical region and since $V_F \approx 10^8$ cm/sec and $\omega_o \approx 10^{13}$ sec^{-1}, we have $\hbar/\delta > q_o = \hbar\omega_o/V_F$. Therefore, both the adiabatic and nonadiabatic regions of the optical vibration spectrum make a contribution to the scattering. For $\hbar/\delta >> q_o$ the contribution of the adiabatic region should dominate. The large temperature independent Raman scattering line-width, which is experimentally observable in a series of metals, can be explained by strong Landau damping for the optical branch in the region $q \gtrsim \hbar\omega_o/V_F$. The comparison

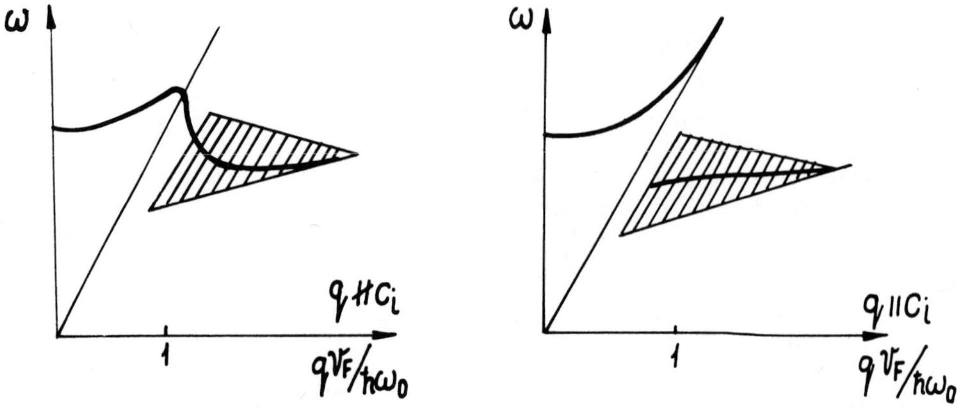

Fig. 2

of the theory and experimental data, obtained by Grant et al[8] is given in Table 1. In case of Be the cross-section is defined by the nonadiabatic region. There is no Landau damping and the width is small. Our theory predicts an asymmetric line with a tail on the short-wave side, due to the strong dispersion of the branch in the nonadiabatic region.[9]

Studies of scattering by lattice vibrations in heavily doped semiconductors are of great interest.[2,3] The electronic properties of these materials are in many respects analogous to those of metals. An important point is that the electron concentration in them is a variable parameter and we can investigate the concentration dependences. In the light frequency region in n-Ge with $n \approx 7 \times 10^{20}$ cm^{-3} and $V_F \approx 10^7$ cm/sec, $\delta \approx 10^{-4}$ cm, the optical phonon frequency $\omega_0 \approx 10^{14}$ sec^{-1} and thus $\hbar/\delta < q_0$. Under these conditions only the nonadiabatic region contributes to the scattering. The interaction with the optical phonons that has been

Table 1

M	ω_o, cm^{-1}	$\frac{q_o}{\hbar} = \frac{\omega_o}{V_F}$, cm^{-1}	$\frac{1}{\delta}$, cm^{-1}	$\Gamma = \begin{cases} \pi\zeta & \omega_o \delta / V_F \\ 0 & q > \frac{1}{\delta} \end{cases}$	
					exp
Cd	44	1.3×10^4	10^6	0.01	0.05^+
Zn	70	3×10^4	10^6	0.03	0.1^+ 0.02^{++}
Bi	100	3×10^4	0.6×10^6	0.2	0.1^{++}
Mg	120	6×10^4	0.7×10^6	0.09	0.1^{++}
Be	460	1.8×10^5	3×10^5	0	0.03^{++}

\+ experimental data of Ref. 1.
++ experimental data of Ref. 8.

studied in this paper occurs in Ge.[5] Light scattering experiments in heavily doped n-Ge have not exhibited an appreciable concentration shift or asymmetry of the Raman-scattering line. It may be thought, therefore, that the case $\hbar/\delta << q_o$ is realized in n-Ge experiments, i.e., the extreme long-wave part of the nonadiabatic region makes a contribution to the scattering.

INTERFERENCE RAMAN SCATTERING IN THE CASE OF DEGENERATE ELECTRON BANDS

The effect of electrons on Raman scattering becomes more complicated when there is band degeneracy and the separation between them is of the order of a phonon frequency. The Raman scattering line exhibits in this case considerable asymmetry and the shape of the spectrum depends on the frequency of the exciting light. Cardona et al[10] have shown that qualitative identification of the spectra can be based on the Fano theory of interference scattering.[11] The quantitative theory of the Raman spectra for the degenerate vibrational state and threefold degenerate electronic excitation spectrum can be formulated in terms of the standard Green function technique.[12]

The diagram technique was formerly applied by Wendin[13] to the problem of photoabsorbtion by heavy atoms. The cross-section of the Raman scattering due to band-to-band electron transitions is equal to[18]

$$\frac{\partial^2 \Sigma(k,k')}{\partial\omega\partial\theta} = 2\frac{\omega^4}{c^4}\, e_\alpha(k) e^*_\beta(k)\, \mathrm{Im}\, G_{\alpha\beta,\gamma\delta}(\omega,\omega') e^*_\gamma(k') e_\delta(k') \tag{7}$$

The imaginary part of the function $G_{\alpha\beta,\gamma\delta}(\omega,\omega')$ is equal to the Raman scattering tensor $i_{\alpha\beta,\gamma\delta}$. The analysis of the diagram representation of the perturbation theory series for $G_{\alpha\beta,\gamma\delta}$ is given in the Appendix. It leads to the unique decomposition of $G_{\alpha\beta,\gamma\delta}$ into resonant and non-resonant parts:

$$G_{\alpha\gamma,\beta\delta}(\omega,\omega') = G^{(0)}_{\alpha\gamma,\beta\delta}(\Delta\omega) + \sum_j \frac{R^{(j)}_{\alpha\gamma}(\omega,\omega')\, R^{(j)}_{\beta\delta}(\omega,\omega')}{\Delta\omega - \omega_{oj} - \pi_j(\Delta\omega)} \tag{8}$$

where $\Delta\omega = \omega - \omega'$; ω_o is the long-wavelength limit frequency of optical phonons with polarization j; $G^{(0)}_{\alpha\gamma,\beta\delta}(\Delta\omega)$, $R^{(j)}_{\alpha\gamma}$, $\pi_j(\Delta\omega)$ are slowly varying functions of $\Delta\omega$. In the case of weak electron-phonon interaction they could be treated by perturbation theory.

In the lowest approximation of perturbation theory $G^{(0)}_{\alpha\gamma,\beta\delta}(\omega,\omega')$ is the electron loop shown in Fig. 3a. The solid line represents the Green function of the band electron, the dot corresponds to the vertex of the electron-phonon interaction. An ordinary diagram calculation gives[14]

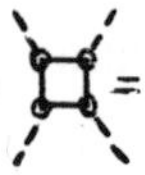

$$= \mathrm{Im}\, G^{(0)}_{\alpha\gamma,\beta\delta} = \pi \sum_{i,n,m,f} \frac{f^\alpha_{im} f^\gamma_{mf} f^\beta_{im} f^\delta_{nf}}{(\hbar\omega - \epsilon_m + \epsilon_f)(\hbar\omega - \epsilon_n + \epsilon_i)} (n_i - n_f) \tag{9}$$

$$\times\ \delta(\hbar\Delta\omega - \epsilon_f + \epsilon_i)\ .$$

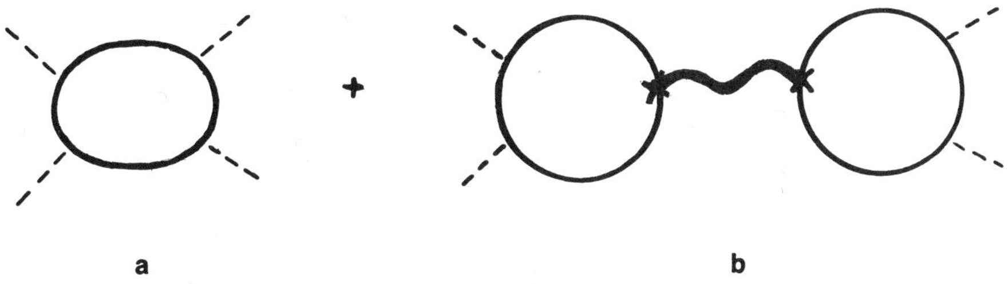

Fig. 3

The spectrum (9) is a continuous band within the interval from ω_{min} to ω_{max}, as shown in Fig. 4.

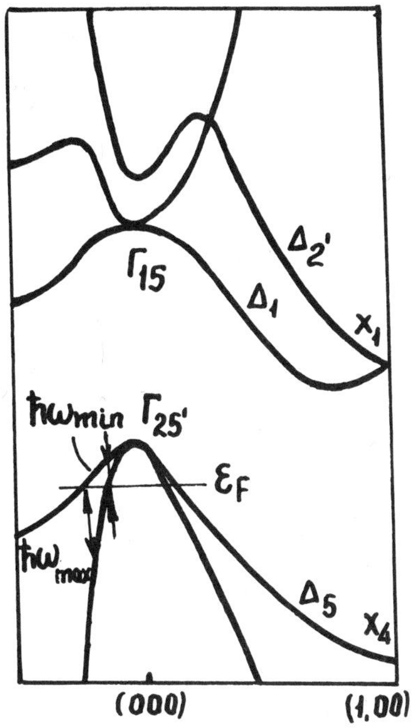

Fig. 4

In the lowest approximation the phonon polarization operator $\pi_j(\Delta\omega)$ equals

$$= \pi_j(\Delta\omega) = \sum_{m,n,p} |\langle n\vec{p}|V_j|m\vec{p}\rangle|^2 \frac{n(\epsilon_{n\vec{p}}) - n(\epsilon_{m\vec{p}})}{\hbar\Delta\omega - \epsilon_{m\vec{p}} - \epsilon_{n\vec{p}} + i\delta} \tag{10}$$

here "x" corresponds to the vertex of the electron-phonon interaction. The quantities $R^{(j)}_{\alpha\gamma}$ are complex electron polarizabilities calculated to first order in the electron-phonon interaction,

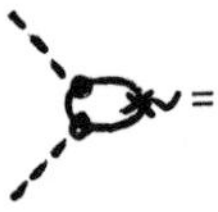

$$= R^{(j)}_{\alpha\gamma}(\Delta\omega) = \sum_{mnp} \frac{f^{\alpha}_{im} f^{\gamma}_{mf}}{\hbar\omega - \epsilon_m + \epsilon_f} \langle i\vec{p}|V_j|\vec{p}f\rangle \frac{n(\epsilon_{i\vec{p}}) - n(\epsilon_{f\vec{p}})}{\hbar\Delta\omega - \epsilon_{f\vec{p}} + \epsilon_{i\vec{p}} + i\delta} . \tag{11}$$

It is seen from (11) that $R^{(j)}_{\alpha\gamma}(\Delta\omega)$ is complex when band-to-band transitions contribute to Raman scattering.

The lattice Raman scattering cross-section is defined by the imaginary part of the diagram in Fig. 3b. It consists of two contributions. The first

$$\mathrm{Re}[R^{(j)}_{\alpha\gamma} R^{(j)}_{\beta\delta}]\ \mathrm{Im}\,[\Delta\omega - \omega_{oj} - \pi_j]^{-1}$$

leads to a resonant peak. The second

$$\mathrm{Im}\,[R^{(j)}_{\alpha\gamma} R^{(j)}_{\beta\delta}]\,\mathrm{Re}[\Delta\omega - \omega_{oj} - \pi_j]^{-1}$$

does not vanish within the interval $[\omega_{min}, \omega_{max}]$ and has an anti-resonant form. If the phonon energy $h\omega_o$ is of the order of the interval $[\omega_{max} - \omega_{min}]$ then overlapping of these two spectra occurs. The result is shown in Fig. 5.

As an example we consider the back-scattering geometry where the incident and the scattered light are polarized along OY and OZ correspondingly. Then the cross-section (7) contains only $G_{yz,yz}$. Just the phonon of OX polarization contributes to $R^{(j)}_{yz}$ in crystals of diamond symmetry.[15]

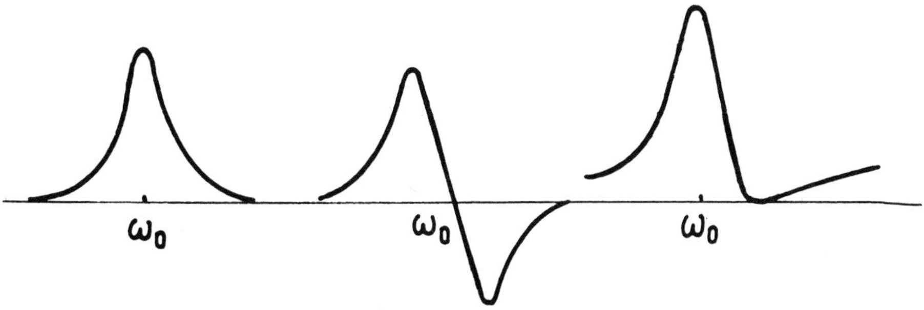

Fig. 5

Substituting (8) - (11) in (7) one gets the cross-section in the form

$$\frac{d^2\Sigma}{d\omega d\theta} = \sigma_o(\Delta\omega) + \sigma_1(\Delta\omega)\frac{(\epsilon + Q)^2}{\epsilon^2 + 1} \quad . \tag{12}$$

Here $\sigma_o(\Delta\omega)$ is the cross-section of nonresonant scattering

$$\sigma_o(\Delta\omega) = \frac{2\omega^4}{c^4} \text{ Im } G^{(0)}_{yz,yz} - \sigma_1(\Delta\omega) \, . \tag{13}$$

The first term is due to band-to-band nonresonant electron scattering. The second one is the correction to it due to the electron-phonon interaction

$$\sigma_1(\Delta\omega) = \frac{2\omega^4}{c^4} \frac{[\mathrm{Im}\, R_{yz}^{(x)}(\Delta\omega)]^2}{\mathrm{Im}\, \pi^{(x)}(\Delta\omega)} \tag{14}$$

and

$$\epsilon = \frac{\Delta\omega - \omega_o - \mathrm{Re}\, \pi^{(x)}(\Delta\omega)}{\mathrm{Im}\, \pi^{(x)}(\Delta\omega)} . \tag{15}$$

The most important quantity here is the Fano profile factor

$$Q = - \frac{\mathrm{Re}\, R_{yz}^{(x)}(\Delta\omega)}{\mathrm{Im}\, R_{yz}^{(x)}(\Delta\omega)} . \tag{16}$$

When $|Q| >> 1$ the line shape is Lorentzian. When $|Q| \approx 1$ we have an asymmetric Fano contour.

Since the Raman scattering by the lattice proceeds through the electrons, both $\mathrm{Re}\, R_{yz}^{(x)}$ and $\mathrm{Im}\, R_{yz}^{(x)}$ are proportional to the electron-phonon interaction. Therefore, Q in (16) is of the order of unity, and the spectrum should be asymmetric.

A similar situation occurs in exciton absorption in indirect gap semiconductors or for excitons near a saddle point where band-to-band absorption interferes with the resonant exciton line. In case of lattice absorption by the crystal with free carriers Q is inversely proportional to the small electron-phonon interaction and is large. Therefore, there should be no interference.

It has been shown by Fano[11] that in the case of discrete level interference with a one dimensional continuum, $\sigma_o = 0$ and the cross-section[14] vanishes at $\epsilon = -Q$. For the discrete level interference with the three-dimensional continuum $\sigma_o \neq 0$ and the cross-section does not vanish for any ϵ. In the back-scattering geometry discussed in our paper only the phonons of OX polarization are created. The electronic excitations are almost one dimensional and $\sigma_o \approx 0$. If the light-scattering leads to simultaneous creation of excitations of different symmetry, then $\sigma_o \neq 0$. For example, the continuous scattering spectrum results in p-Si from electron transitions between light and heavy hole bands. To analyze this we assume $\mathrm{Re}\,R_{yz}^{(x)}$ to be equal to its value in a pure crystal and consider

$$\mathrm{Im}\,R_{yz}^{(x)} = \frac{\pi^2}{m^2}\frac{\hbar}{\omega\omega'}\sum_{\vec{p}}\frac{\langle v_1\vec{p}|\vec{p}_y|c\vec{p}\rangle\langle c\vec{p}|p_z|v_2\vec{p}\rangle}{\epsilon_{c\vec{p}} - \epsilon_{v_1\vec{p}} - \hbar\omega} \tag{17}$$

$$\times \langle v_2\vec{p}|V^{(x)}|v_1\vec{p}\rangle[n(\epsilon_{v_1\vec{p}}) - n(\epsilon_{v_2\vec{p}})]$$

$$\times\, \delta(\epsilon_{v_2\vec{p}} - \epsilon_{v_1\vec{p}} - \hbar\Delta\omega).$$

Here $\langle v_1\vec{p}|p_4|c\vec{p}\rangle$ is the momentum matrix element between v- and c-bands. When the incident light frequency differs from the interband transition frequency $|\epsilon_{c\vec{p}} - \epsilon_{v\vec{p}} - h\omega| >> \Delta\omega \approx \omega_o$ one can neglect the p-dependence of the term in curly brackets. We then have

$$\mathrm{Im}\,R_{yz}^{(x)} = \frac{\pi}{m^2}\frac{\hbar}{\omega\omega'}\frac{\langle v_1|\vec{p}_y|c\rangle\langle c|p_z|v_2\rangle}{\epsilon_c^r - \epsilon_v^r - \hbar\omega}\langle v_1|V^{(x)}|v_2\rangle \tag{18}$$

$$\times \sum_{\vec{p}}[n(\epsilon_{v_1\vec{p}}) - n(\epsilon_{v_2\vec{p}})]\,\delta(\hbar\Delta\omega - \epsilon_{v\vec{p}} - \epsilon_{v_1\vec{p}}).$$

It follows from (14) that in the same approximation $\sigma \approx 0$.

Equations (17) and (18) enable us to find the concentration and temperature dependence of Q. The concentration dependence can be extracted in the simplest way when the electronic contribution to $\mathrm{Re}_{yz}^{(x)}$ is small. According to (10) and (18) the concentration dependence of $\mathrm{Im}\,R_{yz}^{(x)}$ coincides with that of $\mathrm{Im}\,\pi^{(x)}$. The range of integration over p is restricted by the energy conservation law $\hbar\omega_0 = \epsilon_{\vec{p}f} - \epsilon_{\vec{p}i}$ and by the factor $[n(\epsilon_{\vec{p}i}) - n(\epsilon_{\vec{p}f})]$. Therefore, there is no concentration dependence of

$\mathrm{Im}\,\pi^{(x)}$

for an isotropic electron energy spectrum. In the case of considerable anisotropy of $[\epsilon_{\vec{p}f} - \epsilon_{\vec{p}i}]$ the function $\mathrm{Im}\,\pi^{(x)}$ increases with concentration due to expansion of the integration range $[\omega_{max}, \omega_{min}]$. The product should not depend on temperature and concentration. Apparently this agrees with experimental data.[2,10]

Moreover, Eqs. (16), (17) allow one to find the relative sign of Q in n- and p-type silicon. For p-Si the main contribution to $\mathrm{Im}\,R^{(x)}_{yz}$ comes from electronic virtual transition between the Γ'_{25} and Γ_{15} bands. Therefore,

$$\mathrm{Im}\,R^{(x)}_{yz} \sim \frac{\langle v_1;xy|p_y|c;x\rangle\langle c;x|p_z|v_2;xz\rangle}{\epsilon_c^{(\Gamma)} - \epsilon_{v_i}^{(\Gamma)} - \hbar\vec{\omega}} \langle v_2;xz|V_o^{(x)}|v_1;xy\rangle. \tag{19}$$

At the Γ point

$$\langle v_1;xy|p_y|c;x\rangle = \langle c;x|p_z|v_2;xz\rangle$$

so that

$$\mathrm{Im}\,R^{(x)}_{yz} \sim \frac{|\langle v;xy|p_y|c;x\rangle|^2}{\epsilon_c^{(\Gamma)} - \epsilon_v^{(\Gamma)} - \hbar\omega} \langle v_2;xz|V_o^{(x)}|v_1;xy\rangle. \tag{20}$$

In n-type silicon the main contribution to $\mathrm{Im}\,R^{(x)}_{yz}$ is determined by the virtual transition between X_1 and X_4 bands

$$\mathrm{Im}\,R^{(x)}_{yz} \sim \frac{\langle c;X_1|p_y|v;X_4\rangle\langle v;X_4|p_z|c\bar{X}_1\rangle}{\epsilon_c^{(x)} - \epsilon_v^{(x)} - \hbar\omega} \langle c;\bar{X}_1|V_o^{(x)}|c;X_1\rangle. \tag{21}$$

The analysis of the symmetry of the states at the X point for the space group O_h^7, which includes nontrivial translations, gives the relation[16]

$$\langle c;X_1|p_y|v;X_4\rangle = -\langle c;\bar{X}_1|p_z|c;X_4\rangle$$

from which it follows

$$\mathrm{Im}\, R^{(x)}_{yz} \sim -\frac{|\langle c;X_1|p_y|v;X_4\rangle|^2}{(\epsilon^{(x)}_c - \epsilon^{(x)}_v - \hbar\omega)} \langle c\bar{X}_1|V^{(x)}_o|c;X_1\rangle. \tag{22}$$

All the experiments[2,3,10] show that Q does not change its sign with doping. Consequently, $\mathrm{Re}\, R^{x}_{yz}$ does not change its sign with doping either. It follows from Eqs. (20) and (22) that the experimentally established difference between the profile factor sign for p- and n-type silicon can be attributed to the difference between the signs of the electronic Raman tensors for scattering at the Γ point (p-type) and X point (n-type). From this fact we conclude that the sign of the matrix element of the electron-phonon interaction is the same for Γ'_{25} and Γ_{15} states.

According to Eq. (16) the lineshape is a function of the exciting laser energy. The frequency dependence of Q has been analyzed in (17).

APPENDIX

The function $G_{\alpha\gamma,\beta\delta}(\omega\omega')$ in (7) can be represented in the form:[18]

$$G_{\alpha\gamma,\beta\delta}(\omega\omega') = \frac{1}{2\pi i}\int_{-\infty}^{+\infty} dt e^{-i\Delta\omega t}\langle R_{\alpha\gamma}(t)\, R_{\beta\delta}(0)\rangle.$$

The standard diagram representation of $G_{\alpha\gamma,\beta\delta}$ is given in Fig. 6.

The diagrams of c, d type are resonant for $\Delta\omega \to \omega_o$. We select from the perturbation series all the diagrams of nonresonant type. The sum of these diagrams is $G^{(0)}_{\alpha\gamma,\beta\delta}$. The remaining diagrams could be classified by the number of resonant denominators. Every diagram of the set contains $R^{(j)}_{\alpha\gamma}$ and $R^{(j)}_{\beta\delta}$, which correspond to irreducible parts with two electron-photon interaction vertices and one vertex of the electron-phonon interaction with the resonant

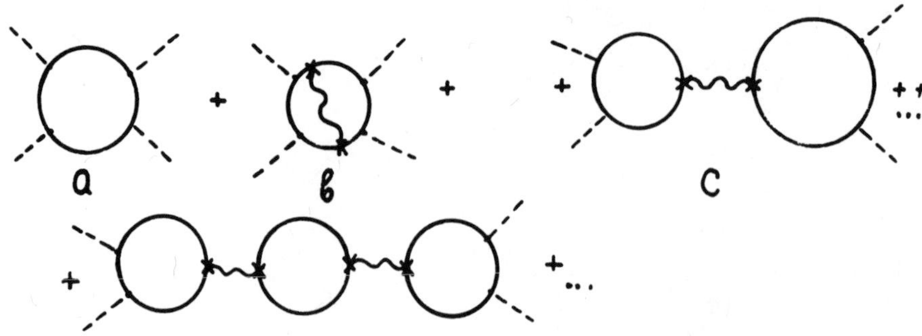

Fig. 6

phonon. The irreducible parts $R^{(j)}_{\alpha\gamma}$ and $R^{(j)}_{\beta\delta}$ are connected by the wavy phonon lines $1/(\omega - \omega_0)$ with an arbitrary number of inserted polarization loops $\pi_{jj'}$ (the diagrams with two vertices of the resonant phonon).

In cubic crystals

$$\pi_{jj'}(q)\big|_{q\to 0} = \pi_j \delta_{jj'} .$$

Eq. (8) is obtained by summation of the resulting progression.

REFERENCES

1. J. H. Parker, D. W. Feldman, M. Ashkin, Light Scattering Spectra of Solids, p. 389, NY, 1969.
2. R. Beserman, M. Jouanne, M. Balkanski, Proc. Intern. Conf. on Semiconductors (Warsaw), p. 1181, 1972.
3. F. Cerdeira, M. Cardona, Phys. Rev. B6, 1440, (1972).
4. I. P. Ipatova, A. V. Subashiev, JETP 66, 722 (1974).
5. G. L. Bir, G. E. Pickus, "Simmetrija i deformatcionnie effecti v poluprovodnicach", M, 1972.
6. A. B. Migdal, JETP 34, 1438 (1958).
7. D. L. Mills, A. A. Maradudin, E. Burstein, Light Scattering in Solids, NY, 1969, p399.
8. W. B. Grant, S. Hüfner, H. Schulz, L. Pelzl, Phys. Stat. Sol. 60, 330 (1973).

9. I. P. Ipatova, A. V. Subashiev, A. A. Maradudin, Light Scattering in Solids, Paris, p86, 1971.
10. F. Cerdeira, I. A. Fjeldly, M. Cardona, Phys. Rev. B8, 4734 (1974).
11. U. Fano, Phys. Rev. 124, 1866 (1961).
12. A. A. Abrikosov, L. P. Gorkov, Dsjaloshinskij "Metodi kvantovoj teorii polja v statisticheskoj phisike", M (1963).
13. G. Wendin, J. Phys. B (At. and Mol. Phys.), 3, 455 (1970).
14. D. Mills, R. F. Wallis, E. Burstein, Light Scattering in Solids, Paris, p107, 1971.
15. M. Poulet, J. P. Mathieu "Spectres de vibration et symmetrie des cristaux", Paris, 1970.
16. J. C. Hancel, H. Hasegawa, M. Nakayama, Phys. Rev. 138, A, 225 (1965).
17. M. Louanne, R. Beserman, I. P. Ipatova, A. V. Subashiev, Sol. State Commun. 16, 1077 (1975).
18. A. A. Maradudin, Sol. State Phys. 19, 1 (1966).

Section IV

High Intensity and Non-Linear Effects

INTERACTION OF ELECTROMAGNETIC WAVES WITH FREE ELECTRONS

Ya. B. Zeldovich

The Institute for Space Investigation
Academy of Sciences of the USSR
Moscow, USSR

In astrophysics one often has to deal with radiation-dominated plasmas.

Its characteristics are that the photon density is much greater than electron and nuclei densities, and also the probability of photon scattering by free electrons is much greater than the probability of photon absorption by a bound electron or an electron which is in the vicinity of a Coulombic centre (an ion).

The most rapid process is the isotropisation of the radiation field due to scattering.

In the second approximation the change of the photon spectrum during scattering must be accounted for. There are two superimposed contributions: the quantum recoil during scattering from electrons initially at rest,

$$\frac{\Delta\nu}{\nu} = -\frac{\hbar\nu}{m_e c^2}(1 - \cos\theta)$$

and the Doppler effect due to thermal motion of the electrons.

In this approximation the spectrum of radiation is readjusted. But as long as the bremsstrahlung and corresponding absorption are neglected, the photon number is conserved during pure scattering. The readjustment of the spectrum with additional condition of a given photon number leads after many collisions to a photon distribution which is given by the Bose-Einstein spectrum with a definite chemical potential of photons μ, with occupation number $[\exp(\hbar\nu + \mu/kT) - 1]^{-1}$.

The most interesting case is the universe as a whole described by a hot Big-Bang theory with strong hydrodynamics and perturbations. The perturbations (at least some of them, with the appropriate wavelength) are damped by radiative viscosity and heat conduction of the radiation-dominated plasma. But these dissipative processes are a source of input into the plasma. The energy density is greater than that corresponding to a Planckian equilibrium with given photon density. The final adjusted spectrum must be Bose-Einstein with positive $\mu > 0$.

Even small energy input leads (after readjustment) to strong effects in the longwave range:

$$F_\nu \sim \hbar \nu^3 (\exp(\mu/kT) - 1)^{-1}$$

instead of

$$F_\nu \sim \nu^2 kT .$$

The observations do not show deviations of the well-known "2.7° blackbody cosmic relic radiation" from the Rayleigh Jeans $\nu^2 kT$ law up to 20 cm. This yields an important upper limit to the amplitude of perturbations of the early universe.

Another field of application of scattering theory is to a rarefied plasma under the action of intense longwave radiation of quasars and pulsars. The radiation has an excessively high brightness temperature, $kT_b >> m_e c^2$, in the longwave range of the spectrum (T_b up to 10^{30} in pulsars!), but the spectrum is cut off at $\nu_c << kT_b/h$ so that the integrated energy density is $\epsilon << \sigma T_b^4$. In this case the induced scattering corresponds to the quadratic $n_1 n_2$ part of the familiar Bose expression $n_1(1 + n_2)$ for the probability of $1 \to 2$ scattering. It is not compensated by the $n_2 n_1$ factor for the $2 \to 1$ reverse process when the change of frequency for the recoil is accounted for. This means that the net effect has an extra small factor $\hbar\nu / m_e c^2$ when $kT_b >> m_e c^2$, and the occupation numbers $n_1, n_2 \sim kT_b/\hbar\nu$ are so great that induced scattering is more important than spontaneous scattering. The physical meaning of the process has been discussed and the evolution of the electron temperature and electromagnetic spectrum were studied. In this case Bose-condensation of photons is predicted, but it does not go smoothly. Theory predicts the possibility of shock wave formation in the spectrum, and the shock is accompanied by an oscillatory structure of the spectrum with quasilines whose width is of the order of the Doppler shift,

$$\Delta\nu/\nu \sim (kT_e/m_e c^2)^{\frac{1}{2}}.$$

The problems mentioned above are treated in a review article of the author, Soviet Physics Uspekhi, Vol. 18, No. 2, pp. 79-98 with 64 references (translation from Uspekhi Fizicheskich Nauk, Vol. 115, No. 2, 161, 1975).

RELATION BETWEEN LIGHT-SCATTERING CROSS-SECTIONS AND THE NONLINEAR OPTICAL SUSCEPTIBILITIES OF LIQUIDS AND SOLIDS

Robert W. Hellwarth

Dept. of Physics and Electrical Engineering
University of Southern California
Los Angeles, California 90007 USA

We establish the theoretical connection between the light scattering spectrum of a medium and its (third-order) nonlinear optical susceptibility. This connection is used to obtain the nonlinear susceptibility spectra.

INTRODUCTION

It has long been appreciated that the light scattering spectrum exhibited by a material is related to the same physical processes that give rise to the dc Kerr effect in nonpolar materials, and to the optical Kerr effect in all materials.[1] With the advent of lasers (quantum generators), many new nonlinear optical effects, closely related to the Kerr effect, have been discovered. These include:

(1) self-focusing, self-phase modulation, and the self-induced polarization changes of a monochromatic incident beam,
(2) stimulated Raman and combinational scattering,
(3) changes in refraction induced in a probe beam by the presence of a pump beam,
(4) electric-field-induced optical second harmonic generation and rectification,

(5) three-wave mixing and coherent anti-Stokes Raman spectroscopy (CARS),
(6) electric-field induced absorption.

All effects such as these are caused by a nonlinear term $\vec{p}^{(3)}$ in the electric polarization density which is third-order in the macroscopic electric field $\vec{E}$.

In this paper we develope the intimate connection of the differential light scattering cross sections of a material to the nonlinear susceptibility functions that govern the aforementioned nonlinear effects in the same material. We consider only effects involving electric fields and waves whose frequencies are well below any electronic excitation frequencies of the material. In such circumstances the Born-Oppenheimer (BO) approximation may be used. It yields relations which are valid even for liquids and solids whose quantum states are not known or well-understood. The plan of the paper is as follows. First we recall an old theorem relating the simple stimulated gain (Raman, combinational, Mandelstam-Brillouin, etc.) in any transparent medium to its differential light-scattering cross-sections. (This theorem is true even when the BO approximation is not.) Then we use the BO approximation to write a general expression for the nonlinear polarization density $\vec{p}^{(3)}(\vec{r}t)$ for any time-and space-dependent macroscopic electric electric field $\vec{E}(\vec{r}t)$. Next we calculate the gain stimulated in a probe beam at ω_s by a strong pump beam at ω_0 in terms of susceptibility functions appearing in the general expression. Then we are able to use the original gain-scattering theorem to relate the light-scattering cross-sections to these susceptibility functions that completely determine all the third-order nonlinear 'optical' effects in the material (except for a contribution by the purely electronic 'hyperpolarizability' which, if relevant, is determined by a separate measurement). By 'optical' we mean effects in which the frequencies involved are above those at which the vibrations and other movements of nuclei cause linear absorption.

To illustrate the above application of light-scattering we shall use the observed scattering spectra of several glasses to determine their nonlinear susceptibilities. For simplicity, we will limit most of our discussion to isotropic media, i.e., fluids and glasses. The treatment of crystals having lower symmetry follows closely that which we give here.[2]

THEORY

When a strong electric-field wave of angular frequency ω_o, (unit) polarization $\hat{e}_o$ and amplitude E_o propagates in a medium, it causes an exponential intensity gain g_{os} cm^{-1} for a second probe wave of ω_s and $\hat{e}_s$ (in addition to any losses that the wave might ordinarily experience). This gain is related to the cross-section $d^2\sigma_{os}/d\Omega d\omega_s$ (cm^{-1} sec) per unit volume to scatter a photon from $(\hat{e}_o, \omega_o)$ into a differential solid angle $d\Omega$ and frequency range $d\omega_s$ around the $(\hat{e}_s, \omega_s)$ mode. For a medium in thermal equilibrium at temperature T this is[3]

$$g_{os} = \frac{E_o^2 \pi^2 c^3 n_o (1-e^{-\hbar\Delta/kT})}{\hbar\omega_o \omega_s^2 n_s^2} \frac{d^2\sigma_{os}}{d\Omega d\omega_s} , \tag{1}$$

where n_o and n_s are the refractive indices for the incident and scattered waves and $\Delta \equiv \omega_o - \omega_s$. This gain is of course greatest at the peaks of resonances in the cross-section, falling to zero at large Δ and at $\Delta = 0$. In the thermal medium, (1) shows the gains to be positive when $\Delta > 0$, and negative (indicating stimulated loss) when $\Delta < 0$.

For the nonlinear effects we shall consider, where all of the frequencies involved are much lower than electronic frequencies, then there are two distinct physical contributions to $\vec{p}^{(3)}(\vec{r}t)$ (apart from electrostriction-induced strains, which we ignore). First, there is an "electronic" contribution from the nonlinear distortion of the electron orbits around the average positions of the nuclei. This polarization responds rapidly to field changes, within a few electronic cycles ($\sim 10^{-16}$ sec). In isotropic materials it contributes a term to $\vec{p}^{(3)}$ well approximated by the instantaneous form

$$\tfrac{1}{2}\vec{\sigma}E(\vec{r}t)E^2(\vec{r}t) \tag{2}$$

for frequencies well below the electronic band gap. The scalar coefficient σ is independent of temperature at fixed density. It may be called the 'electronic bulk hyperpolarizability'.

A second, "nuclear", contribution arises from an optical-field induced change in the motions of the nuclei; about these changed motions the electronic currents respond linearly to the optical fields. After the sudden impression of a field this contribution can be observed only following a time lapse of the order of the time ($\sim 10^{-12}$ sec) required for a nucleus to execute a vibrational or rotational cycle. In isotropic materials this contribution to $\vec{p}^{(3)}$ must be of the form (at some position $\vec{r}$)

$$\vec{E}(t) \int_{-\infty}^{\infty} ds\ a(t-s)E^2(s) + \int_{-\infty}^{\infty} ds\ \vec{E}(s)b(t-s)\vec{E}(t)\cdot\vec{E}(s) \tag{3}$$

for frequencies well below the electronic band gap. The nuclear response kernals, a(t) and b(t), vanish for $t < 0$ by causality, and generally exhibit a marked temperature dependence at constant density. The form (3) follows from the BO approximation by calculating the average over all states of the nuclei (in the ground electronic state) of the optical polarization density $\overleftrightarrow{\chi}(\{R_n\})\cdot\vec{E}(t)$, expanded to first order in the BO interaction Hamiltonian $-\frac{1}{2}\vec{E}(s)\cdot\overleftrightarrow{\chi}\cdot\vec{E}(s)$. Here, $\overleftrightarrow{\chi}(\{R_n\})$ is the local optical susceptibility operator for any configuration $\{R_n\}$ of the nuclei in some microscopic region (much smaller than optical wavelengths) around $\vec{r}$. The superposition of (2) and (3) is a much simpler form for $\vec{p}^{(3)}$ than is required generally outside the optical Born-Oppenheimer (BO) regime considered here. We now proceed to use this simplification to relate the Raman spectrum to the parameters a(t) and b(t). The electronic bulk hyperpolarizability can be determined by one extra measurement, or by a comparison of measurements of mixed nuclear and electronic origin, as we note later.[4]

Now we may try a field in (2) and (3) of the form

$$\vec{E}(t) = \mathrm{Re}(\hat{e}_o E_o e^{i\vec{k}_o\cdot\vec{r}-i\omega_o t} + \hat{e}_s E_s e^{i\vec{k}_s\cdot\vec{r}-i\omega_s t}) \tag{4}$$

which approximates the waves present in simple stimulated scattering if ω_o, ω_s, and $\vec{k}_o$ are real, but $\vec{k}_s$ is the complex wavevector $\hat{s}(k'+\frac{1}{2}ig_{os})$. One finds that the resulting $\vec{p}^{(3)}$ allows Maxwell's wave equation to be satisfied (for negligible pump wave depletion) for an isotropic medium if the gain is given by

$$g_{xx} = \frac{2\pi\omega_s E_o^2}{nc} \operatorname{Im}(A_\Delta + B_\Delta) \tag{5}$$

when $\hat{e}_o$ and $\hat{e}_s$ represent linear (x) polarizations that are parallel. If $\hat{e}_o$ and $\hat{e}_s$ represent perpendicular linear polarizations (x, y) then Maxwell's equations require the gain to be

$$g_{xy} \equiv \frac{\pi\omega_s E_o^2}{nc} \operatorname{Im}B_\Delta . \tag{6}$$

Other cases may be derived from these. Here n is the refractive index, and

$$A_\Delta \equiv \int_{-\infty}^{\infty} e^{i\Delta t} a(t)dt, \tag{7}$$

and similarly for B_Δ. Clearly, from (2) and (3), a(t) and b(t) must vanish for $t < 0$. Therefore,

$$\operatorname{Re}A_\Delta = \frac{1}{\pi} \fint_{-\infty}^{\infty} \frac{d\nu}{\nu - \Delta} \operatorname{Im}A_\Delta , \tag{8}$$

and similarly for B_Δ.

Comparing (5) and (6) with (1) gives at last the relation between the light-scattering spectrum and the nonlinear susceptibility functions A_Δ and B_Δ which we sought:

$$\operatorname{Im}B_\Delta = \frac{\pi c^4}{\hbar\omega_o \omega_s^3} \frac{d^2\sigma_{xy}}{d\Omega d\omega_s} (1 - e^{-\hbar\Delta/KT}) \tag{9}$$

and a similar relation with A_Δ with $\frac{1}{2}\sigma_{xx} - \sigma_{xy}$ substituted for B_Δ and σ_{xy} respectively. It is seen from (9) that polarized and depolarized light scattering spectra, together with the one real electronic hyperpolarizability parameter σ, determine all nonlinear optical effects in the Born-Oppenheimer regime.

APPLICATION TO GLASSES

The absolute values of the polarized (xx) and depolarized (xy) light-scattering cross-sections have been measured for a number of glasses.[4] The results for five of these glasses are shown in figure 1. The imaginary parts of their nonlinear susceptibility functions A_Δ and B_Δ follow directly from these data via the relations (7) and are shown in figure 2 along with their real parts, derived from them by a numerical integration of the relations (8).

The values for the electronic bulk hyperpolarizability σ for these glasses have been determined by comparing these results from light scattering data with Owyoung's measurements of self-induced polarization changes in the same glasses.[5] This latter effect determines $\sigma + 2B_0$ and B_0 can be read in each case from figure 2.

As a sample application, the use of (2) and (3) in Maxwell's equations gives an index change δn for linearly polarized monochromatic light of field amplitude E that is given by

$$\delta n = \frac{\pi E^2}{2n}\left(\frac{3}{2}\sigma + 2A_0 + 2B_0\right). \qquad (10)$$

This change leads to self-focusing whose threshold and other behavior can be calculated from (10) in the steady state, or from (2) and (3) in transient regimes. Values for δn derived from measurements of light-scattering and self-induced polarization changes agree well with measurements by interferometric methods, when available.[4]

Relations similar to (10) that govern all of the nonlinear optical effects mentioned in our introduction have been reviewed elsewhere.[2] Many other uses of light scattering in predicting and interpreting measurements of these effects have been, and are being, made in many laboratories.[2] It is by now clear that the disciplines of light scattering and nonlinear optical measurements have become quite complimentary.

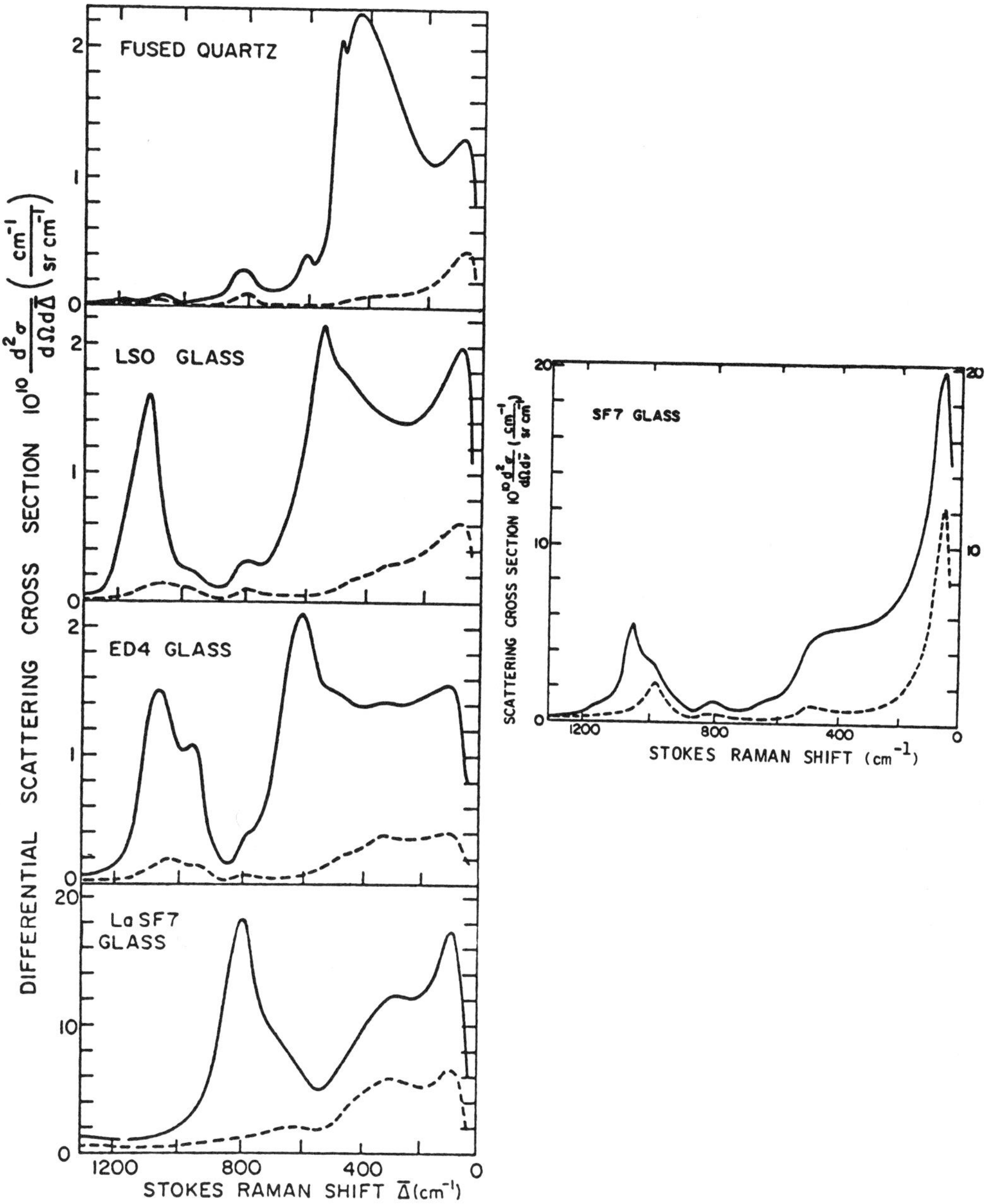

Fig. 1. Differential Raman Scattering cross-sections vs. frequency shift $\overline{\Delta}$ from 514 nm incident beam for five glasses ($\overline{\Delta} = \Delta/(2\pi c)\text{cm}^{-1}$). Solid lines are polarized (xx) and dashed lines are depolarized (xy) cross-sections.

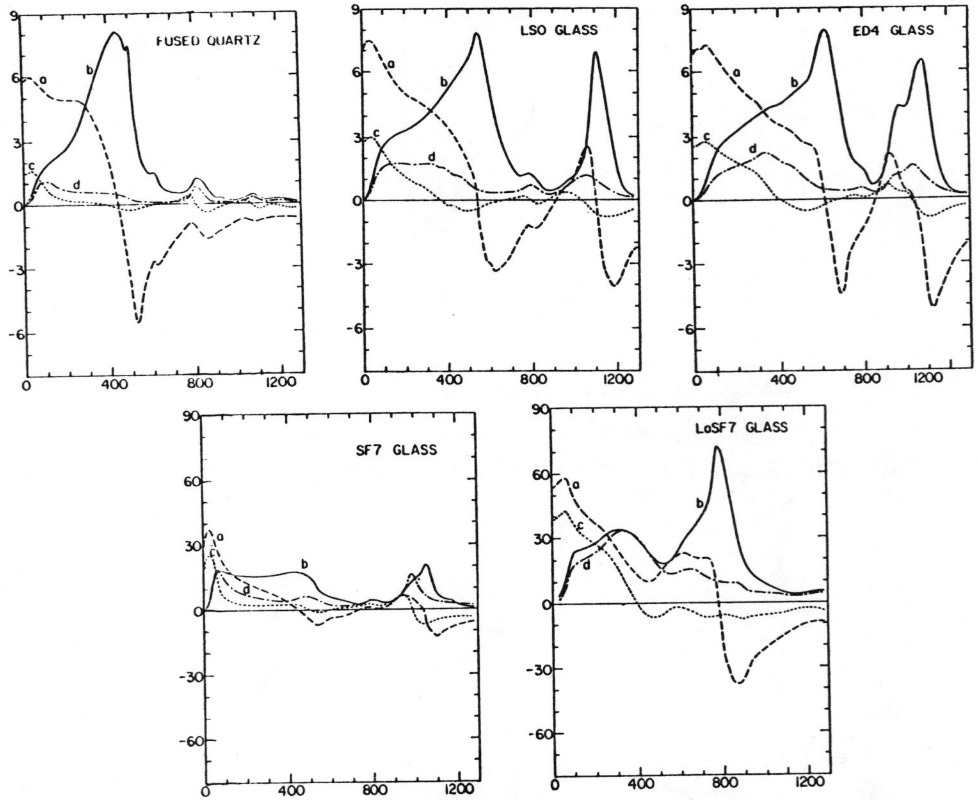

Fig. 2. The Fourier transforms A_Δ and B_Δ of the time-dependent nuclear response functions a(t) and b(t) for five glasses, in terms of which the nonlinear susceptibilities are expressed in equation (3). The curves represent a) $Re(A_\Delta + B_\Delta)$, b) $Im(A_\Delta + B_\Delta)$, c) ReB_Δ, and d) ImB_Δ. The ordinates are 10^{15} times the values in esu. The abscissae give the frequency Δ in wavenumbers (cm^{-1}).

The author would like to thank the Academy of Sciences of the U.S.S.R. and the National Science Foundation of the U.S.A. for making possible the preparation of this paper and its inclusion in an exchange of many stimulating papers and esteemed colleages from both countries, in the First U.S.A.-U.S.S.R. Symposium-Seminar on Light Scattering Theory that was held in Moscow, May 23 to 27, 1975.

REFERENCES

1. I. L. Fabelinskii, Molecular Scattering of Light, (Plenum Press, Inc., New York, 1968), gives an excellent review of light-scattering from molecules and its interpretation.
2. R. W. Hellwarth, Nonlinear Optical Susceptibilities in Liquids and Solids, (Pergamon Press) to be published.
3. R. W. Hellwarth, Phys. Rev. 130, 1850 (1963).
4. R. W. Hellwarth, J. Cherlow, and T. T. Yang, Phys. Rev. B11, 964 (1975); and in Laser-Induced Damage in Optical Materials: 1974, ed. by A. J. Glass and A. H. Guenther (National Bureau of Standards Special Publication 414, 1974) p.207.
5. A. Owyoung, IEEE J. Quantum Electron. 9, 1064 (1973); A. Owyoung, Nat. Bur. Stds. Spec. Publ. No. 387 (U.S. GPO, Washington, D.C., 1973); A. Owyoung, R. W. Hellwarth, and N. George, Phys. Rev. B5, 628 (1972).

COMMENTS

A. S. Davydov: "If, as for light-scattering from polaritons, the scattering is wavevector-dependent, then the relations shown here to nonlinear effects must be altered to include this dependence."

R. Hellwarth: "Although the equation (1) is true even for wavevector-dependent scattering, the remaining analysis must indeed be altered in this case."

P. Platzman: "The simplification of the general nonlinear polarization density to the forms (2) and (3) ought to arise from a selection of a certain subset of diagrams in a diagrammatic perturbation theory. Have you found this diagrammatic expression?"

R. Hellwarth: "No."

THEORY OF LIGHT SCATTERING FROM LIGHT IN MOLECULAR CRYSTALS

L. N. Ovander and Yu. D. Zavorotnev

Donetsk Physico-Technical Institute of the

Academy of Sciences of the Ukranian SSR USSR

The process of light scattering from light (LSL) is as follows. Two photons propagating in a crystal interact and change into two other photons which differ from the first ones both in frequencies and polarizations.

The LSL phenomenon, which is called four-photon scattering, was studied experimentally[1,4] and theoretically. The power of scattered emission is about 0.5 μW at the pumping intensity of approximately 9 μW. Theoretical papers[5-8] deal with the development of the phenomenon through a semiphenomenological theory which allows investigation of the spectral form of the scattering line, taking into account specific features connected with nonmonochromatic pumping, and investigation of the dependence of the frequency of resulting emission on observation direction.[5] This dependence was actually observed in experimental investigations carried out on CdS,[1] calcite[2] crystals, and in water.[3]

In the papers mentioned, however, there is no obvious expression for the scattering tensor expressed in terms of microscopic characteristics of a crystal. The analysis of such an expression allows one to separate different mechanisms which contribute to the scattering probability. It is also possible to find the indicatrix, depolarization power, and some other characteristics of scattered emission. This paper deals with an attempt to make up for this deficiency.

The Hamiltonian of the crystal may be taken as

$$H = \sum_{p\vec{k}} \epsilon_\rho(\vec{k})\,\xi_\rho^+(\vec{k})\,\xi_\rho(\vec{k})$$

$$+ \sum_{k_1 k_2 k_3 k_4} \sum_{\rho_1\rho_2\rho_3\rho_4} W_{\rho_1\rho_2\rho_3\rho_4}(\vec{k}_1, \vec{k}_2, \vec{k}_3, \vec{k}_4)\,\xi_{\rho_1}(\vec{k}_1)\,\xi_{\rho_2}(\vec{k}_2)$$

$$\times\, \xi^+_{\rho_3}(\vec{k}_3)\,\xi^+_{\rho_4}(\vec{k}_4) + \sum_{\vec{k}_1\vec{k}_2\vec{k}_3} \sum_{\rho_1\rho_2\rho_3} [Q_1\,\xi^+_{\rho_1}(\vec{k}_1)\,\xi^+_{\rho_2}(\vec{k}_2)$$

$$\times\, \xi_{\rho_3}(\vec{k}_3) + Q_2\,\xi_{\rho_1}(\vec{k}_1)\,\xi_{\rho_2}(\vec{k}_2)\,\xi_{\rho_3}(\vec{k}_3) + \text{h.c.}] \qquad (1)$$

Here $\epsilon_\rho(\vec{k})$ is the spectrum of photons (polaritons) in the crystal, $\vec{k}$ is the wave vector, ρ is the branch number, ξ^+, ξ are the operators of polariton generation and annihilation. The first sum in (1) is the Hamiltonian in zeroth approximation which describes the behavior of noninteracting polaritons. The subsequent three sums describe the interaction between polaritons. The second sum (four-particle interaction) contributes to the probability of LSL in the first order of perturbation theory. As to the third and fourth sums, together with the corresponding Hermitian conjugated ones, they contribute to the second order of perturbation theory. Contribution to the interaction between polaritons will give the following mechanisms: 1) interaction of a transverse electromagnetic field with charges in the crystal; 2) intermolecular interaction; 3) kinematic interaction which occurs due to the fact that the Hamiltonian of a molecular crystal is initially given by Pauli operators of generation and annihilation. However, these operators become unsuitable in further calculations and it is reasonable to consider Bose-operators. In this case the additional summands which describe interaction between quasiparticles (which is called the kinematic one) appear in the expression for the crystal Hamiltonian. It is necessary to take into account the kinematic interaction when considering the fourth and higher order effects.

One may find the values of W, Q_1 and Q_2 transition probabilities, and the scattering intensity in the framework of a molecular crystal model. The formula for energy flow of scattered emission into the spectral interval $d\omega$ and into the solid angle $d\Omega$ is given as

$$S_4(\omega\Omega m) = B\sum_{ij} X_{ijm} I_{1i} I_{2j} \int \delta(\epsilon_1 + \epsilon_2 - \epsilon_3 - \epsilon_4) |f(\Delta\vec{k})|^2 d\vec{k}_3 \tag{2}$$

where

$$f(\vec{k}) = \frac{8}{L_x L_y L_z K_x K_y K_z} \sin\left(\frac{K_x L_x}{2}\right) \sin\left(\frac{K_y L_y}{2}\right) \sin\left(\frac{K_z L_z}{2}\right), \tag{3}$$

$$\Delta\vec{k} = \vec{k}_1 + \vec{k}_2 - \vec{k}_3 - \vec{k}_4$$

Here I_{1i}, I_{2j} are the pumping intensities; i, j are polarization numbers; B is some coefficient; indices 1, 2 characterize incident emission and 3, 4 characterize scattered emission; $\epsilon_\ell (\ell = 1-4)$ are the photon energies.

The value X_{ijm} is defined by the formula

$$X_{ijm} = \sum_\ell a_{ij\ell m} \ell_i \ell_j \ell_\ell \ell_m, \tag{4}$$

where $\ell_i, \ell_j, \ell_\ell, \ell_m$ are unit vectors determining polarization; $a_{ij\ell m}$ is the tensor of LSL. The explicit expression for $a_{ij\ell m}$ is given below.

As it has been shown, scattering is realized by three mechanisms: by interaction between a transverse field and crystal charges, by molecular and by kinematic interactions. If one takes into account the first mechanism only, the scattering tensor has the form

$$a_{ij\ell m} = \sum_{\mu_1\mu_2} \frac{\rho^{i}_{0\mu_1}\rho^{j}_{\mu_1\mu_2}\rho^{\ell}_{\mu_2\mu_3}\rho^{m}_{\mu_3 0}}{(\epsilon_{\mu_1} - \hbar\omega_1)(\epsilon_{\mu_2} - 2\hbar\omega_1)(\epsilon_{\mu_3} - \hbar\omega_3)}$$

$$+ \sum_{\mu_1\mu_2\mu_3\mu_4} \frac{\rho^{i}_{0\mu_1}\rho^{j}_{\mu_1\mu_2}\rho^{s}_{\mu_2 0}\ell_s\rho^{\ell}_{0\mu_4}\rho^{m}_{\mu_4\mu_3}}{(\epsilon_{\mu_1} - \hbar\omega_1)(\epsilon_{\mu_2} - 2\hbar\omega_1)(\epsilon_{\mu_4} - \hbar\omega_3)} \tag{5}$$

$$\times \frac{\rho^{s}_{\mu_3 0}\ell_s}{(\epsilon_{\mu_3} - 2\hbar\omega_1)} \frac{2\pi}{v_0\epsilon_s(2\vec{k}_1)[\epsilon_s(2\vec{k}_1) - 2\hbar\omega_1]} + \ldots,$$

where v_0 is the unit cell volume, S is the virtual phonon polarization, ϵ_μ is the energy of excitons belonging to the branch μ, B is the molecular dipole momentum. In the case where only intermolecular interactions contribute, the scattering tensor is given by the following formula

$$a^{II}_{ij\ell m} = \sum_{\mu_1\mu_2\mu_3\mu_4} \sum_{x_1y_1x_2y_2} A_{x_1y_1}(\vec{k})A_{x_2y_2}(\vec{k}_4)$$

$$\times \frac{\rho^{j}_{0\mu_1}\rho^{x_1}_{\mu_1 0}\rho^{m}_{0\mu_2}\rho^{x_2}_{\mu_2 0}\rho^{i}_{0\mu}\rho^{y_1}_{\mu_3\mu_4}\rho^{y_2}_{\mu_4\mu_5}\rho^{\ell}_{\mu_5 0}}{(\epsilon_{\mu_1} - \hbar\omega_2)(\epsilon_{\mu_2} - \hbar\omega_4)(\epsilon_{\mu_3} - \hbar\omega_1)(\epsilon_{\mu_5} - \hbar\omega_3)}$$

$$\times \frac{1}{(\epsilon_{\mu_4} - 2\hbar\omega_1)} + \ldots, \tag{6}$$

where

$$A_{xy}(\vec{k}) = \sum_n \frac{\ell^{i\vec{k}\vec{n}}}{|\vec{R}_{0n}|^5} \{ |\vec{R}_{0n}|^2 \delta_{xy} - 3(\vec{R}_{0n})_x (\vec{R}_{0n})_y \},$$

$\vec{R}_{0n}$ is the distance between the zeroth and n-th cells. With due regard to kinematic interaction the LSL tensor has the form

$$a^{III}_{ij\ell m} = \sum_{\mu_1\mu_2} \frac{\rho^i_{0\mu_1} \rho^m_{\mu_1 0} \rho^j_{0\mu_2} \rho^\ell_{\mu_2 0} \epsilon_{\mu_2}}{(\epsilon_{\mu_1} - \hbar\omega_1)(\epsilon_{\mu_2} - \hbar\omega_3)(\epsilon_{\mu_2} - \hbar\omega_4)(\epsilon_{\mu_1} - \hbar\omega_4)}$$

$$+ \sum_{\mu_1\mu_2} \frac{\rho^i_{0\mu_1} \rho^j_{\mu_1 0} \rho^\ell_{0\mu_2} \rho^m_{\mu_2 0}}{(\epsilon_{\mu_1} - \hbar\omega_1)(\epsilon_{\mu_1} - \hbar\omega_3)(\epsilon_{\mu_2} - \hbar\omega_4)} + \sum_{xy\mu_1\mu_2\mu_3}$$

$$A_{xy}(k_1) \frac{\rho^i_{0\mu_1} \rho^x_{\mu_1 0} \rho^j_{0\mu_2} \rho^\ell_{\mu_2 0} \rho^m_{0\mu_3} \rho^y_{\mu_3 0}}{(\epsilon_{\mu_1} - \hbar\omega_1)(\epsilon_{\mu_2} - \hbar\omega_2)(\epsilon_{\mu_2} - \hbar\omega_3)(\epsilon_{\mu_3} - \hbar\omega_4)}$$

Estimates show that in the general case $a^I_{ij\ell m}$, $a^{II}_{ij\ell m}$, $a^{III}_{ij\ell m}$ are of the same order of magnitude. As is seen from the formulae (5)-(7), the tensors have different frequency dependences.

In addition, it is possible using polarization measurements to isolate the contribution to the effect at the expense of intermolecular interaction. This is possible in crystals belonging to the crystal classes O_h, O, T, T_d, T_h, C_{4v}, D_4, D_{4h}, D_{3h}, C_{6v}, D_6, D_{6h}, C_{2v}, D_2, D_{2h}, and S_{4v} if propagation directions and pumping and scattered emission polarizations are as follows

$$s_1 = s_{1x},\ s_2 = s_{2z},\ s_4 = s_{4y},\ \ell_1 = \ell_{1y},\ \ell_2 = \ell_{2x},\ \ell_4 = \ell_{4z}.$$

Using the obvious form for the tensor one can investigate at small angles the indicatrix of hyper-Raman scattering (HRS) which represents the limiting case of LSL, when one of the scattered photons being on the lower polariton branch becomes "nearly phonon". Figure 1 shows the indicatrix for the cubic crystal GaP in relative units. It should be noted that at angles larger than 2.5^{0}-3^{0} the LSL indicatrix goes into the HRS one.

If both scattered quanta are almost photons one must take into account synchronism (3), as when considering second harmonic generation. We have investigated collinear synchronism in uniaxial crystals. Phase-matching for uniaxial positive crystals is realized in the following two cases: a) when both primary beams are extraordinary; b) when one beam is ordinary and another is extraordinary. The synchronism condition assumes LSL only for definite $\vec{k}$ vector directions which form the cones shown in Fig. 2. Figure 2 shows the situation for the case (a) of a positive crystal. The case (b) will give two other cones differing from (a) by the angle between the axis and generatrix.

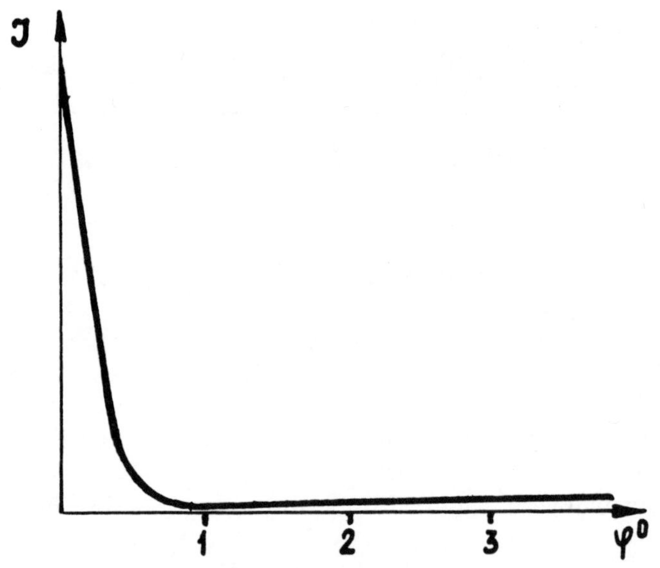

Fig. 1

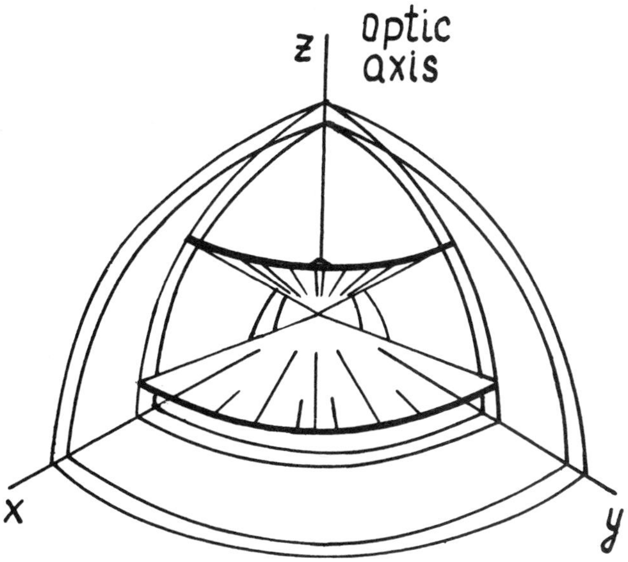

Fig. 2

For the uniaxial negative crystal KDP the synchronism directions are found at frequencies $\omega_1 = \omega_2 = 17264\ cm^{-1}$ (Hg), $\omega_3 = 9843\ cm^{-1}, \omega_4 = 24705\ cm^{-1}$. In this case five cones appear, each of them corresponding to one of the possible types of transformation. If beams of primary emission are ordinary and both scattered quanta are extraordinary (00 → EE), then synchronism occurs at the angle 21°15' relative to the optical axis. In the case of the transformations (00 → 0E), (0E → EE), (0E → 0E), (00 → EO) the angles are 24°33', 30°48', 45°12', 48°28', respectively. For the other interaction types, for example (EE → EE), synchronism is not realized.

The angles at which synchronism is realized were found as solutions of the equation

$$\ell n_4(\varphi) + (i - \ell)\, n_3(\varphi) = \frac{1}{2}\left[n_1(\varphi) + n_2(\varphi)\right],$$

where $n(\varphi)$ is the refraction index as a function of the angle of the beam propagation direction, and $\ell = \omega_4/(\omega_1 + \omega_2)$.

We have considered some other cases realized in uniaxial crystals.

In LSL phenomenon, as well as for second harmonic generation, one gets the ratio $\Delta k \sim (\Delta\varphi)^2$ when 90° synchronism is realized, where $\Delta\vec{k}$ is the angle of deviation from synchronism direction. In the ordinary case $\Delta\vec{k} \sim (\Delta\varphi)$.

REFERENCES

1. A. A. Grinberg, E. M. Rivkin, I. M. Fishman, I. D. Yaroshetskii, Pisma v GETP, 7, 324 (1968).
2. I. G. Meadors, W. T. Kavage, E. K. Damon, Appl. Phys. Letts., 14, 360 (1969).
3. D. L. Weinberg, Appl. Phys. Lett., 14, 32 (1969).
4. Yu. M. Kirin, S. G. Rautian, A. E. Semenov, B. M. Chernobrod, Pisma v GETP, 11, 340 (1970).
5. D. N. Klyshko, GETP, 55, 1006 (1968).
6. B. Ya. Zeldovich, GETP, 58, 1348 (1970).
7. R. I. Sokolovskii, Optika i Spectroskopiya, 33, 586 (1973).
8. R. I. Sokolovskii, GETP, 59, 799 (1970).

COHERENT ACTIVE SPECTROSCOPY OF RAMAN SCATTERING: COMPARISON WITH SPECTROSCOPY OF SPONTANEOUS SCATTERING

S. A. Akhmanov and N. I. Koroteev

Faculty of Physics, Moscow State University
Moscow, USSR

INTRODUCTION

Active spectroscopy of Raman scattering (ASRS) which has been rapidly developing for the last 2-3 years, is an intermediate between the technique of spontaneous and stimulated Raman scattering (RS) of light; it combines wide spectroscopic applicability of the first method with such advantages of the second one as high scattering efficiency, strong excitation and phasing of intramolecular oscillation in large volume, etc.

ASRS has been progressing mainly due to the great success in the development of lasers of tunable frequency.

The use of tunable lasers allowed one to choose a new way of studying RS of light. Conventional study of equilibrium, thermal elementary excitations (intramolecular oscillations of molecules in liquids and gases, phonons, polaritons and magnons in solids) has been replaced by "preparation" of the scattering medium with the help of tunable lasers. Active spectroscopy of RS deals with light scattering just off such nonequilibrium excitations.

In many respects active spectroscopy is advantageous as compared to conventional spectroscopy of spontaneous scattering and it may find some new physical and analytical applications.

ASRS PRINCIPLE

In the ASRS technique a biharmonic laser pumping is used for phasing and strong excitation of combination-active modes (frequencies of the waves used ω_1, ω_2 are usually within the visible or near IR range). A vibrational transition with frequency Ω_R is excited at $\omega_1 - \omega_2 \simeq \Omega_R$; beating of light waves results in the excitation of elementary oscillations due to the coupling between electronic and nuclear motions in a molecule:[1,2]

$$\frac{d^2Q}{dt^2} + 2\Gamma_R \frac{dQ}{dt} + \Omega_R^2 Q = \frac{1}{2M}\left(\frac{\partial\alpha_{ij}}{\partial Q}\right)_0 \epsilon_i^{(1)} \epsilon_j^{(2)*} e^{-i(\omega_1 - \omega_2)t} \quad (1)$$

Q is the normal coordinate of the appropriate oscillation; $\epsilon^{(1)}$, $\epsilon^{(2)}$ are the field amplitudes of the pump waves with frequencies ω_1, ω_2 respectively; $(\partial\alpha_{ij}/\partial Q)_0$ is the derivative of the electronic polarizability of the molecule with respect to the nuclear coordinate; $2\Gamma_R$ is the oscillation damping constant, and M is the reduced molecular mass.

Coherent molecular oscillations are tested with a "probe" ray (frequency ω_L). The intensity of Stokes or anti-Stokes scattering of the probe ray in the case $\omega_1 - \omega_2 = \Omega_R$ highly exceeds that of the corresponding spontaneous scattering. Besides, as the threshold of stimulated R combination scattering (SRS) is not exceeded, there appears no competition between lines and uncontrolled instabilities (see Fig. 1).

In coherent ASRS wave vectors of the pumps $\vec{k}_{1,2}$, probe wave $\vec{k}$ and scattered radiation $\vec{k}_{c,a}$ are related by:[1]

$$\vec{k}_1 - \vec{k}_2 = \vec{k} - \vec{k}_c \; ; \; \vec{k}_1 - \vec{k}_2 = \vec{k}_a - \vec{k}_c \quad (2)$$

Phenomenologically, these interactions may be described in a centrosymmetrical medium by cubic nonlinear susceptibilities χ

$$\vec{E}^{(c)} \sim \chi^{(3)}(\omega_c = \omega_L - \omega_1 + \omega_2)\, \vec{E}\vec{E}^{(1)*}\vec{E}^{(2)} \; ; \; \vec{E}^{(a)} \sim \chi^{(3)}\vec{E}\vec{E}^{(1)}\vec{E}^{(2)*} \quad (2)^*$$

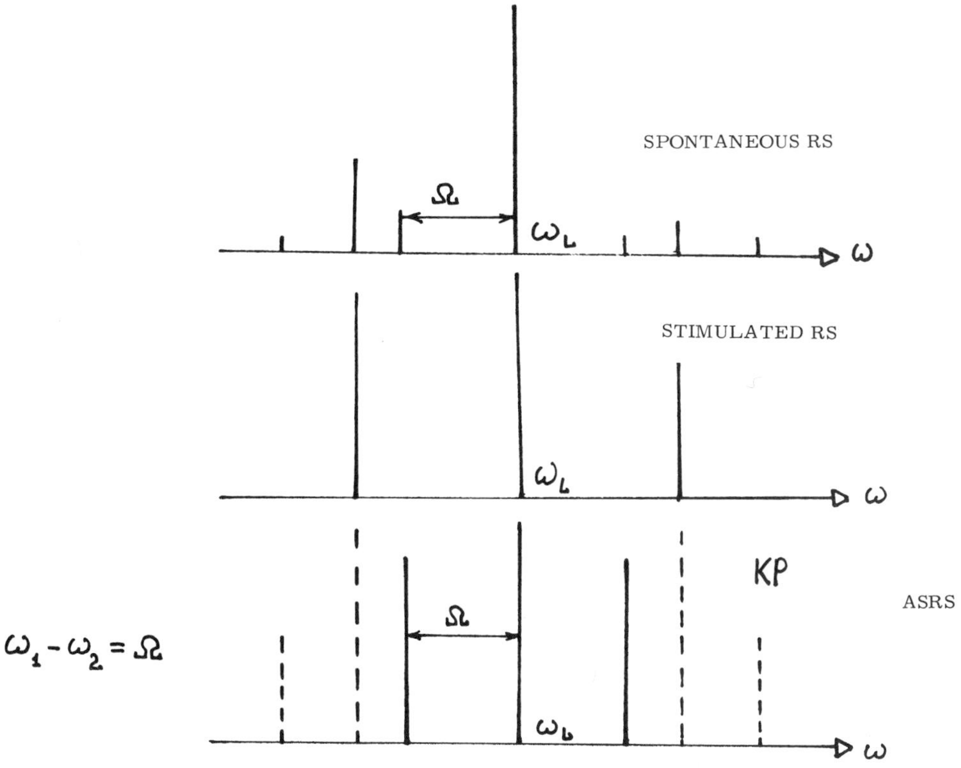

Fig. 1. Qualitative spectral pattern for different variants of Raman scattering (RS). In the scheme of active spectroscopy of RS when the pump frequencies difference $\omega_1 - \omega_2$ is tuned, any line of spontaneous RS can be excited.

As in the experiment intensities $I^{(c,a)} \sim |\vec{E}^{(c,a)}|^2$ are measured, the spectroscopic parameter measured in ASRS by a tunable generator is the dispersion of the square modulus of the appropriate cubic susceptibility $|\chi^{(3)}|^2$, whereas spectra of spontaneous RS correspond to the dispersion of the imaginary part of the susceptibility $\chi^{(3)}$ in the vicinity of vibrational resonances:

$$I_{ckp} \sim I_L \cdot \mathrm{Im}\, \chi^{(3)}(\omega_c = \omega_L - \omega_L + \omega_c)\ ;\ \omega - \omega_c \simeq \Omega_R \tag{3}$$

The nonlinear susceptibility $\chi^{(3)}$ is acted upon by all optical resonances of the medium, including electronic transitions and vibrations of nuclei in the molecule. If the contribution of molecular vibration with frequency Ω_R is isolated, we may write:

$$\chi^{(3)}_{ijk\ell} = \chi^{R}_{ijk\ell} + \chi^{NR}_{ijk\ell} \tag{4}$$

where $\chi^{NR}_{ijk\ell}$ corresponds to the far-apart oscillating and electronic resonances

$$\chi^R \sim \bar{\chi}^R (i - \Delta_1)^{-1} = \left(\frac{\partial \alpha_{ij}}{\partial Q}\right)_0 \left(\frac{\partial \alpha_{kc}}{\partial Q}\right)_0 [\Omega_R^2 - (\omega_1 - \omega_2)^2$$

$$- 2i\Gamma_R(\omega_1 - \omega_2)]^{-1}$$

If none of the frequencies ω_1, ω_2, ω_L falls into the absorption region, then

$$\chi^{NR} = (\chi^{NR})^* \text{ and } |\chi^{(3)}|^2 = (\mathrm{Re}\, \chi^R + \chi^{NP})^2 + (\mathrm{Im}\, \chi^R)^2$$

or

$$\frac{|\chi^{(3)}(\omega_{c,a} = \omega_L \mp \omega_1 \pm \omega_2)|^2}{(\chi^{NR})^2} = 1 + \frac{\alpha_1^2}{1 + \Delta_1^2} - \frac{2\alpha_1 \Delta_1}{1 + \Delta_1^2} \tag{5}$$

where

$$\alpha_1 = \overline{\chi}^R/\chi^{NR}, \quad \Delta_1 = [(\omega_1 - \omega_2) - \Omega_R]/\Gamma_R$$

The "active spectrum" (5) differs from the spontaneous one (4) in the appearance of a "pedestal" conditioned by $\chi^{NR} \neq 0$ and the interference between the "real" and the "imaginary" parts of $\chi^{(3)}$.

ACTIVE SPECTRA INTENSITY

It should be noted that in contrast to spontaneous RS, the indicatrix of coherent scattering is considerably anisotropic: the intensity of Stokes (anti-Stokes) scattering of the probe ray has its maximum in the phase-matched direction (2). In other directions coherent scattering is negligible. As a result, radiation scattered off coherent molecular oscillations is obtained in the form of a well-collimated, practically fully spatially coherent beam. In the case when the energy exchange between pumping waves may be neglected, its intensity may be given by (3):

$$I^{(c,a)} = (\ell n \frac{d\sigma}{d\theta})^2 \, 2^4 c^4 \omega_{c,a}^{-6} (2\hbar\Gamma_R)^{-2} n^{-4} I_L I_1 I_2 \tag{5}$$

where $d\sigma/d\theta$ is the differential cross-section of spontaneous (Stokes) RS; $2\Gamma_R$ is the line width of spontaneous RS, I_L, $I_{1,2}$ are the intensities of the appropriate beams; and $n \simeq n_{1,2c,a}$ are the refractive indices for frequencies ω_1, ω_2, ω_c, ω_a. In deducing (5) only the Raman contribution of χ^R to $\chi^{(3)}$ has been taken into account.

The gain in total power of the signal in ASRS as compared to spontaneous (Stokes) scattering is given by

$$\eta = \frac{P^{(c)}_{ASRS}}{P^{(c)}_{SRS}} \simeq \ell n \frac{d\sigma}{d\theta} \, 2^4 c^4 \, \omega_c^{-6} \, (2\hbar\Gamma_R)^{-2} \, n^{-4} \, I_1 I_2 (\delta\theta)^{-1} \tag{6}$$

where $\delta\theta$ is the solid angle within which Stokes spontaneous Raman scattering (SRS) is measured. The factor η may be very high; it depends on the intensity of the pump waves (in our experiments $\eta \sim 10^3$ to 10^5).

SPECTRAL RESOLUTION OF ASRS

In the case where RS-active oscillations are excited as a result of beating of waves from two narrow-band sources, the spectral resolution of ASRS is determined by the ratios of the spectral width of the Raman line $\Delta\Omega_R$ to the line width of the tunable oscillator. The active spectrum is detected by the frequency tuning of the latter. Tunable parametric oscillators and dye lasers which have lately become available and which have spectrum line widths ($\sim$ 100 Mhz) limited practically by the pulse duration,[4] when used in ASRS scheme will make possible measurement of RS lineshapes with an accuracy that was inaccessible in conventional RS spectroscopy.

Meanwhile, in ASRS wide band tunable oscillators can also be used together with an almost monochromatic generator with fixed frequency without any deterioration of the spectral resolution of ASRS as compared to that for spontaneous RS. In this case for the line shape of the form $|\chi^{(3)}|^2$ to be resolved the active spectrum $S(\omega_{c,a})$ should be normalized with respect to the spectrum of the wide-band pumping component (let it be $S(\omega_2)$):[5]

$$|\chi^{(3)}(\omega_{c,a} = \omega_L \mp (\omega_1 = \omega_2))|^2 \sim S(\omega_{c,a})/S(\omega_2) \quad (7)$$

In such a case the active spectrum is registered at the monochromator scanning, i.e., the same as in spontaneous RS. The resolution is determined by monochromator performance. However, the greater intensity of the scattered light allows one to obtain "active" spectra at very high values of signal-to-noise ratio.

From the point of view of obtaining from ASRS spectra reliable spectroscopic information about the investigated combination-active transition of the optical medium, it is important to require that a strong energy exchange between pumping waves, as well as between the waves of the probe and scattered radiation be absent. In practice, this is equivalent to requiring the intensities of all light beams directed into the medium (frequencies ω_L, ω_1, ω_2) to be lower than the threshold values for the excitation of stimulated

RS or for the effect of inverse RS to be vividly displayed. In the opposite case, distortions in the spectra of the scattered probe ray in the ASRS scheme increase to such a degree that active spectra have no resemblance at all with the contour $|\chi^{(3)}|^2$.[6,7]

It should be noted that the appearance of "a pedestal" in ASRS spectra, which results from the presence of $\chi^{NR} \neq 0$, introduces some peculiarity in the measurement of weak lines in RS by the ASRS technique (especially of the lines of the second and higher orders and overtones): in this case it is necessary to register a slight change of $|\chi^{(3)}|^2$ due to $\chi^R << \chi^{NR}$ on the background with respect to the relatively high "pedestal" $|\chi^{NR}|^2$. It is evident that this peculiarity diminishes the selectivity of the ASRS technique, when weak lines are under investigation. The aforementioned drawback can be overcome either by using tunable oscillators with radiation spectrum scanning, which allows measurement of the derivative $|\chi^{(3)}|^2$ with respect to frequency, or by making use of the special methods of coherent "extraction" of the pedestal (see below).

ACTIVE SPECTROSCOPY OF RESONANT RAMAN SCATTERING

The intensities of Stokes and anti-Stokes components of spontaneous RS are shown to increase considerably when the excitation line is approaching the absorption band. This phenomenon is called resonant RS.[8] The ASRS technique can be modified in such a way that it will help to study resonance RS spectra as well. For this purpose one-photon absorption should be provided at the frequency of one of the pumping waves (e.g., $\omega_1 \simeq \Omega_E$, where Ω_E is the central frequency of the electronic transition). Generally speaking, the probe wave frequency in this case may be chosen in the transparency band of the substance under investigation, so that clear differentiation of proper resonant RS effects from "background" effects of the type of resonant fluorescence and wide-band (Stokes) luminescence becomes possible.

In the case of homogeneous multicomponent media absorption at the pumping frequency ω_1 may be provided either by a resonant electronic transition in the same molecule where vibrational resonance is under investigation in scattering, or by a similar transition in a molecule of another component of the mixture. Here, due to the coherent nature of the scattering process in ASRS, contributions of different components of the mixture into the scattered light signal are not summed up, as was the case in spontaneous RS, but they interfere, thus causing deformations of active spectra typical for resonant RS.

Therefore, it becomes possible to obtain "active" resonant RS spectra in the transparency band of different gases or liquids when some resonance-absorbing additives are mixed up with them. Moreover, it turns out that a proper choice of the concentration of the absorbing additive and detuning from the center of the absorption band give a sharp increase of the "contrast ratio" of the dispersion curve $|\chi^{(3)}|^2$
(contrast ratio $K = \mathrm{Max}|\chi^{(3)}(\Delta_1')|^2/\mathrm{Min}|\chi^{(3)}(\Delta')|^2$)
in the vicinity of any arbitrarily weak RS line. It results from coherent "subtraction" of the active spectrum pedestal in the vicinity of the RS line under investigation at the expense of the real part of the resonance susceptibility connected with electron transition. The condition of maximum increase of the contrast ($K \to \infty$) is as follows:

$$-\alpha_1\alpha_2 = 1 + (\Delta_2 - \alpha_2)^2$$

$$\alpha_2 = N_2\bar{\chi}_2^E/(N_1\chi_1^{NR} + N_2\chi_2^{NR})\Gamma_E;\ \Delta_2 = (\omega_1 - \Omega_E)/\Gamma_E \tag{8}$$

where the expression given below is taken for $\chi^{(3)}$ (intermolecular interaction being neglected):

$$\chi^{(3)} = N_1\chi_1^{NR} + N_2\chi_2^{NR} + \frac{N_1\bar{\chi}_1^R}{i - \Delta_1} + \frac{N_2\bar{\chi}_2^E}{i - \Delta_2} \tag{9}$$

N_1, N_2 are the densities of the number of molecules of scatterer and absorber respectively, and

$$\bar{\chi}^R \simeq \frac{c^4}{\hbar\omega_c^4}(d\sigma/d\theta);$$

$d\sigma/d\theta$ is the cross-section of spontaneous (Stokes) RS,

$\chi^E = 4\sigma_1^2 c^2 \Gamma_E^2 / \hbar\omega_c^2 \omega_1 \Omega_R$; Γ_E is the half-width of electronic transition, and σ_1 is the cross-section of one-photon absorption.

In the case where the one-quantum transition is saturated, the condition of maximum contrast (8) may be transformed into

$$-\alpha_1\alpha_2 = 1 + (\Delta_2 - \alpha_2)^2 + \gamma - \frac{\alpha_2^2 \gamma}{1 + \gamma + \Delta_2^2} \qquad (10)$$

where $\gamma = A|dE^{(1)}|^2/\Gamma_E \hbar^2$ is the saturation parameter of the resonant transition.

Figures 2-4 show calculated curves illustrating active spectra, when condition (8) is followed, and also the relationship between contrast and absorption line parameters.

ACTIVE SPECTROSCOPY: EXPERIMENTAL RESULTS

It has been stated above that, in principle, active spectra contain greater spectroscopic information than spontaneous RS spectra. In particular, they include data concerning both imaginary and real parts of the nonlinear susceptibility of the third order. The interference of different contributions in the cubic susceptibility results in the appearance of more or less pronounced dips accompanying peaks which take the place of RS lines. The ratio of intensities and relative spacing of peaks and dips in active spectra make it possible to determine relative values and phases (i.e., spectra) of Raman and electronic contributions.

Figure 5 shows, as an example, an active spectrum of a calcite crystal (solid line) as compared to a spontaneous RS spectrum (dotted line).

Active spectra were taken with different orientations of polarization vectors of interacting beams. From the spectra given it can be concluded that contributions of the vibration $A_{1g}(\overline{\chi}^R_{1122}(A_{1g}))$ and electronic sub-system of the crystal (χ^{NR}_{1122}) into the component $\chi^{(3)}_{1122}$ are of the same sign, while contributions of E_g vibrations $(\overline{\chi}^R_{1122}(E_g))$ have the opposite sign than to the contribution χ^{NR}_{1122}. Due to the symmetry of vibrations the positive

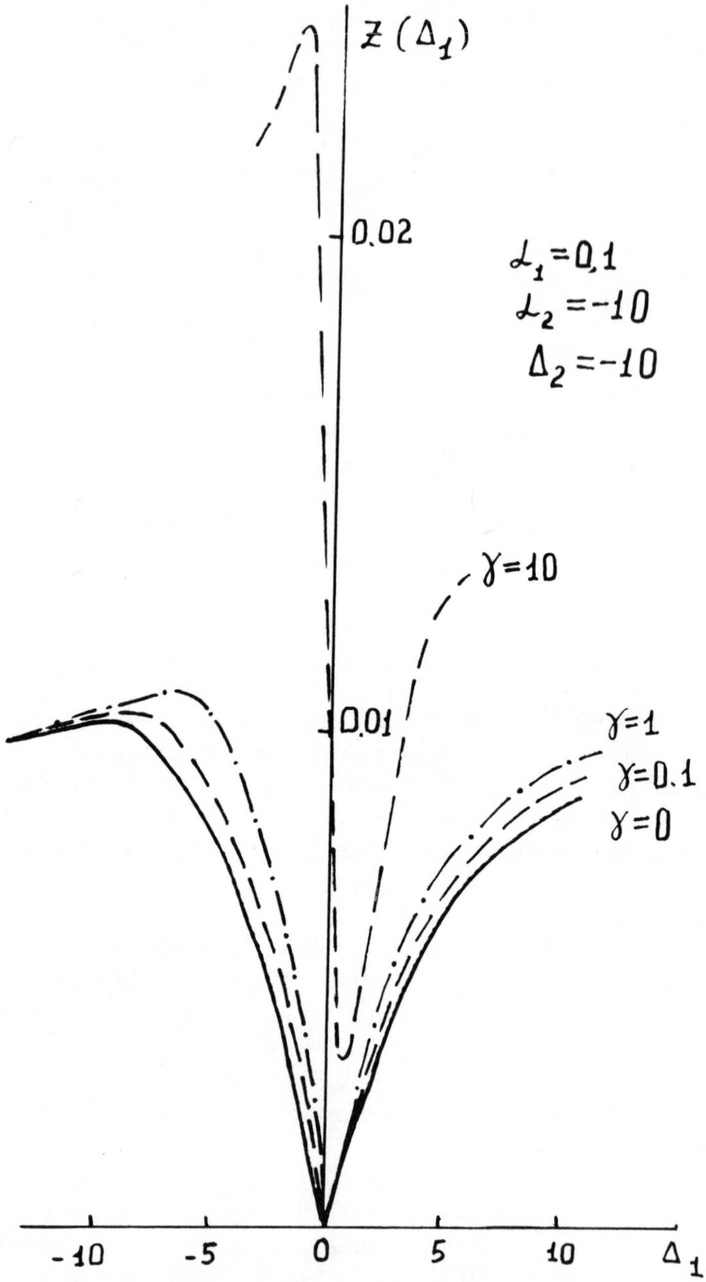

Fig. 2. Dispersion curve for the square modulus of cubic susceptibility ($z(\Delta_1) = |\chi^{(3)}|^2/(\chi_1^{NR} N_1))^2$ for the condition of maximum contrast (8) for different values of the saturation factor of the resonant transition γ.

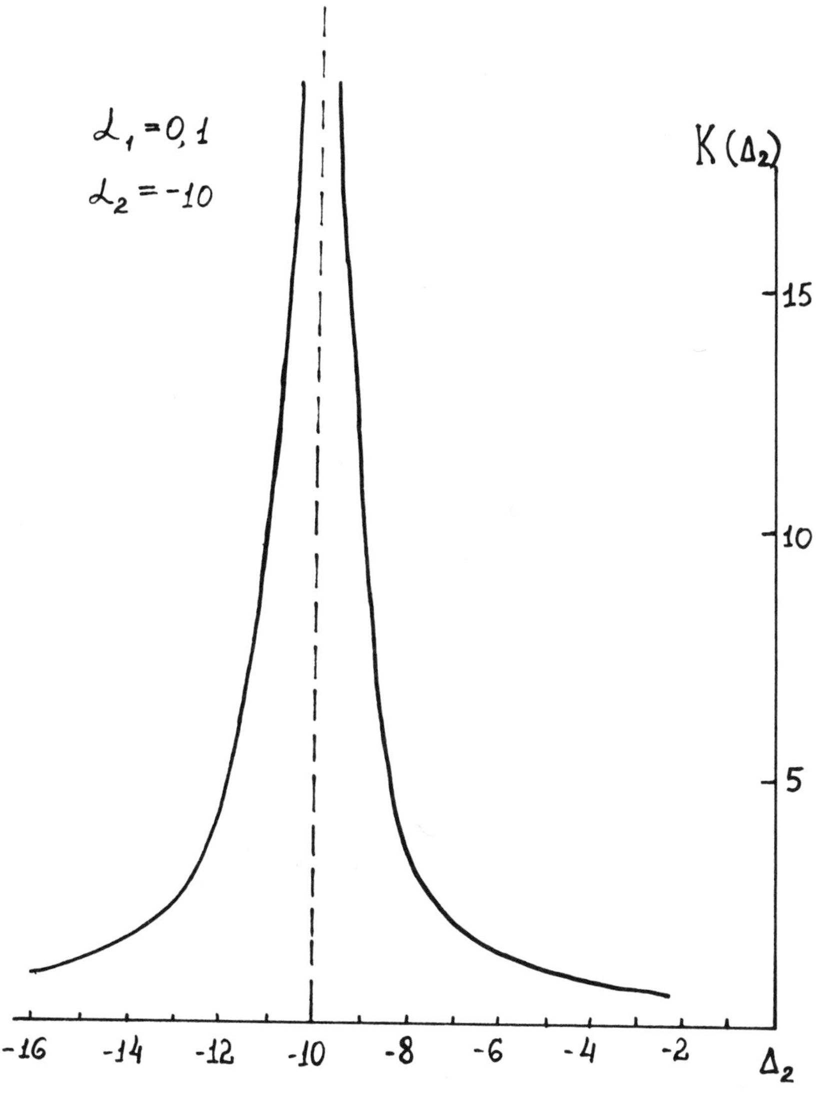

Fig. 3. Normalized contrast of the dispersion curve $|\chi^{(3)}|^2$ for a mixture of two substances $K(\Delta_2) = k(\Delta_2)/k(\alpha_2 = 0)$ vs. frequency detuning Δ_2 from the center of absorption band.

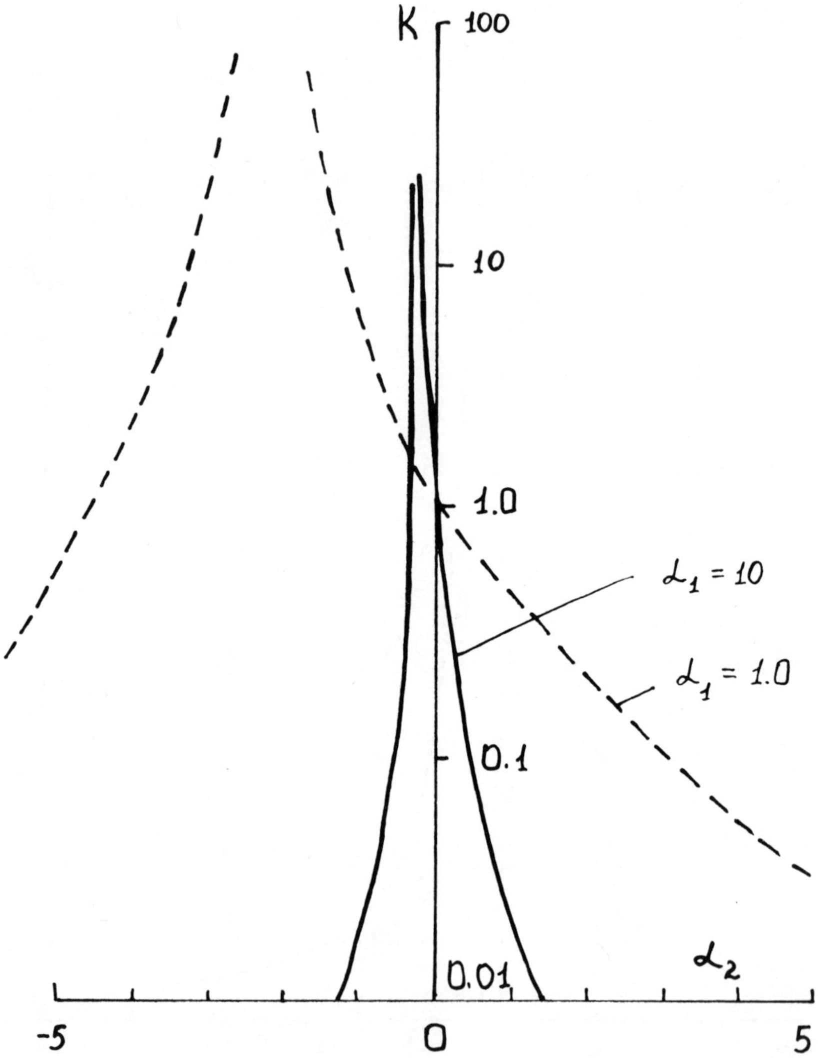

Fig. 4. Normalized contrast K of dispersion curve $|\chi^{(3)}|^2$ for a mixture of two substances vs. α_2 value; the detuning Δ_2 is fixed and equal to zero.

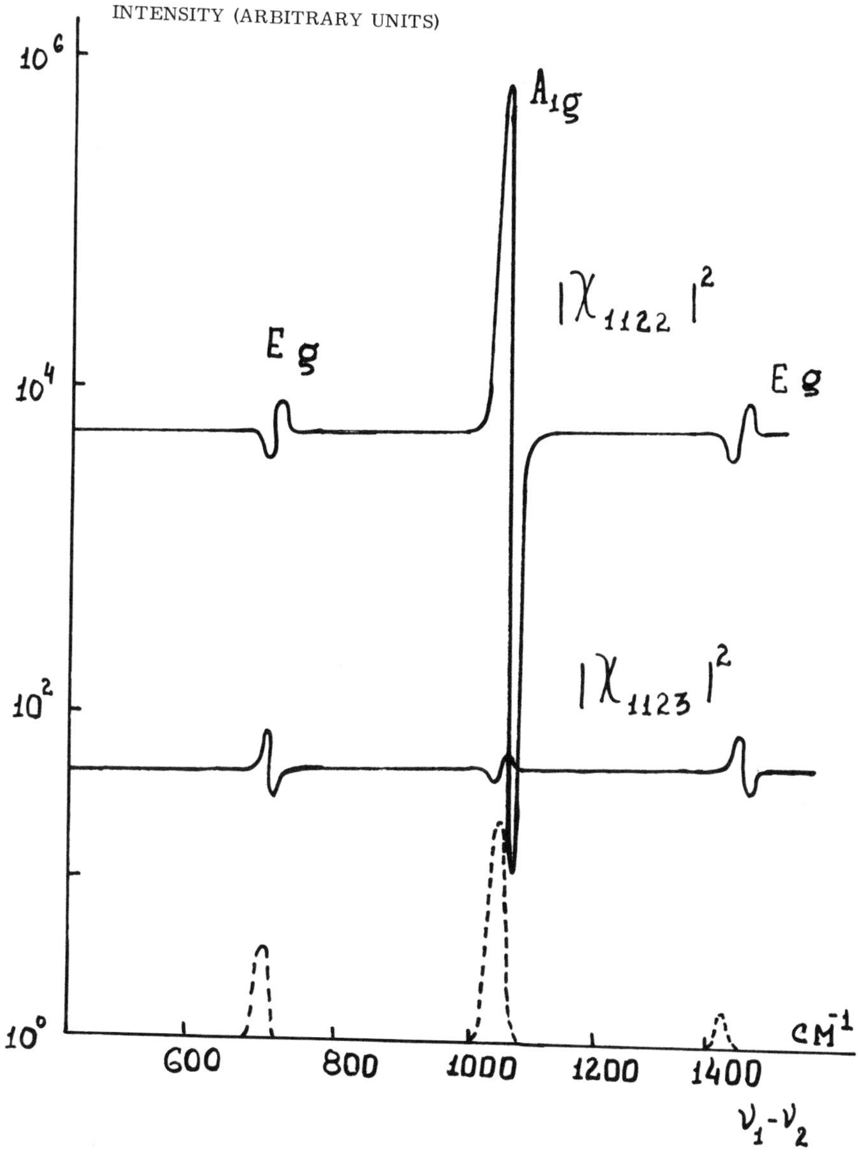

Fig. 5. "Active" (Solid lines) and spontaneous (dotted lines) spectra of a calcite crystal in a region of frequencies of inner vibrations of the ion $CO_3^{=}$.

sign may be attributed to the term $\overline{\chi}^{R}_{1122}(A_{1g})$ and the negative sign to the terms $\overline{\chi}^{R}_{1122}(E_g)$; therefore, one can conclude that the sign of χ^{NR}_{1122} is positive.[5, 9]

From the dispersion curve $|\chi^{(3)}_{1123}|^2$ it is quite evident that nondiagonal elements of the RS tensor for a fully-symetrical vibration A_{1g} of calcite are practically equal to zero and, therefore, high values of these elements, observed in some earlier works,[10] should be attributed to the imperfection of the experimental technique.

Experiments on ASRS of liquids carried out on the experimental set-up shown schematically in Fig. 6, provide data about the value of different invariants of the RS molecular tensor. In particular, measurements on benzene with the use of ASRS technique have shown that, evidently, the RS tensor of a two-fold-degenerate vibration (E_g^+ with the frequency 1178 cm^{-1}) is asymmetric, in contrast to general opinion.[9] It should be noted that quite recently the authors of Ref. 11 came to the similar conclusion on the basis of accurate measurements of the depolarization ratio of the spontaneous RS line corresponding to this vibration.

A unique high level of anti-Stoke signal of ASRS* which coincides in intensity with the Stokes-component of the scattered probe ray, allows to make reliable quantitative measurements of RS spectra for highly luminescent and dyed media and, in particular, for mixtures and solutions with absorbing additives. Thus, using ASRS we could easily detect spectra of benzene, toluene, and other organic liquids with different dyes dissolved in them (rodamine 6G and B, iodine, cobalamine, criptocyanine, etc.) up to concentrations which provide the absorption of one of the pumping waves and a probe wave at a level of 5-10 cm^{-1}. Deformations of the contours of active spectra lines for these organic solvents upon changes in the concentration of absorption additives have been measured. As an example, in Fig. 8 the deformation

---

*When working with the experimental set-up shown shematically in Fig. 6, the signal of anti-Stokes scattering of the probe ray on coherent vibration of benzene molecules ($\Omega_R/2\pi c$ = 992 cm^{-1}) could be observed with the naked eye, on a screen placed behind the cell with the liquid under investigation, when direct pumping rays were blocked.

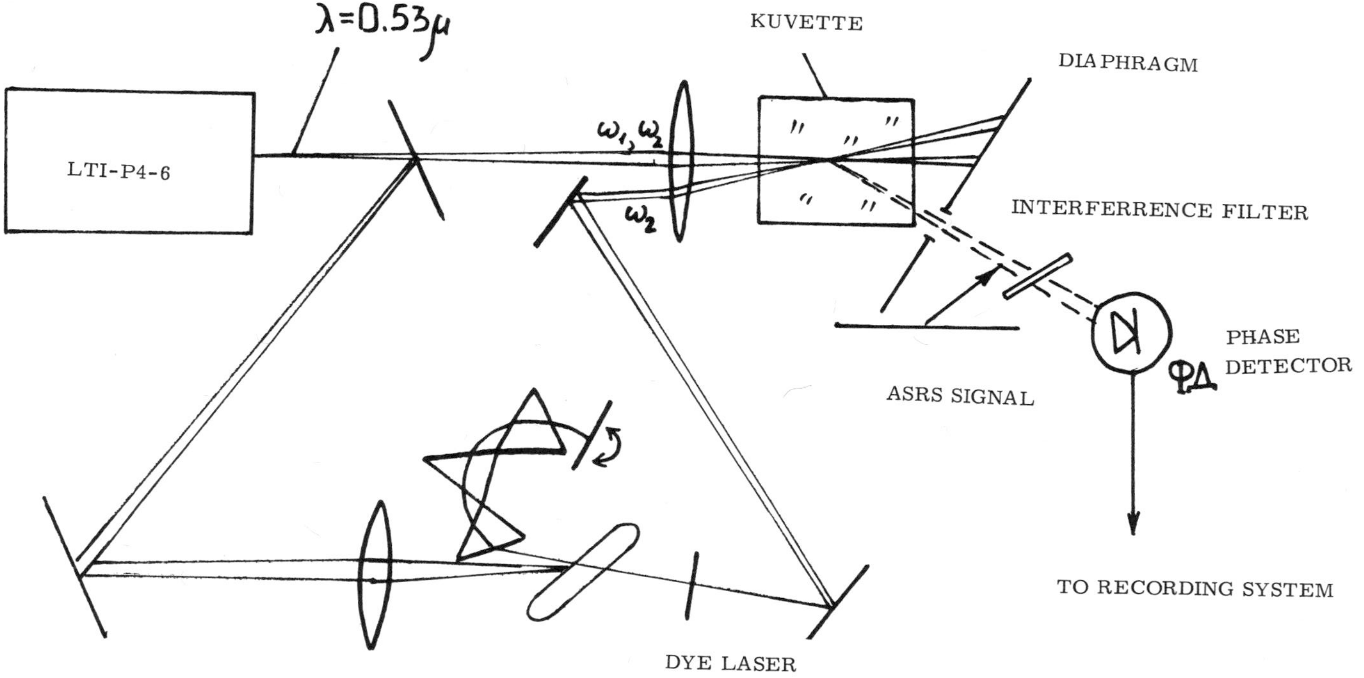

Fig. 6. The scheme of an experimental set-up for studying liquids by ASRS based on a laser of the type LTI-P4-6 with doubled frequency of radiation on YAG-Nd^{37} and a laser on the solution of rodamine 6G or B in ethyl alcohol. The angle between interacting rays is chosen in such a way as to follow the condition of phase-matching (2) (in this scheme $\omega_L = \omega_1$, $\vec{k}_L = \vec{k}_1$).

pattern is given for the A_1-vibration of toluene molecules ($\Omega_R/2\pi c$ = 1208 cm^{-1}) with an increase in concentration of rodamine 6G dissolved in it (absorption band for the dye and relative spacing of pumping lines and probe radiation are given in Fig. 7). A nonmonotonic pattern of the deformation obtained corresponds to the ideas given previously in the present paper concerning coherent subtraction of the pedestal of active spectra and the respective increase in the contrast ratio of the dispersion curve.

SUMMARY

Data given in the present paper alongside with data from some works of foreign authors (e.g., Ref. 12-15) show that active spectroscopy of RS is becoming an effective method for investigating properties of a substance, and that it considerably broadens the analytical applicability of light scattering. Although, naturally, one cannot state that ASRS can fully replace conventional spectroscopical techniques of spontaneous scattering, this method opens some new opportunities.

As far as physical aspects are concerned, these are:

1) uniquely high sensitivity and spectral resolution;
2) the possibility of obtaining data which refer simultaneously to different dynamic sub-systems of the medium (interference of different contributions, interaction of electronic shell and molecular nuclei, etc.);
3) the possibility of a new approach to resonant scattering;
4) obtaining high levels of excitation of molecular vibrations for studying highly anharmonic oscillations;
5) direct measurement of molecular relaxation times; revelation of energy dissipation routes from strongly excited combination-active modes;
6) new possibilities of measuring surface states.

Active spectroscopy also opens some new opportunities in analytical applications of light scattering.

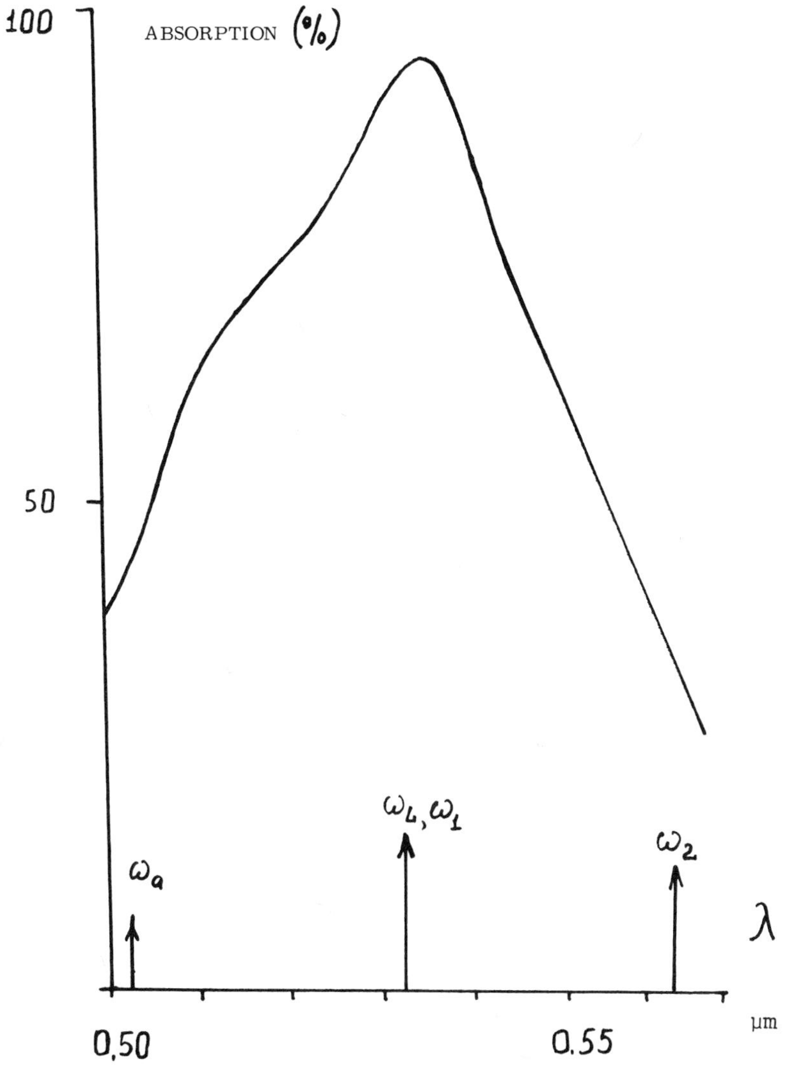

Fig. 7. Absorption band for rodamine 6G in toluene and relative spacing of pump lines, probe and scattered radiations. The cell with the solution is of length $\ell = 5$ mm. Dye concentration: $N_2 = 1 \times 10^{17}$ cm^{-3}.

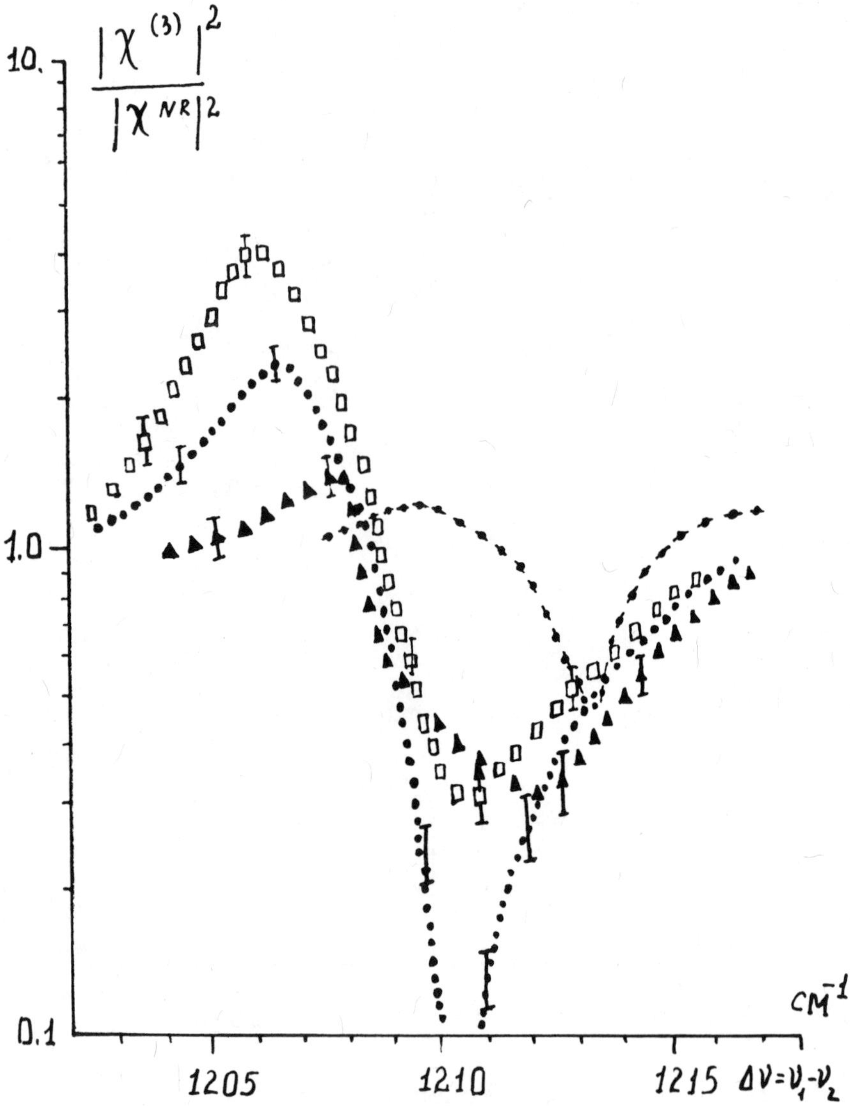

Fig. 8. Dispersion curve $|\chi^{(3)}|^2$ for toluene with a dye additive within the range of the A_1-line ($\Omega_R/2\pi c$ = 1208 cm^{-1} of the toluene molecules vs. dye concentration.

pure toluene - □□□□ N_2 = 5x10^{16} cm^{-3} ▲▲▲▲

N_2 = 1x10^{16} cm^{-3} - ●●●● N_2 = 1x10^{17} cm^{-3} ✦✦✦✦

Further development of the ASRS technique is connected with the improvement of tunable lasers (especially with the development of tunable narrow-band oscillators for IR and UV ranges) and with the use of frequency-tunable pico-second pulses.

REFERENCES

1. J. A. Giordmaine, W. Kaiser, Phys. Rev. 144, 676 (1966).
2. R. W. Terhune, P. D. Maker, Phys. Rev. 137, A801 (1965).
3. S. A. Akhmanov, V. G. Dmitriev, A. I. Kovrigin, N. I. Koroteev, V. G. Tunking, A. I. Kholodnykh, Pis'ma v JETPh, 15, 600 (1972).
4. T. W. Hänsch, Appl. Opt. 11, 895 (1972); M. Levenson, N. Bloembergen, PRL 32, 645 (1974).
5. S. A. Akhmanov, N. I. Koroteev, A. I. Kholodnykh, J. Raman Spectr. 2, 239 (1974).
6. N. I. Koroteev, I. L. Shumai, Quant. Electronica 2, 2489 (1974).
7. N. I. Koroteev, Opt. Spectr. 29, 543 (1970).
8. P. P. Shorygin, Usp. Fis. Nauk 109, 293 (1973).
9. S. A. Akhmanov, N. I. Koroteev, JETPh 67, 1306 (1974).
10. L. Coutur, Ann. de Phys. 2, 5 (1974); S. Bhagavantam, Proc. Ind. Acad. Sci. A11, 62 (1940).
11. I. V. Aleksandrov, Ya. S. Bobovich, A. V. Bortkevich, JETPh 68, N°4 (1975).
12. M. D. Levenson, N. B. Bloembergen, Phys. Rev. B6, 3962 (1972).
13. M. D. Levenson, N. B. Bloembergen, Phys. Rev. B10, 4447 (1974).
14. R. F. Begley, A. B. Harvey, R. L. Byer, Appl. Phys. Lett. 25, 387 (1974); J. Chem. Phys. 61, 2466 (1974).
15. P. R. Regnier, J. P.-E. Taran, Appl. Phys. Lett. 23, 5 (1973).

RAMAN SCATTERING INDUCED BY STRONGLY NONEQUILIBRIUM PHONONS

S. A. Bulgadaev and I. B. Levinson

L. D. Landau Institute for Theoretical Physics
Academy of Sciences of the USSR
Moscow, USSR

This paper is connected with the theory of Raman scattering in two ways. First of all, we shall consider a new type of elementary excitation closely related to longwave optical phonons, and the best way to observe these excitations is to perform a Raman scattering experiment. Secondly, these elementary excitations are collective excitations in a nonequilibrium phonon system, and the best way to achieve the nonequilibrium state is to excite the phonon system by some Raman-type process.

The theoretical work was stimulated by the experiment of Colles and Giordmain, performed in 1971 [1]. In this experiment a diamond sample was irradiated by a mercury lamp, and the hot luminescence from some impurity centre was monitored. Then the sample was simultaneously pumped by two beams. The difference of the beam frequencies was in resonance with the longwave optical phonon frequency: $\nu_1 - \nu_2 = \Omega_0$. At low pumping intensities the hot luminescence displays some structure with the optical phonon frequency Ω_0, but at higher pumping intensities there also appears a structure with a subharmonic frequency $\frac{1}{2}\Omega_0$. The authors' interpretation is in line with Orbach's ideas [2]. Due to the beating of laser beams optical phonons with k = 0 are generated. These long-wave phonons decay by a parametric anharmonic process into two acoustical phonons with equal and opposite momenta: $k = 0 \rightarrow (+\vec{q}_0) + (-\vec{q}_0)$. The frequencies of these phonons are obviously equal one to another, and equal to one half of the optical phonon frequency: $\omega_0 = \frac{1}{2}\Omega_0$. Now it is possible to give a tentative interpretation of the observed hot luminescence structure at $\frac{1}{2}\Omega_0$. It appears because acoustical phonons with the frequency ω_o change the rate of some selected radiationless transitions of the impurity centre.

As a result of the laser pumping the acoustical phonon system is highly excited. But in contrast with thermal excitation only a small part of the acoustic modes is excited. The frequency ω_o determines the magnitude of the momentum $\vec{q}$: it is equal to some q_0. And now it is evident that in $\vec{q}$-space only those modes are excited which are near the sphere $q = q_0$. The excited acoustical phonon distribution is displayed in Fig. 1, where N_q are the occupation numbers for various phonon modes. The width of the distribution Δq is very small, because it is due to the uncertainties of the energy-momentum conservation laws only [3]. This means that the width of the phonon distribution is determined by finite phonon life-times, or finite spectral width of the laser beams, or non-plane-wave structure of these beams, or other similar spectral uncertainties. Perhaps the most important factor is the laser spectral width, or, maybe, the phonon life-time. In both cases it implies that the phonon width in the frequency scale $\Delta\omega q \simeq 1\ \mathrm{cm}^{-1}$, while the central phonon frequency $\omega_o \simeq 500\ \mathrm{cm}^{-1}$.

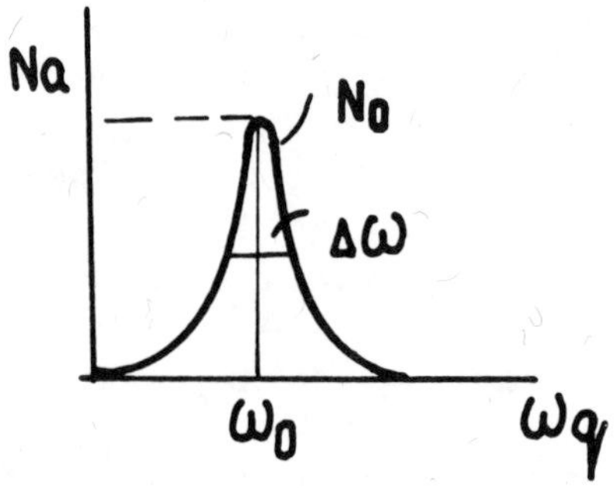

Fig. 1. Acoustic phonon distribution.

It is best at this point to formulate the theoretical problem. Suppose a dielectric crystal is given at zero temperature, in which the external sources maintain a narrow phonon distribution, as has been displayed. What are the collective excitations in such a nonequilibrium phonon system? What is the Raman-scattering caused by these excitations?

When the phonon distribution is smooth, like the thermal one, then second sound is the collective excitation.

Our result is that in the case of a narrow distribution there exist new types of collective excitations, similar to zero sound in a Fermi liquid. The existence of zero sound in a Fermi system is connected with the singularity of the distribution function near the Fermi surface. The zero sound excitations may be visualized as oscillations of the Fermi surface, that is, as oscillations of the singularity position. A narrow phonon distribution has a singularity at ω_o, slightly smoothed by $\Delta\omega q$. The singularity is the cause of the new type of collective excitations we found. Since the singularity is smoothed, the excitations have finite lifetimes.

To find elementary excitations we calculate the two-particle acoustic-phonon Green function (Fig. 2). The interaction is the anharmonic vertex shown in the same figure. The strength of this interaction is measured by the optical phonon spontaneous decay width F_o. The acoustical phonon distribution is assumed to be of Lorentzian form. The Lorentzian is described by two parameters: the width of the distribution $\Delta\omega$ and the occupation number N_o of the central mode q_o.

Since we consider a situation far from equilibrium, the ordinary Green function methods are inapplicable. The most convenient method is that developed by Keldysh [4]. In this method the one-particle Green function is a matrix (2x2). Among four Green functions only two are independent. These independent functions are the retarded Green function G_r and the ordinary correlator G_s.

When we are interested in collective excitations, we must investigate the properties of the two particle Green function with respect to the total momentum $\vec{k} = \vec{q}_1 + \vec{q}_2$ and the total energy $\Omega = \omega_1 + \omega_2$. For this reason we can join the acoustical lines by

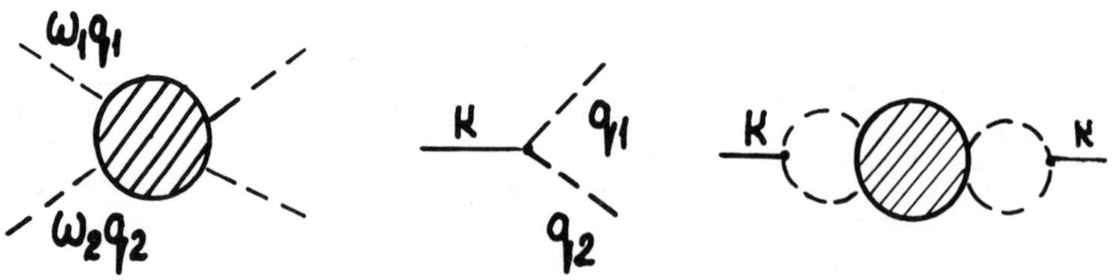

Fig. 2. Two-particle acoustic-phonon Green function, interaction vertex, one-particle optical-phonon Green function.

the interaction vertex (see Fig. 2), and this means that we can calculate the one-particle optical phonon Green function. Its polarization operator π may be calculated in the simplest approximation. In an ordinary case of a smooth distribution the appropriate small parameter is the adiabatic parameter F_0/Ω_0. In our case of a narrow distribution this parameter is not effective, and the simplest approximation is based on another parameter $N_0 \Delta\omega / \omega_0$, which is equal to the number of acoustical phonons per unit cell. This parameter may be assumed to be small.

Now we can calculate the retarded optical-phonon Green function $G_r(\vec{k}, \Omega)$. For simplicity we consider the case of long-wave excitations, $\vec{k} = 0$. Then the poles of the Green function can be found explicitly. They are depicted in the complex Ω-plane in Fig. 3. When there are no acoustic phonons, (this implies $N_0 = 0$), we have one pole at $\Omega_0 - i\frac{1}{2}F_0$. It is the usual pole of the optical phonon. When acoustic phonons are present (this implies $N_0 \neq 0$), two poles exist. For small N_0 the new pole is located at $\Omega_0 - i\Delta\omega$ (Fig. 3, which is for the case $\Delta\omega > \frac{1}{2}F_0$). When N_0 grows, the poles shift to meet one another in a vertical direction. At some

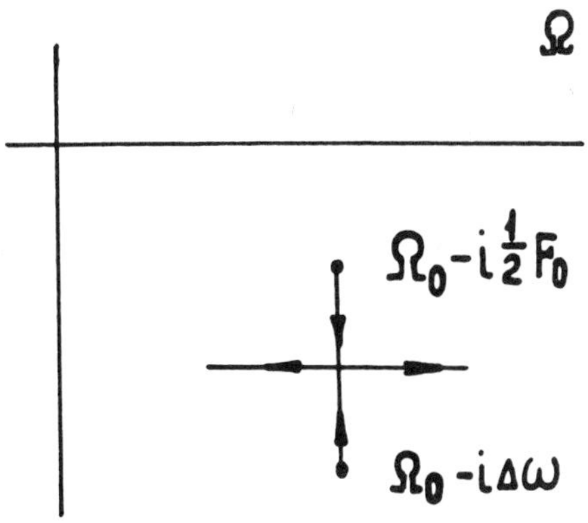

Fig. 3. Poles of the optical-phonon retarded Green function.

critical value of the acoustic phonon concentration the poles meet half way, and then diverge in a horizontal direction. The critical value of N_o is

$$N_o^* = (\alpha - 1)/8\alpha \quad , \quad \alpha = 2\Delta\omega/F_o \, .$$

Now it is evident that new excitations are important only when the acoustic phonon distribution is narrow enough. If $\Delta\omega$ is not small, the lifetime of a new excitation is short. It can be seen from the location of poles, that the lifetime of the excitation is shorter than the spontaneous lifetime of the optical phonon. It is in agreement with the usual point of view that the present acoustic phonons stimulate the decay of optical phonons, that is, F_o must be replaced by $F_o(1+2N_o)$.

But let us now consider the case, when $\Delta\omega \lesssim \frac{1}{2}F_o$. Then the picture in the Ω-plane is the same, only the poles are interchanged. In this case the lifetime of the excitation is longer than the spontaneous lifetime of the optical phonon. This means that a narrow distribution of acoustic phonons destimulates the optical

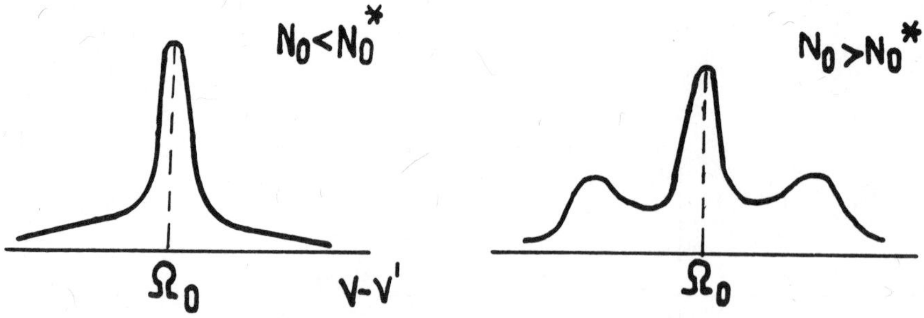

Fig. 4. Stokes scattering spectral dependence.

phonon decay, in contrast to the effect of a broad distribution. In experiments of Alfano and Shapiro [5] the lifetime of the optical phonon in calcite was measured under high excitation, and it was found to be twice as long as the spontaneous time. Maybe this result has some connection with our theory.*

The last point we want to consider is the Raman-scattering from the elementary excitations we have discussed. Or, in other words, how does the optical-phonon Raman scattering change when a narrow distribution of subharmonic acoustic phonons is pumped in the crystal?

In ordinary theories of Raman scattering the scattering probability is given in the form

$$w(\nu') \sim \left\{ {n+1 \atop n} \right\} \operatorname{Im} G_r(\nu - \nu')$$

Here ν is the frequency of the incident light, ν' is the frequency of the scattered light, n are the occupation numbers of the optical phonons, G_r is their retarded Green function. For smooth distributions the occupation number factor gives no spectral de-

---

*It must be noted, however, that in similar recent experiments of Laubereau, Wochner and Kaiser [6] the lifetime of the optical phonon in calcite was found to be the same as the spontaneous one.

pendence. In our situation, when distributions are narrow, the result for the scattering amplitude may be written in the same form, but instead of the occupation numbers we have some correlators G_s with an essential spectral dependence.

Now the form of the Raman line becomes very complicated and dependent on the concentration of the acoustic phonons. Let us consider the Stokes line at $\nu' = \nu - \Omega_0$ (Fig. 4). Then for acoustical phonon concentrations below the critical one we have a superposition of two lines centered at Ω_0, one with the width $\Delta\omega$, and the other with the width F_0. For concentrations above the critical one the Raman line looks like a triplet. The width of the central peak is $\Delta\omega$, and the width of every satellite is F_0.

REFERENCES

1. M. J. Colles and J. A. Giordmain, Phys. Rev. Lett., 27, 670 (1971).
2. R. Orbach, Phys. Rev. Lett., 16, 15 (1966).
3. I. B. Levinson, ZhETF, 65, 331 (1973) [Sov. Phys. - JETP 38, 167 (1974)].
4. L. V. Keldysh, ZhETF, 47, 1515 (1964) [Sov. Phys. - JETP, 20, 1918 (1965)].
5. R. R. Alfano and S. L. Shapiro, Phys. Rev. Lett., 26, 1247 (1971).
6. A. Laubereau, G. Wochner and W. Kaiser, Optics Comm., 14, 75 (1975).

MULTIPHONON RAMAN SCATTERING IN SEMICONDUCTING CRYSTALS

Bernard Bendow

Deputy for Electronic Technology/RADC
Solid State Sciences Division
Hanscom AFB MA 01731 USA

A critical review of recent theoretical investigations of multiphonon Raman scattering in semiconducting crystals is presented. Although various aspects of available experimental data can be understood on the basis of existing theories, a quantitative and comprehensive account of many-phonon phenomena is still lacking.

INTRODUCTION

There has been considerable interest in recent years in multiphonon Raman scattering (henceforth, MRS) in semiconducting crystals, especially under resonance conditions.[1-5] The objective of this paper is to present a critical overview of the highlights of theoretical work relevant to MRS, and to provide a status report of research in the field. Our attention here will be restricted to many-phonon ($n \geqslant 3$) scattering. An extensive literature exists on the two phonon scattering problem; the interested reader is directed to Refs. 6-11, for example.

Although many-phonon phenomena have been widely investigated with regard to absorption wings (Urbach edges,[12] impurity sidebands[13] and high frequency infrared wings,[14] for example) RS offers the prospect of studying new features not generally present in absorption. For the most part, these are associated with frequencies near to or above resonance with the fundamental electronic transitions of the crystal. Probably the most spectacular aspect of scattering above the gap in semiconductors is the appearance of a series of sharp (nearly intrinsic anharmonic widths) multiphonon lines with relatively slowly decreasing strengths in the scattered spectrum. Results from the pacesetting experiment of Leite et al[1] are indicated in Fig. 1. This behavior contrasts markedly with scattering below the gap, where sharp multiphonon features are not generally encountered. Rather, the spectrum is smooth and continuous, and decreases rapidly with increasing order (typically by one to two orders of magnitude per phonon). Part of the attraction in investigating resonance MRS is the capability of separately tracing distinct multiphonon lines as a function of incident photon frequency, which provides information about exciton and/or phonon dispersions. Yu, Shen and co-workers,[5,8] for example, have utilized data such as that illustrated

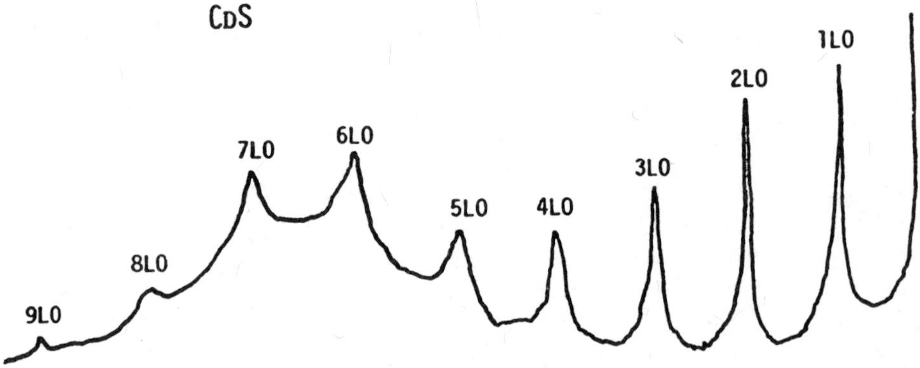

Fig. 1. Uncorrected scattered spectrum of CdS at ~ 300°K with laser excitation at 4579 Å (from Ref. 1).

in Fig. 2 for this purpose. Generally, one requires only fairly simple models of the frequency dependence of RS to extract information about quasiparticle dispersions.

One advantage of MRS as a probe of many-phonon effects is the selective enhancement which highlights various phonons or combinations of phonons in the spectrum. While such selectivity may result in part from the dependence of the scattering on the geometrical configuration, it stems principally from the particulars of the exciton and phonon dynamics which characterize the MRS. For example, spectra stemming from exciton-phonon interactions characterized by couplings which are restricted to a narrow range of wavevectors are expected to display correspondingly narrow lines. Another consideration is that phonons

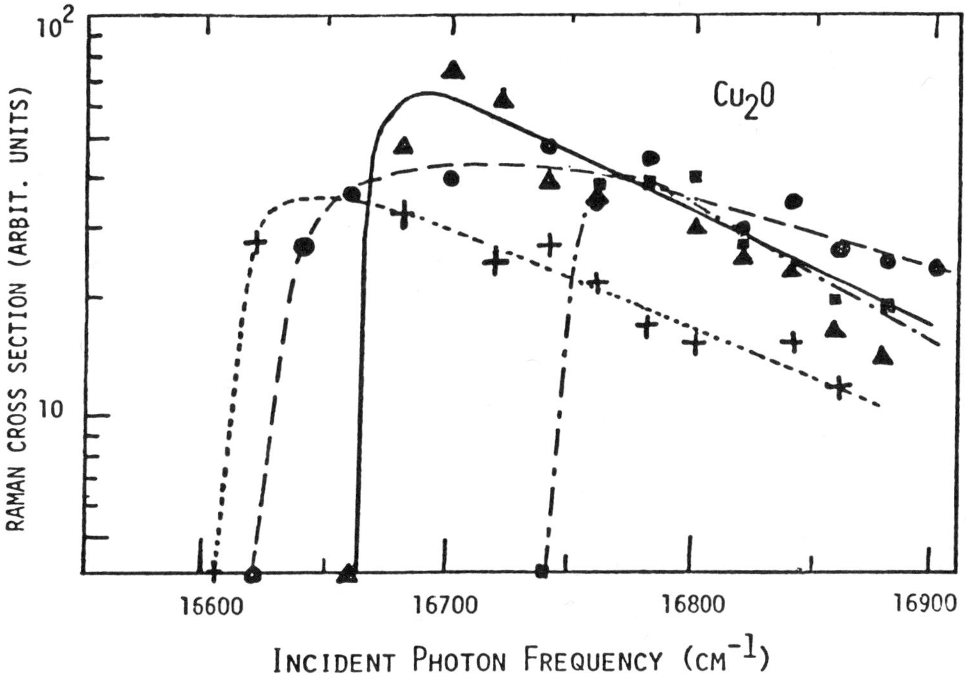

Fig. 2. Raman cross section vs. incident photon frequency for various three and four phonon modes of Cu_2O. Pluses are identified as $2\Gamma_{12}^- + \Gamma_{25}^-$, triangles as $2\Gamma_{12}^- + \Gamma_{15}^{-(1)}$, circles as $3\Gamma_{12}^-$ and squares as $4\Gamma_{12}^-$; solid lines are theoretical curves. (From P.Y.Yu and Y. R. Shen, Phys. Rev. B12, 1377 (1975)).

corresponding to the appearance of multiple poles in the scattering probability will be selectively enhanced. We elaborate on these and related factors further on in this paper.

MRS BELOW THE GAP

Multiphonon Raman scattering below the gap is adequately described in terms of fluctuations in the polarizability[15] $P(\{\underset{\sim}{u}\})$, where $\{\underset{\sim}{u}\}$ is the set of lattice displacements of the crystal. For conceptual simplicity let us restrict attention to a single displacement $\underset{\sim}{u}$. Then making some additional assumptions about isotropy, one can show that the efficiency R may be approximated as

$$R \sim \sum_n f_n \rho_n(\Delta\omega)$$
$$f_n = \frac{1}{n!}\left(\frac{\partial^n P}{\partial u^n}\right)_o^2, \quad \rho_1(\omega) = \langle u(t)u(o)\rangle_\omega \qquad (1)$$

where $\Delta\omega$ is the frequency shift, and ρ_n are convolutions of the (anharmonic) displacement correlation function[16] in frequency space, $\rho_1(\omega)$. The above expansion corresponds to keeping just the quadratic term in a cumulant expansion[17] for R. For the present problem we are interested in $R(\Delta\omega)$ for values $\Delta\omega/\omega_o > 3$, where ω_o is the phonon frequency. The behavior of the ρ's and sums such as that appearing in Eq. 1 have been studied extensively with respect to multiphonon infrared absorption (MIRA).[14,18] The derivaties f_n tend to decrease nearly exponentially,[14,18] while the ρ_n's tend for large n toward increasingly broad Gaussians centered near $n\omega_o$,[19] i.e., the structure in ρ_1 is progressively smoothed out. Since the ρ_n's are essentially the nth order phonon densities of state,[20] one may say that the scattering is dominated by density of states effects below the gap. The degree of structure in the spectrum depends to a large extent on the particular crystal studied and the selection rules governing the scattering configuration under consideration. Studies of MIRA reveal very little structure for $n \gtrsim 3$ at room temperature in ionic solids, but indicate the persistence of structure in covalent solids such as the zinc-blende semiconductors;[21] similar trends may be expected for MRS as well. Methods utilized in Sec. IVB of Ref. 22 may be taken over to the present problem, from which one concludes that the spectrum will decrease nearly exponentially with frequency shift for $3 \lesssim \omega \lesssim 7$. The typical decrease predicted is about one to two orders of magnitude per phonon at room temperature, and is greater for the semiconductors than for the highly ionic materials. The temperature dependence is formally of the usual multiphonon type, but the T-dependence of the phonon spectrum, which may substantially suppress the T-dependence of R for $\Delta\omega \gg \omega_0$, must in general be taken into account.[23]

The above discussion has omitted the frequency dependence of P, which will be instrumental in the resonance regime. Nevertheless, it is easily seen that the continuous scattered spectrum given by Eq. (1) will, to an excellent approximation, simply be enhanced uniformly as resonance is approached from below, in a fashion similar to one-phonon RS, say. Thus, resonance conditions offer the opportunity of detecting the very rapidly decreasing portion of the MRS spectrum which is not normally observable in the non-resonant regime. In addition to the continuous portion, however, we may expect various quasi-discrete features to arise in the spectrum. These could be the result of either the wave vector dependence of the coupling, or of selective resonance enhancement, a topic to be discussed further in the following sections. The quasi-discrete features are distinguishable from the continuous background because their widths and peak positions depend on the incident photon frequency, and display a different (generally stronger) resonance enhancement from the background. Thus an analysis of such features provides information about couplings and crystalline band structure. To this date, however, the bulk of experimental and theoretical work on MRS has concentrated on the range of frequencies very close to or above resonance. We therefore emphasize this regime throughout the remainder of this paper.

RS ABOVE THE GAP: GENERAL CONSIDERATIONS

As remarked previously, the most striking many-phonon spectra are observed above the gap in II-VI semiconductors,[1-4,6] in the form of series of narrow lines at frequencies nearly equal to multiples of the LO mode at $\underset{\sim}{k} = 0$. For example, as many as nine overtones in CdS, eight in ZnO and five in ZnSe, ZnTe and ZnS are observed, with these numbers being roughly proportional to the size of the polaron coupling constant. The line intensities, which may appear atop of a luminescent background, first decrease and then increase through a maximum with increasing order. The intensity is enhanced as a line approaches resonance with excitons from above, and lines coincident with frequencies below the gap become broadened and decreased in intensity. The nature of the spectrum suggests that the scattering is due to the intraband Frohlich interaction, which couples to just LO phonons and strongly weights $\underset{\sim}{k} = 0$ phonons because of its long range. We note that sharp overtone lines have not been observed in Si and Ge, or III-V's such as GaP, InAs and InSb. A useful theory of MRS must account for the number of lines observed in different materials, the slow decrease in the intensities of the lines, and the frequency dependence of the intensity and width of the lines when tuned through the gap.

A principal factor in the selective enhancement of many-phonon lines above the gap is the possibility of vanishing energy denominators in the perturbation theory expression for the scattering efficiency. Thus, although experiments in Cu_2O involve scattering attributable to short-range deformation potentials, they nevertheless display a variety of distinct many-phonon spectral features.[4, 8] When only weakly dispersive optical phonons are involved, then the relative positions and widths of the peaks depend weakly on incident frequency, while they vary with frequency when one or more acoustic phonons participate in the scattering.

To investigate the frequency dependence of MRS we follow a development due to Zeyher,[24] which shows that the "most resonant" term in the n-phonon contribution to R takes the form

$$R^{(n)} \sim \sum_{\underset{\sim}{q}_i, \alpha_i} F(\underset{\sim}{q}_1 j_1 \cdots \underset{\sim}{q}_n j_n) F^*(\underset{\sim}{q}_{\alpha_1} j_{\alpha_1} \cdots \underset{\sim}{q}_{\alpha_n} j_{\alpha_n}) \delta(\sum_i \underset{\sim}{q}_i)$$

$$F(\underset{\sim}{q}_i j_i \cdots \underset{\sim}{q}_n j_n)$$

$$= \sum_{\gamma_1 \cdots \gamma_{n+1}} \frac{g^*(\gamma_1) V(\gamma_1 \gamma_2; \underset{\sim}{q}_1 j_1) \cdots\cdots V(\gamma_n \gamma_{n+1}; \underset{\sim}{q}_n j_n) g(\gamma_{n+1})}{D(\gamma_1) D(\gamma_2 \underset{\sim}{q}_1) \cdots\cdots\cdots D(\gamma_{n+1}, \underset{\sim}{q}_1 \cdots \underset{\sim}{q}_n)} \tag{2}$$

$$D(\gamma, \underset{\sim}{q}_1 \cdots \underset{\sim}{q}_{\ell-1}) = \omega - \omega_\gamma(\sum_{i=i}^{\ell-1} \underset{\sim}{q}_i) - \sum_{i=i}^{\ell-1} \omega_o(\underset{\sim}{q}_i j_i) + i\Gamma(\omega - \sum_{i=i}^{\ell-1} \omega_o(\underset{\sim}{q}_i j_i))$$

where $\omega_o(j)$ and ω_γ are phonons and excitons of type j and γ, respectively, g is the photon-exciton coupling, Γ the exciton damping function and $V(\gamma\gamma')$ is the matrix element of the single phonon-exciton interaction; photon polarization indices have been suppressed for notational simplicity. It has been assumed that V depends only on the transferred momentum $\tilde{q}$, and that Γ depends only on frequency. Equation (2) immediately reveals that the exciton and phonon dispersions and couplings will significantly affect the frequency dependence of R. The possibility of vanishing D's above the gap will clearly influence the frequency dependence for lines tuned through the gap. We will investigate and discuss MRS in more detail on the basis of Eq. (2) in the following section.

CALCULATIONS OF MULTIPHONON RS

Although a microscopic treatment based on expressions such as in Eq. (2) would be highly desirable, from a practical point of view it is very difficult to evaluate such expressions. This situation has spurred the development of simplified models to interpret MRS spectra. We here briefly describe two such models, cascade and configuration coordinate, after which we return to a further inspection of the microscopic theory.

Configuration coordinate models have been utilized extensively in a variety of other connections, such as impurity sidebands,[13] Urbach tails[25] and RS from gases.[26] Essentially, the ground and excited electronic state energies are assumed to depend adiabatically on the phonon displacement u, so that for unit mass

$$E_o = \tfrac{1}{2}\omega_o^2 u^2$$
$$E_i = E_i^{(o)} + \tfrac{1}{2}\omega_o^2 u^2 - \gamma\omega_o^{3/2} u \qquad (3)$$

where γ is a polaron coupling constant. One obtains for the nth order contribution to R,

$$R^{(n)} \sim \sum_{im} \frac{f_i e^{-\gamma^2/2}}{\omega^2 - \omega_i^2} J_{om} J_{mn}$$

$$J_{mn} = \frac{m!}{n!} \left(\frac{\gamma^2}{2}\right)^{n-m} \left[L_m^{n-m}\left(\frac{\gamma^2}{2}\right)\right]^2 \qquad (4)$$

where f_i incorporates electronic overlap, and L is an associated Laguerre polynomial; the above expression has assumed that the electronic and vibrational overlaps decouple completely. Calculated results from Ref. 27 based on this model are illustrated in Fig. 3. Similar calculations employing a Gaussian distribution of impurity states below the gap are given in Ref. 28. Another treatment of MRS analogous to the one discussed here has been presented by Mulazzi.[29] Although explicit results were not displayed in the latter, the appearance of an imaginary part in the

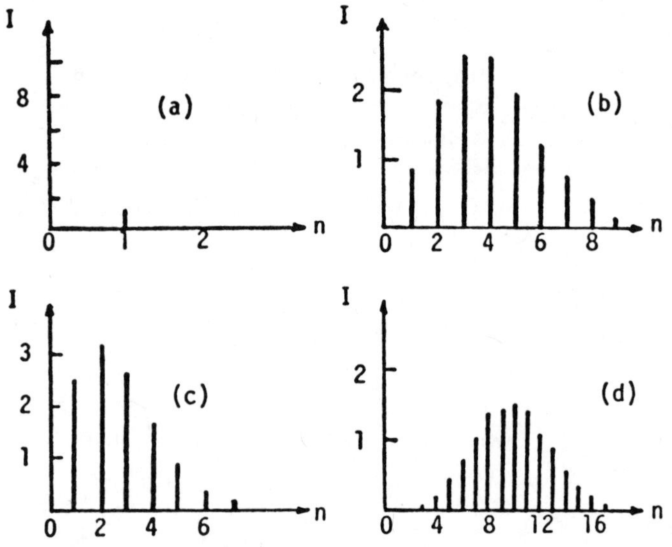

Fig. 3. Calculated relative intensities I of scattered Raman lines vs order n for various values of coupling constant $\gamma^2/2$: (a) 0.1, (b) 2.5, (c) 4.0 and (d) 10.0 (from Ref. 27).

expressions for R may enable one to account for the frequency dependence of MRS as one tunes through the gap.

The advantage of the configuration coordinate model is that it is non-perturbative. However, such a model would appear most appropriate for intermediate or strong polaron couplings, while for most materials of interest here the couplings are weak (at least weak enough to justify perturbation theory). Unfortunately, the relationship between γ and the usual Frohlich polaron coupling constant is not clear. Other disadvantages in the model are that wave vector effects are suppressed, and that it is uncertain whether differences between above and below gap scattering can be properly taken into account.

We now turn attention to the cascade model[30] of MRS. Inspection of Eq. (2) reveals that, in general, the real part of n of the n+1 energy denominators D will vanish simultaneously for a special set of $\underset{\sim}{q}$'s. In the zero-damping limit vanishing D's result in δ-functions representative of the creation of real (rather than virtual) electronic intermediate states, which subsequently interact with other states through the electron-phonon interaction V. The cascade picture of Martin and Varma[30] accounts for just processes consisting of n real transitions (vanishing D's) and a single virtual transition. The incident photon leads to creation of a real exciton which may decay via a variety of channels, including optic and acoustic phonon scattering and radiative recombination. Since LO phonon scattering is typically the fastest decay mechanism, the exciton is successively scattered through real states of energy $\omega - m\omega_{LO}$, with a small but finite probability of radiative recombination (creating a photon of frequency $\omega - m\omega_{LO}$) during the mth step. The probability per photon of such recombination at step m is $S_m = \Gamma_R(m)/\Gamma_{LO}(m)$, the ratio of the radiative and LO-scattering probabilities per unit time. The cascade may continue until the exciton frequency falls below the gap, in which case LO-phonon scattering is precluded by energy conservation. The Raman lines are pictured as broadening below the gap due to interactions with acoustic phonons (thermalization). Although calculated values of Ref. 30 provide a reasonably good account of the original CdS observations,[1] they are in poorer agreement with subsequent measurements of Damen et al[31] on crystals with improved surface quality.

We now return to an inspection of microscopic theory as embodied by Eq. (2). Two simple limits of interest are that of discrete, well-separated exciton levels, and that of dense or essentially continuous exciton levels. We will conclude that under certain conditions in the former instance results similar to the cascade approach are applicable. However, in general, interference effects arising from a continuous spectrum of electronic levels will not be negligible. In particular, this appears to be the case for the tetrahedral semiconductors considered here.

Consider the evaluation of Eq. (2) when the sum over intermediate states (γ_2 to γ_n) is restricted to a single exciton $\bar{\gamma}$, but full sums over initial and final excitons γ_1 and γ_n are retained. Such a model is appropriate for certain of the experimental conditions in Cu_2O, considering the large binding energy of the 1s-exciton and the dipole-forbidden character of the yellow series. Ignoring the $\tilde{q}$-dependence of $V_{\gamma\gamma'}$ and employing a dispersionless phonon ω_o and an effective mass dispersion for the exciton, the $\tilde{q}_i$ integrations in the term $(\alpha_1 \ldots \alpha_n) = (1 \ldots . n)$ take the form

$$\int \frac{d\tilde{q}_n |V|^2}{|\omega - \omega_1(\tilde{q}_n) - n\omega_o + i\Gamma|^2}$$

$$\sim \begin{cases} \dfrac{\bar{\Gamma}(\omega - n\omega_o)}{\Gamma(\omega - n\omega_o)} & \text{for } \omega > \omega_1(0) + n\omega_o \\[2ex] \dfrac{|V|^2}{\mathrm{Im}\sqrt{Z_n}} & \text{for } \omega < \omega_1(0) + n\omega_o \end{cases} \tag{5}$$

$$Z_n = \omega - \omega_1(0) - n\omega_o + i\Gamma(\omega - n\omega_o)$$

where $\bar{\Gamma}$ is the decay rate of exciton ω_1 due to scattering with the creation of a phonon,

$$\bar{\Gamma} \propto |V|^2 \mathrm{Re}\sqrt{Z_n} \; , \tag{6}$$

in contrast to the total damping Γ. The entire contribution to R arising in this way will contain (n-1) second order poles ("double resonances") in the limit of zero damping and is the most divergent of the n! possible contributions stemming from permutation of indices in Eq. (2). Thus, for typical values of Γ just the single most divergent term need be considered above the gap, while below the gap all n! terms need be retained in general. The above-gap efficiency may then be written in the physically more suggestive form

$$R^{(n)}(\omega) \propto P_A(\omega) \left[\prod_{m=1}^{n-2} \frac{P_S(\bar{\omega})}{P(\bar{\omega})} \right] \frac{P_E(\omega-(n-1)\omega_o)}{P(\omega-(n-1)\omega_o)}, \qquad \bar{\omega} = \omega - m\omega_o \tag{7}$$

where P_A, P_E, P_S and P denote the probabilities per unit time for phonon-assisted absorption and emission, scattering via ω_o, and decay via all possible mechanisms, respectively. Note that energy and momentum are conserved in each step; thus, for example, the absolute values of the exciton and phonon momenta are fixed in the first step, and similarly restricted in higher steps. Spectral features will shift and change shape because of the dependence of the allowed $\underset{\sim}{q}$ values on incident frequency. Such considerations play a principal role in the deduction of dispersive properties from the frequency dependence of MRS in Cu_2O.

Equation (2) may be reduced to single non-trivial integration for the case of a non-interacting electron—hole contimuum and constant $V(q) \propto C$, as described in Ref. 32. In general, interference effects between channels turn out to be non-negligible, and the efficiency does not factonize in the manner indicated by Eq. (7). One can show that the lines in the continuum vary in the weak coupling limit as

$$R^{(n)} \sim \begin{cases} C^{2n} & n = 1, 2 \\ C^6 |\log C| & n = 3 \\ C^6 & n \geq 4 \end{cases} \tag{8}$$

where C is the polaron coupling strength. This result contrasts with that of the cascade model, where $R^{(n)} \sim C^2$ for all n. Although we have considered just constant electron-phonon coupling in obtaining Eq. (8), the results are expected to be qualitatively analogous for the case of peaked $V(\underset{\sim}{q})$'s. The reason is that the effect of peaked $V(\underset{\sim}{q})$'s is essentially to restrict the spectrum to a series of narrow, evenly spaced lines, which is the same result obtained by utilizing dispersionless phonons and constant coupling. The C-dependence displayed by Eq. (8) explains why more lines are observed in crystals with larger polaron coupling constants, and why none have been observed in those with very small couplings, such as Si and Ge.

Numerical evaluation of the efficiency indicates that the relative intensities of the overtones depend strongly on the frequency dependence of Γ, which is a sum of the optic phonon decay Γ_o and a usually small background decay Γ_B due to all other mechanisms. The intensity will be enhanced for overtones lying near to the gap, since Γ will typically be large and nearly constant far above the gap, but will decrease sharply to $\Gamma \approx \Gamma_B$ as the gap is approached from above. Below the gap the real parts of various D's can no longer vanish, and the line intensities rapidly decrease. These considerations clearly explain many of the observed features of MRS spectra. Unfortunately, the sensitivity of R on $\Gamma(\omega)$ implies that impurities and surface conditions may strongly influence the resonance properties of MRS. Model calculations carried out for the present case are indicated in Fig. 4, and display the slow decrease of successive lines observed experimentally. Computations for CdS with $n = 2$ results normalized to experiment predict values which are 1.05, 1.22 and 0.98 times the experimentally observed values[31] for $n = 3$, 4 and 5, respectively.

A final matter to which we allude is the temperature (T) dependence of MRS above the gap. While an $(n(\omega_o) + 1)^m$ dependence would be predicted by (undamped) perturbation theory for an mth order line, and an $(n(\omega_o) + 1)$ dependence for all orders by the cascade model, the model calculations described in the previous paragraph lead to a T-dependence which is intermediate to these two extremes.[32] Thus, the T-dependence of R could well serve as a check on the validity of the various predictions. Unfortunately, appropriate data of this sort does not appear to be presently available.

CONCLUDING REMARKS

In the above we have described various aspects of theoretical interpretations of MRS in crystals. Not only is the phenomenon of MRS of interest in its own right, but it serves as a potentially

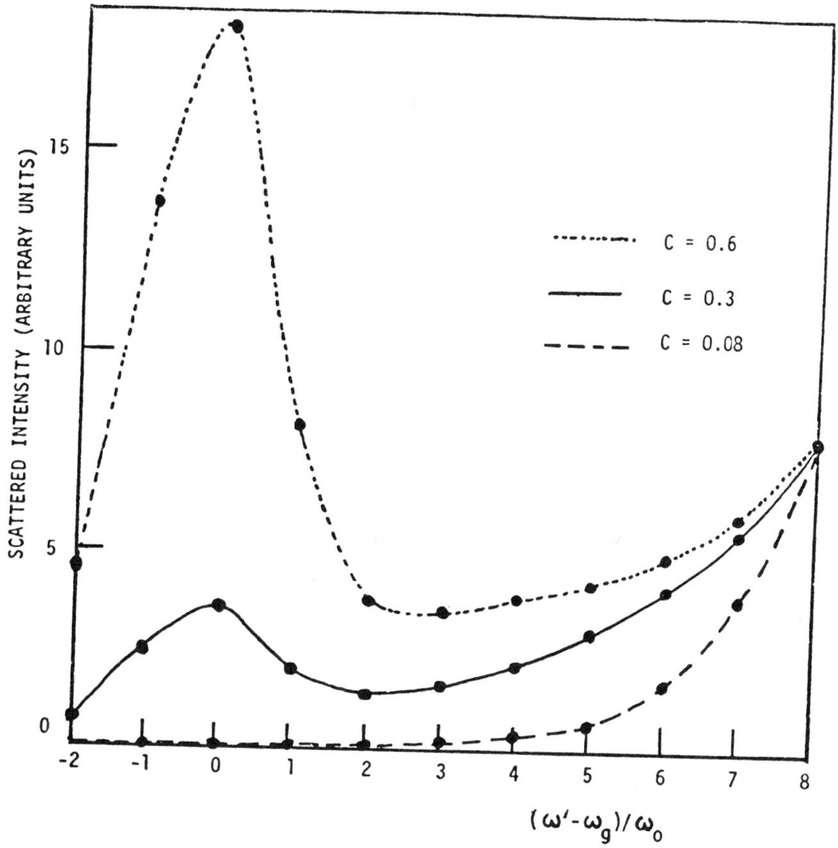

Fig. 4. Calculated intensity distribution of overtones for various coupling strengths C. The incident photon frequency $\omega = \omega_g + 10\,\omega_0$, where ω_g is the gap frequency; ω' is the scattered photon frequency. The lines have been interpolated to connect the calculated points (dark dots). (From Ref. 32).

powerful tool for investigating quasiparticle couplings and dispersion in crystals. Moreover, the fundamental nature of resonant scattering above the gap may be revealed from a theoretical analysis of MRS.

The investigations to date provide the following picture of MRS: Below the gap, the spectrum is dominated by density of states effects, and is therefore nearly continuous and rapidly decreasing with increasing order. As resonance is approached from below, various energy denominators in the perturbation expression for the efficiency can become small and selective enhancement of various spectral features results. Above the gap many denominators can resonate simultaneously, producing highly selective enhancement of certain phonons, and resulting in long sequences of slowly decreasing overtones. Were it not for interference between decay channels, the spectrum could be described as arising from the cascade of an exciton, which is successively scattered by LO phonons until the gap frequency is reached, with a finite probability of radiative emission during each step. Although such a description is never exact, it might be expected to provide a rough account of MRS behavior in certain instances.[33]

The configuration coordinate picture correctly predicts the observed slow decrease of overtone intensities in MRS above the gap. It is not certain whether the model is capable of correctly accounting for the dependence of the number of observed lines on the size of the polaron coupling constant, or for the frequency dependence of the spectrum as lines are tuned through the gap. Moreover, wave-vector effects, which are known to influence the frequency dependence of certain spectral features, are not explicitly taken into account. The cascade model shares some of the same deficiencies; in addition, interference between channels is ignored. The dependence of the number of observed lines is incorrectly predicted in the cascade model. It is also deficient, perhaps, from an aesthetic point of view, since scattering phenomena below and above the gap are not unified by a single expression for the efficiency covering both regimes.

The evaluation of special terms in the microscopic theory under a variety of simplifying assumptions has contributed significantly to the interpretation of observed MRS spectra. Features such as the dependence of the number of observed lines on the polaron coupling constant, and the frequency dependence of the spectrum as one tunes through the gap, are well explained on the basis of such calculations. Interference effects, which are influential for densely packed electronic levels, can be taken account in a rigorous manner. For well-separated levels under certain conditions, a cascade-like representation of the scattering as a product of step probabilities emerges. The wider applicability of the results of the microscopic calculations is uncertain because of the simplifying assumptions, such as $\tilde{q}$-independent couplings, which were utilized in their derivation.

We conclude that although many general features of MRS spectra can be understood on the basis of existing theories, a detailed account of observed line shapes and positions, as well as their frequency dependence, is not as yet available. To properly determine the origin of the MRS spectra in II-VI's, for example, it may well be necessary to calculate the contributions of diagrams containing less "double resonances" than those which have been considered to the present. It will also be necessary to determine more fully the influence of the electron-hole interaction and the $\tilde{q}$-dependence of the electron-phonon coupling function on the predicted efficiency. More extensive experimental data, especially of the frequency and temperature dependence of MRS above and below the gap, would be highly desirable to provide comparison with theoretical predictions.

The author thanks Dr. R. Zeyher of the Max Planck Institute, Stuttgart, for extensive discussions; the present paper has also drawn heavily on material supplied by him to the author.

REFERENCES

1. R. C. Leite, J. F. Scott and T. C. Damen, Phys. Rev. Lett. 22, 780 (1969).
2. J. F. Scott, R. C. Leite and T. C. Damen, Phys. Rev. 188, 1285 (1969); J. F. Scott, ibid, B2, 1209 (1970); J. F. Scott, T. C. Damen, W. T. Silfvast, R. C. Leite and L. E. Cheesman, Optics Comm. 1, 397 (1970); T. C. Damen, R. C. Leite and J. Shah, in "Proc. 10th Conf. Phys. of Semicon." (US Atomic Energy Comm., 1970).
3. M. V. Klein and S. P. Porto, Phys. Rev. Lett. 22, 782 (1969).
4. E. Gross, S. Permagorov, V. Travnikov and A. Selkin, J. Phys. Chem. Solids 31, 2595 (1969); E. Gross, S. Permagorov, Ya. Morozenko and B. Kharlamov, Phys. Stat. Sol. (b)59, 551 (1973).
5. P. Y. Yu and Y. R. Shen, Phys. Rev. Lett. 32, 939 (1974); P. Y. Yu, in "Proc. 3rd Conf. Light Scattering in Solids," M. Balkanski, ed. (Flammanon, Paris, in press). P. Y. Yu and Y. R. Shen, Phys. Rev. B12, 1377 (1975).
6. R. H. Callender, S. S. Sussman, M. Selders and R. K. Chang, Phys. Rev. B7, 3788 (1973); P. B. Klein, J. J. Song, R. K. Chang and R. H. Callender, in Proceedings, op cit, Ref. 4.
7. B. A. Weinstein and M. Cardona, Phys. Rev. B8, 2795 (1973); M. A. Renucci, J. B. Renucci, and M. Cardona,

Sol. State Comm. 14, 1229 (1974); M. A. Renucci, J. B. Renucci, R. Zeyher and M. Cardona, Phys. Rev. B10, 4309 (1974); R. L. Schmidt, B. D. McCombe and M. Cardona, Phys. Rev. B11, 746 (1975); W. Kiefer, B. D. McCombe, W. Richter, R. L. Schmidt and M. Cardona, in Procs., op cit, Ref. 4.

8. P. Y. Yu, Y. R. Shen, Y. Petroff and L. M. Falicov, Phys. Rev. Lett. 30, 283 (1973); P. Y. Yu, Y. R. Shen, and Y. Petroff, Sol. State Comm. 12, 973 (1973).
9. J. Oka and T. Kushida, J. Phys. Soc. Japan 33, 1372 (1972).
10. P. B. Klein, H. Masui, J. J. Song and R. K. Chang, Sol. State Comm. 14, 1163 (1974).
11. R. Zeyher, Phys. Rev. B9, 4439 (1974).
12. See, for example, J. D. Dow and D. Redfield, Phys. Rev. B5, 594 (1972).
13. See, for example, M. H. Pryce, in "Phonons", R. W. Stevenson, ed. (Plenum, NY, 1966).
14. See, for example, B. Bendow, S. C. Ying and S. P. Yukon, Phys. Rev. B8, 1679 (1973); M. Sparks and L. J. Sham, ibid, 3037 (1973); D. L. Mills and A. A. Maradudin, ibid, 1617 (1973); T. C. McGill, R. W. Hellwarth, M. Mangir and H. V. Winston, J. Phys. Chem. Solids 34, 2105 (1973).
15. M. Born and K. Huang, "Dynamical Theory of Crystal Lattices" (Oxford U.P., London, 1956).
16. See, for example, R. A. Cowley in "Phonons", R. W. Stevenson, ed (Plenum, NY, 1966).
17. B. Bendow and S. P. Yukon, in "Optical Props. Highly Transparent Solids", S. S. Mitra and B. Bendow, eds (Plenum, NY, 1975).
18. Some relevant papers include B. Bendow et al, op. cit. Ref. 14; K. V. Namjoshi and S. S. Mitra, Phys. Rev. B9, 815 (1974); B. Bendow, S. P. Yukon and S. C. Ying, ibid., B10, 2286 (1974).
19. A. Sjolander, Ark. Fyski 14, 315 (1958).
20. See, for example, A. A. Maradudin, E. W. Montroll, G. H. Weiss and I. P. Ipatora, "Theory of Lattice Dynamics in the Harmonic Approximation" (Academic, NY, 1971).
21. T. F. Deutsch, J. Phys. Chem. Solids 34, 2091 (1973).
22. B. Bendow, S. P. Yukon and S. C. Ying, Phys. Rev. B10, 2286 (1974).
23. See, for example, B. Bendow, Appl. Phys. Lett. 23, 133 (1973); M. Sparks and L. J. Sham, Phys. Rev. Lett. 31, 714 (1973).

24. R. Zeyher (unpublished material).
25. T. H. Keil, Phys. Rev. 144, 582 (1966).
26. W. Kiefer, Appl. Spectr. 28, 115 (1974).
27. H. Malm and R. R. Haering, Can. J. Phys. 49, 1823 (1971).
28. M. L. Williams and J. Smit, Sol. State Comm. 8, 2009 (1970).
29. E. Mulazzi, Phys. Rev. Lett. 25, 228 (1970).
30. R. M. Martin and C. M. Varma, Phys. Rev. Lett. 26, 1241 (1971).
31. T. C. Damen, R. C. Leite and J. Shah, in "Proc. 10th Conf. Phys. of Semicon." (U.S. Atomic Energy Comm., 1970).
32. R. Zeyher, Sol. State Comm. 16, 49 (1975).
33. R. M. Martin, Phys. Rev. B10, 2620 (1974).

LIGHT SCATTERING AS A METHOD OF STUDYING THE TRANSFORMATION OF INCOHERENT PHONON FLUX INTO A COHERENT ACOUSTICAL SIGNAL

S. V. Gantsevich, V. L. Gurevich, V. D. Kagan
and R. Katilius

A. F. Ioffe Physico-Technical Institute of the
Academy of Sciences of the USSR, Leningrad, USSR

The purpose of the present paper is to discuss the physical situation in which some light scattering experiments can give deep insight into the interesting physical phenomenon of the coherentization of an acoustical flux in the course of its amplification in a semiconductor.

As is well known [1], if one applies to a semiconductor a dc electric field which is sufficiently high to cause the supersonic drift of carriers, then the amplification of thermal noise takes place along the crystal provided the electron-phonon coupling is not too small. This is the case, for instance, in piezoelectric semiconductors, such as CdS, ZnO, and others.

At first the noise intensity grows exponentially along the crystal. The equation for the average phonon occupation number $\bar{N}_q$ is linear, and its solution is exponential (see, e.g., Ref. [2]):

$$v_x \frac{d\bar{N}_{\vec{q}}}{dx} - \gamma^0_{\vec{q}} \bar{N}_{\vec{q}} = 0 , \quad \bar{N}_{\vec{q}}(x) = \bar{N}_{\vec{q}}(0) \exp (\gamma^0_{\vec{q}} x / v_x) \tag{1}$$

Here v_x is the sound velocity, x is the distance from the crystal boundary, and $N_{\vec{q}}$ is the thermal equilibrium average occupation number which determines the initial noise intensity. The amplification constant $\gamma^0_{\vec{q}}$ as a function of the wave vector has its maximum in the long-wave region, so the whole problem, in fact, is classical rather than quantum-mechanical. The above-mentioned linear theory of amplification can be illustrated by a set of diagrams of the Wyld [3] type (see Fig. 1). The two-phonon vertices in these diagrams describe the phonon interaction with the active medium, i.e., with the drifting electron system. (The derivation of the Wyld-type diagram technique for long-wave phonons interacting with an electron system can be found in Ref. 4).

In Fig. 2 the spectral, i.e., the frequency dependence of the amplification constant, is shown for the wave vector parallel to the drift direction $\vec{x}$. For the phonon wave vector non-parallel to the drift direction the curves are of the same type but go lower. So the wings of the angular and spectral distribution of noise at large x are amplified much less than the center of the distribution. As a result, the angular and spectral interval of noise is effectively narrowing in the course of its propagation. But the intensity fluctuations are relatively large in the amplified flux, as they are in thermal equilibrium. The narrowing cannot transform the noncoherent noise into a coherent signal since linear amplification, of course, cannot change statistical properties of noise.

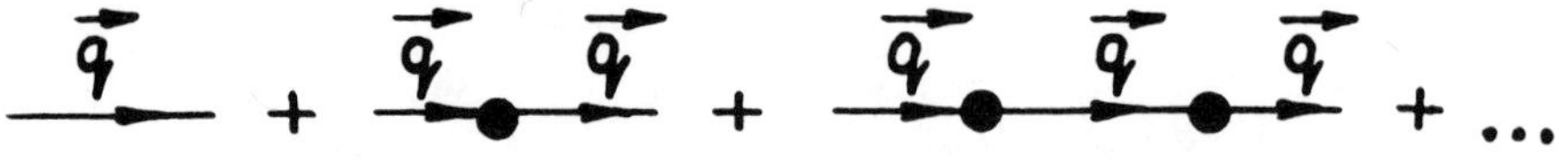

Fig. 1. A set of diagrams of the linear theory.

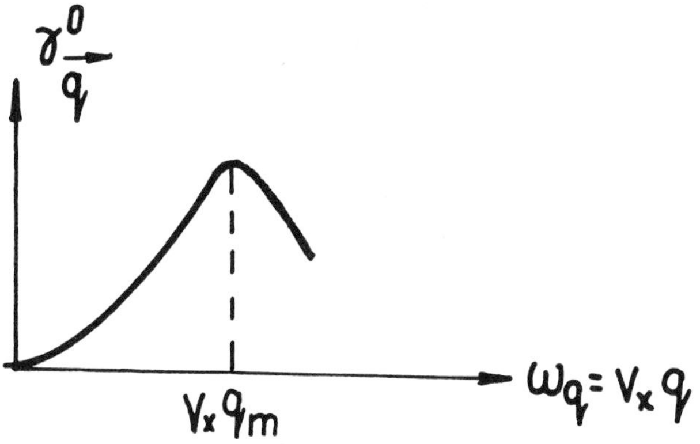

Fig. 2. The typical frequency dependence of the linear amplification constant $\gamma^0_{\vec{q}}$.

So, let us consider nonlinear effects which become essential at larger distances. First, the amplification constant becomes dependent on the phonon intensity:

$$\gamma_{\vec{q}} = \gamma^0_{\vec{q}} - \sum_{\vec{q}\,'} W_{\vec{q}\vec{q}\,'} \overline{N}_{\vec{q}\,'} \tag{2}$$

If the kernel $W_{\vec{q}\vec{q}\,'}$ is a smooth function of wave vectors, then the nonlinear contribution is approximately proportional to the integral phonon intensity

$$\overline{N}(x) = \sum_{\vec{q}} \overline{N}_{\vec{q}}(x) .$$

For the positive kernel W it leads to the attenuation of distribution wings, because for these wings the nonlinear term predominates. At the same time, the waves near the center of distribution are still being amplified, because near the maximum of the linear amplification constant the linear term predominates.

Intensity fluctuations in such a self-narrowing flux are still large and the statistics remains unchanged because the above-mentioned nonlinear effect is simply the renormalization of the linear amplification constant. The corresponding equation for the average phonon intensity,

$$v_x \frac{d\overline{N}_{\vec{q}}}{dx} - \gamma_{\vec{q}} \overline{N}_{\vec{q}} = 0 \qquad (3)$$

can be represented by a set of the self-consistent-field type diagrams depicted in Fig. 3. It should be mentioned here that in semiconductors the main role is played by the indirect four-phonon interaction via the electron system [4]. This is why the vertices are four-point in these diagrams. Note that these diagrams describe consecutive events of interaction and closely resemble the diagrams of the linear theory depicted in Fig. 1.

But there is also another type of diagram which can be constructed with the four-phonon vertices. An example is shown in Fig. 4. This diagram describes the simultaneous rather than consecutive interaction of phonons. The frequencies of the phonons taking part in this process should obey the energy conservation law:

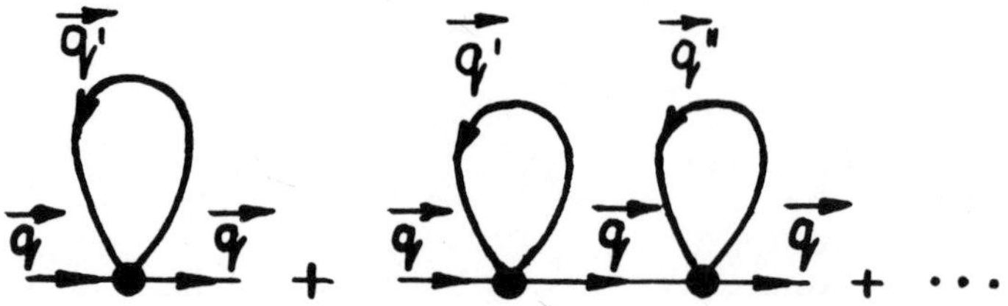

Fig. 3. A set of diagrams describing the consecutive non-linear phonon-phonon interaction.

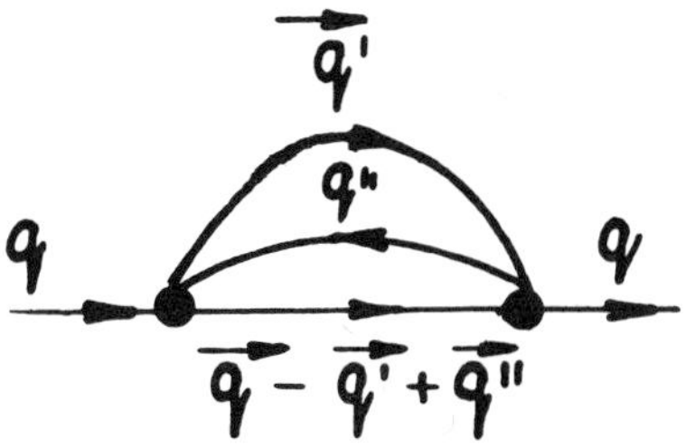

Fig. 4. An example of simultaneous multiphonon process.

$$\omega_{\vec{q}} - \omega_{\vec{q}'} - \omega_{\vec{q} - \vec{q}' + \vec{q}''} + \omega_{\vec{q}''} = 0 \tag{4}$$

So, four phonons chosen at random from the flux cannot, in general, take part in this simultaneous process, while for the consecutive interaction no restriction of this sort exists. It should be noted that the energy conservation law would be automatically followed also in the simultaneous interaction for the phonons with the same wave vectors (i.e., if $\vec{q}'' = \vec{q}' = \vec{q}$). So in such a situation both types of processes would be essential. Now let us remember that in our phonon system the phonon frequency has the natural width of the order of magnitude of the linear amplification constant γ^0. This constant determines the accuracy of the energy conservation. Therefore, simultaneous interaction becomes essential, when the spectral width of our self-narrowing flux becomes of the order of magnitude of this constant:

$$\Delta\omega \lesssim \gamma^0 \tag{5}$$

By the "spectral width of the flux" we mean the order of magnitude of the left-hand side in Eq. (4) for the randomly chosen wave vectors $\vec{q}$, $\vec{q}'$, $\vec{q}''$, belonging to the flux:

$$\Delta\omega = \left| \omega_{\vec{q}} - \omega_{\vec{q}'} - \omega_{\vec{q}-\vec{q}'+\vec{q}''} + \omega_{\vec{q}''} \right| \tag{6}$$

The direct comparison of the analytical expressions for the diagrams confirms these simple arguments, showing the significance of simultaneous interaction in the flux of the small spectral width. One can see that no equation for the average intensities can exist in this case due to the very complicated character of simultaneous processes. Just these processes lead to the change of the flux statistics. This can be seen from the direct summation of all diagrams (consecutive and simultaneous). The summation is possible for the extremely narrow flux, when $\Delta\omega$ becomes much less than γ^0. The procedure of summation will be described elsewhere. The result of the summation shows the exponential decrease of the level of intensity fluctuations with the increase of the coordinate x, i.e., the appearance of an acoustic wave with a fixed amplitude. The intensity of the wave is*

$$N = 2\gamma^0_{\vec{q}_m} / W_{\vec{q}_m \vec{q}_m} \tag{7}$$

The factor "2" in this formula is connected with the fundamental difference of the statistical properties of noise and of a signal. For the noise we have:

---

*It should be noted [4] that three-phonon interactions (being of resonant nature) can become essential in a narrow flux together with four-phonon interactions. The effective enhancement of the three-phonon interaction in a narrow flux can cause an additional difference (apart from the factor "2" discussed above) between the non-linear attenuation constant of noise, W, and that of the signal. The discussion of this interesting phenomenon, however, complicates the physical picture which is already rather complicated. This is why we do not consider it here. Besides, three-phonon processes remain negligible in a narrow flux, if a large sound velocity dispersion exists in a semiconductor, so that

$$\left| \omega_{2\vec{q}} - 2\omega_{\vec{q}} \right| >> \gamma^0$$

$$\overline{N_{\vec{q}}^{m}} = m!\ \overline{N}_{\vec{q}}^{m} \tag{8}$$

whereas for a signal, of course

$$\overline{N_{\vec{q}}^{m}} = \overline{N}_{\vec{q}}^{m} \tag{9}$$

Here $N_{\vec{q}}$ is the "momentary", i.e., unaveraged intensity. The difference in the methods of decoupling of the higher distribution functions for the noise and for the signal makes impossible any decoupling procedure in the intermediate case. This is why any nonlinear equation for the averaged intensities cannot survive in the intermediate stage of flux evolution. And this is why all the diagrams become essential in the description of a narrow flux. Preventing any decoupling of many-phonon distribution functions, simultaneous processes provide the continuous conversion of a narrow noise flux into a wave with a fixed, nonfluctuating amplitude.*

So we have demonstrated the possibility of the generation of an acoustic signal with a stable amplitude from incoherent noise in the course of its propagation along a sufficiently long crystal. It would be extremely interesting to observe this transformation. The Brillouin scattering of light from the amplified acoustic flux is now the usual technique for the investigation of acoustic flux in semiconductors (see, e.g., [5]). But usual Brillouin scattering experiments give information only about the averaged phonon intensities and are insufficient for the investigation of the flux statistics. The change of the flux statistics can be checked by a photocount investigation of the scattered light [6, 7]. If gradual transformation of the acoustic noise into a signal takes place along the crystal, the light scattered from different regions of the crystal ought to have different properties. The corresponding photocount distribution can be changed gradually from the

*The conclusion about the complete suppression of the amplitude fluctuations is valid only as long as we neglect random forces describing the stochastic interaction of the amplified flux with the active medium. This interaction secures a small but finite level of signal intensity fluctuations.

Bose-Einstein distribution to the Poisson distribution [7]. Some change, perhaps of this nature, was observed by Siebert and Wonnenberger [6]. In their experiment the Poisson distribution maximum appeared on the initially monotonically decreasing photocount curve.

There was no evidence of the spectral width narrowing in their experiment, so we reserve our judgement about the applicability of our theory to this experiment. It would be extremely interesting to carry out photocount experiments with longer samples in the frequency region near the maximum amplification frequency. The suitable material may be the one in which (like in ZnO at certain temperature regions [8]) the nonlinearity associated with the electron trapping effects [9] is essential. The nonlinearity caused by trapping of electrons can assure the smooth nonlinear attenuation which, as we have seen, can create the conditions for self-narrowing (in the wave vector space) and subsequence self-coherentization of the amplified acoustic flux.

REFERENCES

1. A. R. Hutson, J. H. McFee, D. L. White, Phys. Rev. Lett. 7, 237 (1961).
2. V. L. Gurevich, in Proceedings of the 9th Intern. Conf. on Phys. of Semicond., Moscow, July 23-29, 1968, Nauka, Leningrad, 1968, p.899-900.
3. H. W. Wyld, Ann. of Phys. 14, 143 (1961).
4. S. V. Gantsevich, V. D. Kagan, R. Katilius, Zh. exper. teor. fiz. 67, 1765 (1974).
5. H. Kuzmany, Phys. Stat. Sol. (a) 25, 9 (1974).
6. F. Siebert, W. Wonnenberger, Phys. Lett. 37A, 367 (1971).
7. M. Lax, Fluctuation and Coherence Phenomena in Classical and Quantum Physics. Brandeis Univ. Summer Institute Lectures in Theor. Physics: Statistical Physics, Phase Transitions and Superfluidity. Eds. M. Chrétien, E. P. Gross and S. Deser. Gordon and Breach, N.Y., 1968.
8. E. Mosekilde, Phys. Rev. B2, 3234 (1970).
9. R. Katilius, Fiz. tverd. tela, 10, 458 (1968). Yu. V. Gulyaev. Fiz. Tekhn. Poluprovodnikov 2, 628 (1968).

Section V

Bulk and Surface Polaritons

EXCITONS, POLARITONS AND LIGHT ABSORPTION IN CRYSTALS

A. S. Davydov

Institute for Theoretical Physics
Academy of Sciences of the Ukrainian SSR
Kiev, USSR

INTRODUCTION

Many optical and spectral properties of solid dielectrics are connected with the formation of excitons - currentless collective electronic states characterized by a wave vector $\vec{k}$ and some set of other discrete quantum numbers. In particular, the propagation of light through a crystal is caused by the transformation of photons into excitons followed by the excitation energy transfer to other photons (light scattering) and crystal lattice vibration phonons (light absorption).

This brief review will be concerned with the interaction of light with a crystal using the phenomenological and quantum-statistical theories developed by Eremko, Myasnikov, Serikov and the author.[1-4]

PHENOMENOLOGICAL THEORY

The dielectric permeability tensor $\epsilon(\omega, \vec{k})$ is a macroscopic characteristic of a crystal which defines its interaction with a long-wave electromagnetic field of real frequency ω and wave vector $\vec{k}$. For optically isotropic crystals that we are interested in this tensor degenerates into a scalar.

If in the range of frequencies ω the contribution to $\epsilon(\omega, \vec{k})$ is made by an isolated band of electronic excited states with frequencies $\Omega(\vec{k})$ and the dipole character of absorption,* the dielectric permeability is represented by an approximate relation

$$\epsilon(\omega, \vec{k}) = \epsilon_0 - f^2 \{[\omega + i\gamma(\omega, T)]^2 - \Omega^2(\vec{k})\}^{-1}, \qquad (1)$$

where ϵ_0 is the contribution to the dielectric permeability from all the other elementary excitations, except for $\Omega(\vec{k})$; $f^2 = 4\pi e^2 F(mv)^{-1}$ is the parameter characterizing the intensity of interaction of photons with the electronic excitation $\Omega(\vec{k})$; F is the oscillator strength of a given electronic excitation; v is the volume of a unit cell of the crystal; m and e are the mass and charge of an electron.

The function (1) may be analytically continued into the region of complex values of ω. At $\Omega(\vec{k}) >> \gamma(\Omega_k, T) > 0$ its poles $\Omega(\vec{k}) - i\gamma(\Omega_k, T)$ in the lower half-plane of complex ω with fixed real $\vec{k}$ determine the frequencies $\Omega(\vec{k})$ and the average lifetime of electronic elementary excitations. These excitations are called the excitons.

When calculating theoretically the dispersion law of $\Omega(\vec{k})$ one must take into account a complete Coulomb (without retardation) interaction between electric charges of a crystal. For long-wave electronic excitations we can write

$$\Omega^2(\vec{k}) = \Omega_0^2 + \alpha k^2 c^2, \qquad (ka << 1), \qquad (2)$$

Here $\alpha = \hbar\Omega_0(2m^*c)^{-1}$ is the exciton effective mass, and c is the light velocity.

The damping $\gamma(\omega, T)$ contained in (1) is defined by the interaction of excitons with crystal lattice vibrations. It is essentially dependent on the frequency ω and the temperature T of a crystal. The qualitative character of this dependence is represented in Fig. 1 for excitons with a positive effective mass in the neighborhood of the frequency Ω_0, which corresponds to the bot-

*The case of a quadrupole absorption was considered by Eremko and the author in Ref. 2.

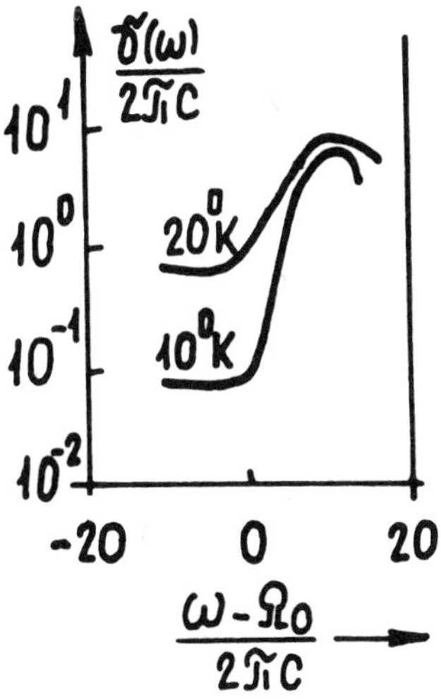

Fig. 1. The temperature dependence of the damping coefficient in the resonance frequency region.

tom of the exciton band. The Fourier transforms $\vec{A}(\omega, \vec{k})$ of the vector potential

$$\vec{A}(\vec{r}, t) = \int d\omega \int d^3k \vec{A}(\omega, \vec{k}) \exp\{i(\vec{k}\vec{r} - \omega t\}$$

of a long-wave electromagnetic field in a crystal are determined by the Maxwell equations

$$[c^2k^2 - \omega^2\epsilon(\omega, \vec{k})]\vec{A}(\omega, \vec{k}) = 4\pi c\vec{j}(\omega, \vec{k}) \tag{3}$$

where $\vec{j}(\omega, \vec{k})$ are the Fourier transforms of the densities of external currents. A free electromagnetic field in a crystal (without external currents) is defined by the equation

$$[c^2\vec{k}^2 - \omega^2\epsilon(\omega, \vec{k})]\vec{A}(\omega, \vec{k}) = 0 \tag{4}$$

provided that

$$c^2\vec{k}^2 - \omega^2 \epsilon(\omega, \vec{k}) = 0 \tag{5}$$

With fixed real values of the wave vectors $\vec{k} = \vec{Q}$ the solutions of equation (4)

$$\vec{A}(\vec{r}, t) = \vec{A}_0 \exp\{i[\vec{Q}\vec{r} - \omega_\ell(\vec{Q})t]\} \tag{6}$$

characterize the "mixed" exciton-photon elementary excitations called polaritons. Their dispersion laws $\omega_\ell(\vec{Q})$ are defined by equation (5). In the general case $\omega_\ell(\vec{Q})$ are complex functions of the real vector $\vec{Q}$. Under the condition (2) there are two branches of polariton states.[5,6] The real parts $\mathrm{Re}\,\omega_\ell(\vec{Q})$ characterize the dispersion of polaritons, and the imaginary parts $1/\tau_\ell = -\mathrm{Im}\,\omega_\ell(\vec{Q})$ define their damping.

At $t >> \tau_\ell$ the polariton states are damped and there remains in a crystal only a forced electromagnetic field

$$\vec{A}(\vec{r}, t) = c \int d\omega \int d^3k\, D(\omega, \vec{k}) \vec{j}(\omega, \vec{k}) \exp\{i[\vec{k}\vec{r} - \omega t]\} \tag{7}$$

caused by external currents;

$$D(\omega, \vec{k}) = 4\pi [c^2 k^2 - \omega^2 \epsilon(\omega, \vec{k})]^{-1} \tag{8}$$

is the photon Green's function in the crystal.

We assume that in a crystal of infinite dimensions the densities of external currents are defined by

$$j_x(\vec{r}, t) = j_0 \delta(t) \exp(iQz), \quad j_y(\vec{r}, t) = j_z(\vec{r}, t) = 0 \tag{9}$$

By substituting (9) into (7) and calculating the integral by means of residue theory, we can show that the vector potential of a forced spatially homogeneous field in a crystal is represented by the superposition of two polariton states

$$A_x(z,t) = \frac{c j_0}{2\pi} e^{iQz} \int d\omega D(\omega, Q) e^{-i\omega t}$$
$$= j_0 \sum_{\ell=1}^{2} B_\ell \exp\{i[Qz - \omega_\ell(Q)t]\}, \quad (t \geq 0) \tag{10}$$

where $\omega_\ell(Q)$ are the complex poles of the photon Green's function (8), i.e., the solution of equation (5) at $k = Q$, and B_ℓ are the coefficients proportional to the residues of the integrand in (10). Expressions (10) were analyzed by Eremko and the author in Ref. 1.

We now define the resonance frequency Ω_r under the condition that the frequencies and the wave vectors of an exciton and a phonon in a crystal are equal. Then with the resonance value of the vector

$$Q_r = \Omega_r n_0 c^{-1}$$

and the inequalities $n_0\Omega_r >> f_0, \Omega_2 >> \gamma_2$ which are always valid the roots of the equation

$$c^2 Q_r^2 - \omega^2 \epsilon(\omega, Q_r) = 0$$

are equal to

$$\omega_{1,2}^2 = \Omega_r^2 - i\gamma_r \Omega_r \pm \Omega_r \sqrt{f_0^2 \epsilon_0^{-1} - \gamma_0^2 - i\gamma_r f_0^2 (\epsilon_0 \Omega_r)^{-1}} \tag{11}$$

where γ_0 and f_0 are the values of the relaxation parameter and f at the resonance frequency Ω_r.

We consider the following two limiting cases:

(A) The condition of a strong coupling of excitons with phonons is satisfied:

$$f_0^2 >> \epsilon_0 \gamma_0^2 \tag{12}$$

Then (11) transforms to

$$\omega_{1,2} \approx \Omega_r \pm f_0 (2n_0)^{-1} - i\gamma_0/2 \; ; \; n_0 = \sqrt{\epsilon_0}$$

and the vector potential (10) takes the form

$$A_x(z,t) = \frac{ic\pi}{\epsilon_0} e^{iQz} \left\{ \frac{e^{-i\omega_1 t}}{\omega_1} + \frac{e^{-i\omega_2 t}}{\omega_2} \right\}$$

Thus the energy density averaged over a time comparable with a period $2\pi/\Omega_r$ of fast field oscillations changes according to the law

$$W(t) \approx W(0) e^{-\gamma_0 t} \cos^2 \left(\frac{f_0 t}{2n_0}\right), \quad t \to 0 \tag{13}$$

So, in the case of a strong exciton-photon coupling (as compared to relaxation processes) the energy density decreases with time, oscillating with a frequency f_0/n_0. This result is also valid for local excitations, since the presence of spatial dispersion was not taken into account explicitly when it was derived. The frequency of oscillations of the energy density is proportional to the difference between the frequencies of the two polariton branches at $Q = Q_r$.

We now introduce the effective time-damping coefficient γ_{eff} through the relation

$$\gamma_{eff} = W(0) / \int_0^\infty W(t)\, dt$$

Then it follows from (12) that if

$$\gamma_{eff} = 2\gamma_0$$

the curve

$$W_{eff}(t) = W_0 \exp(-\gamma_{eff} t)$$

restricts the same area as the curve (13).

(B) The condition of a weak exciton-photon coupling is satisfied (relaxation processes play an important role):

$$f_0^2 << \epsilon_0 \gamma_0^2$$

In this case the density of the electromagnetic energy is a crystal varies according to the exponential law

$$W(t) = W(0) \exp\left(-\frac{f_0^2 t}{2\epsilon_0 \gamma_0}\right) \qquad (14)$$

Here the index of the exponent is proportional to f_0^2, i.e., the oscillator strength F of the quantum transition. The same result may also be obtained using perturbation theory.

Of much more interest are the forced spatially inhomogeneous waves which arise in a crystal under the action of monochromatic light. In this case the wave frequency ω is real, and the wave vector $\vec{k}$ has no defined value (spatial damping). In order to investigate the passage of such waves through a crystal, we suppose that in a crystal in the plane z = 0 there are external currents such as

$$j_x(\vec{r}, t) = j_0 \delta(z) e^{-i\omega t}, \quad j_y(\vec{r}, t) = j_z(\vec{r}, t) = 0$$

The nonzero Fourier components of these currents

$$j_x(\omega', \vec{k}) = j_0 \delta(\omega' - \omega)\, \delta(k_x)\, \delta(k_y)/2\pi$$

excite in a crystal, according to (7), a stationary field with vector potential

$$A_x(z, t) = \frac{2e^{-i\omega t}}{\omega} j_0 \int \frac{dN \exp\{i\frac{\omega}{c}Nz\}}{N^2 - \epsilon(\omega, N)} \tag{15}$$

where

$$N \equiv ck/\omega \tag{16}$$

Taking into account (2) and (15), one can transform (1) in the frequency range $\omega = \Omega_0$ to

$$\epsilon(\omega, N) \approx \epsilon_0 - \frac{f^2/2\omega}{\omega - \Omega_0 + i\gamma - v\omega N^2 (2cn_0)^{-1}} \tag{17}$$

where

$$v = \hbar n_0 \omega / m^* c$$

is the velocity of an exciton with effective mass m^* and wave vector $n_0\omega/c$.

Using the analytic continuation of the dielectric permeability $\epsilon(\omega, N)$ into the region of complex values N and applying the residue theory method, we can transform (10) in the region $z \geq 0$ to

$$A_x(z, t) = j_0 \sum_{\ell=1}^{2} B_\ell \exp\{i\omega[N_\ell z/c - t]\} \tag{18}$$

where

$$N_\ell = n_\ell + i\kappa_\ell \tag{19}$$

are the complex roots of the equation

$$N^2 - \epsilon(\omega, N) = 0 \tag{20}$$

Furthermore,

$$\frac{B_1}{B_2} = \frac{N_2(N_2^2 - \epsilon_0)}{N_1(N_1^2 - \epsilon_0)} \approx \left(\frac{N_2}{N_1}\right)^3 \tag{21}$$

The individual terms

$$B_\ell e^{-\kappa_\ell \omega z/c} \exp\{i\omega[n_\ell z/c - t]\} \tag{22}$$

in the sum (18) will be called normal electromagnetic waves in a crystal. The amplitude of these waves decreases exponentially with increasing z.

It was shown by Myasnikov and the author in Ref. 3 that both normal waves in the sum (14) are the same, i.e., $N_1 = N_2$ and $B_1 = B_2$ at the frequency

$$\omega_0 = \sqrt{(\Omega_0^2 + \alpha f^2)(1 - \alpha\epsilon_0)} \approx \Omega_0$$

provided there is a temperature T_0 at which the equality

$$\gamma(\omega_0, T_0) = f\sqrt{\alpha} \tag{23}$$

is valid. If the equality (23) is valid, T_0 is called the critical temperature. For the model crystal (like anthracene) introduced by Myasnikov and the author in Ref. 7, the critical temperature is 15°K. In real crystals the critical temperature may take values higher or lower than this one. Figure 2 represents the values $N_1(\omega)$ and $N_2(\omega)$ calculated in Ref. 7 for a model crystal at T = 20°, which exceeds the critical value.

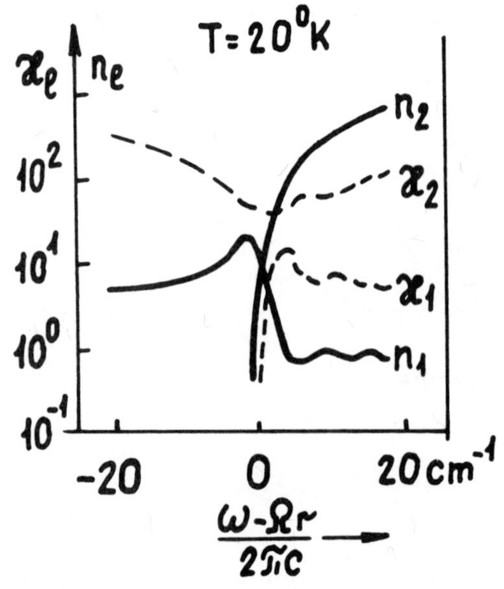

Fig. 2. The refractive indices n_1 and n_2 and the absorption coefficients κ_1 and κ_2 for two normal electromagnetic waves propagating in a crystal at a temperature above the critical one.

The solid lines in Fig. 2 represent the real (n_1, n_2), and the dashed lines the imaginary (κ_1, κ_2) parts of the complex refractive indices. The scale of ordinates is logarithmic. The figure shows that the inequality $|N_2| >> |N_1|$ is valid in the frequency range in point. Therefore, according to (21), a normal wave with refractive index N_2 occurs in a crystal with a very small amplitude; moreover, for this wave $\kappa_2 >> \kappa_1$ and so it is fast damping. Thus the light propagating in a crystal coincides, in practice, with one normal wave with refractive index n_1 and absorption coefficient κ_1.

At a temperature (10^0K) less than the critical one the frequency dependence of N_1 and N_2 is represented in Fig. 3. We can see that the inequality $|N_2| < |N_1|$ is valid in the frequency range $\omega < \Omega_0$, so that only a wave with refractive index $n_2 \neq 0$ and absorption coefficient $\kappa_2 \approx 0$ contributes to the superposition (18). In the frequency range $\omega > \Omega_0$ the inequality $|N_1| < |N_2|$ obtains, so that the dominant contribution to the light propagation in a

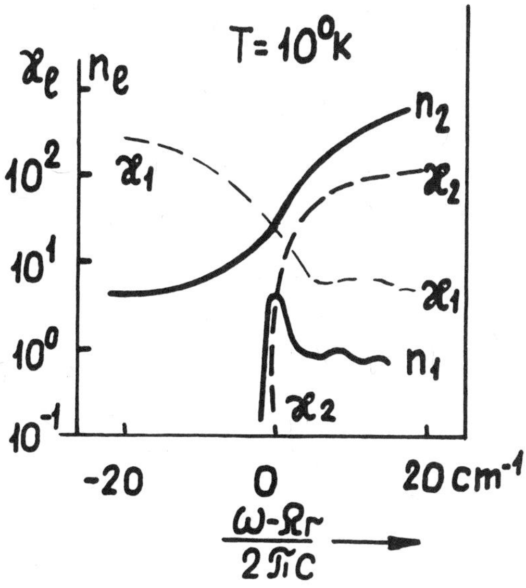

Fig. 3. The refractive indices n_1 and n_2 and the absorption coefficients for two normal electromagnetic waves propagating in a crystal at a temperature below the critical one.

crystal is made only by the first wave with refractive index $n_1 < n_2$ and absorption coefficient $\kappa_1 << \kappa_2$. Thus the dispersion curve $n(\omega)$ at a temperature below the critical one is not a smoothly varying function of frequency. With the frequency increasing to the left of Ω_0 the refractive index increases, and then in going through a certain region in the neighborhood of Ω_0 there takes place a "skip" from the curve $n_2(\omega)$ to the curve $n_1(\omega)$.

In the "skipping" region both normal waves with refractive indices $n_1(\omega)$ and $n_2(\omega)$ make comparable contributions to (18), provided their absorption coefficients κ_1 and κ_2 differ little. Consequently, in this range of frequencies and temperatures the refractive index of a crystal has no definite value and the beam of light from a crystal loses its coherence (the superposition of waves with unequal difference of phases). The loss of coherence seems to be very important at frequencies and temperatures of a crystal which are close to the "critical" ones, since when approaching the critical values of ω_0 and T_0, the contributions of both waves to the superposition (18) become equal. As already mentioned above, both normal waves are identically equal at the critical point itself (ω_0, T_0). This case is hardly realizable in reality, since we always deal with quasi-monochromatic light.

The results of the theory[3] were qualitatively confirmed by Brodin, N. Davydova and M. Strashnikova[8] who investigated the refraction in CdS in the exciton band region $A_{n=1}$ over the temperature range 4-80°K. It was established that the critical temperature lies in the range 25-50°K.

A more detailed investigation of the passage of light through a crystal in the range of frequencies coinciding resonantly with the frequency of dipole excitons was performed by Eremko and the author.[1] The frequencies $\omega = \Omega_{k_0} = \Omega_r$ and the wave vectors of photons and excitons are the same at the resonance. Consequently, if the frequencies of long-wave ($ka \ll 1$) excitons are given by

$$\Omega_{\vec{k}} = \Omega_0 + \hbar \vec{k}^2 (2m^*)^{-1}$$

the resonance frequency Ω_r is determined by the equation

$$\Omega_r = \Omega_0 + v_0 n_0 \Omega_r (2c)^{-1} \tag{24}$$

where

$$v_0 \equiv \hbar \Omega_r n_0 (m^* c)^{-1}$$

is the velocity of excitons corresponding to the resonance frequency.

In the resonance region the roots of equation (20) are determined by the equation

$$N^2_{1,2}(\Omega_r) \approx n_0^2 + \frac{cn_0\gamma_0}{v_0\Omega_r}\left(i \pm \sqrt{\frac{f_0^2 v_0}{cn_0\gamma_0^2} - 1}\right), \tag{25}$$

We consider the following two limiting cases:

A) Spatial dispersion is very important. In this case the inequality

$$v_0 f_0^2 >> c n_0 \gamma_0^2 \tag{26}$$

holds and (25) is replaced by an approximate expression

$$N_{1,2}^2(\Omega_r) \approx n_0^2 \pm \frac{f_0}{\Omega_r}\sqrt{\frac{cn_0}{v_0}} + i\frac{\gamma_0 c n_0}{v_0 \Omega_r} \tag{27}$$

If in addition the inequality (low velocity of excitons at large f_0)

$$\sqrt{\frac{cn_0}{v_0}} >> \frac{n_0^2 \Omega_r}{f_0} \tag{28}$$

holds, then

$$N_1 = n + i\kappa \ , \ N_2 = \kappa + in,$$

where

$$n = \left[\frac{f_0}{\Omega_r}\sqrt{\frac{cn_0}{v_0}}\right]^{\frac{1}{2}}, \qquad \kappa = \frac{\gamma_0}{2\sqrt{f_0\Omega_r}}\left[\frac{cn_0}{v_0}\right]^{\frac{3}{4}}.$$

Thus, $\kappa << n$ and the flow of the energy of the electromagnetic field in a crystal in the region $z \geq 0$ is determined by

$$S(z) = \tfrac{1}{2}S(0)\{ne^{-2(\Omega_r/c)\kappa z} + \sqrt{2}\, e^{-(\Omega_r n/c)z} \cos(\tfrac{\pi}{4} + \tfrac{\Omega_r n}{c} z)\} \tag{29}$$

Since there are two terms in the curly brackets in (29), the dependence of the energy flux density in the range of small z does not correspond to a simple exponential law. The z-dependence of

$-\ell n S(z)$ is represented by a broken curve with a steep initial section.

The nonexponential law of the change in intensity of polarized light (which passes through a paradichlorbenzol crystal) depending on the crystal width at 4.2°K was observed by Delyukov and Klimusheva[9, 10] in the range of frequencies of 35650 cm^{-1} and 35660 cm^{-1} (see Fig. 4). The same figure also represents the theoretical curve corresponding to (29).

In the case when the velocity of excitons is very high at small f and instead of (28) a reciprocal inequality holds and the parameter γ_0 is so small that the inequality (26) is valid, we can obtain from (27) the approximate values

$$N_1(\Omega_r) = n_1 + i\kappa , \quad N_2(\Omega_r) = n_2 + i\kappa ,$$

where

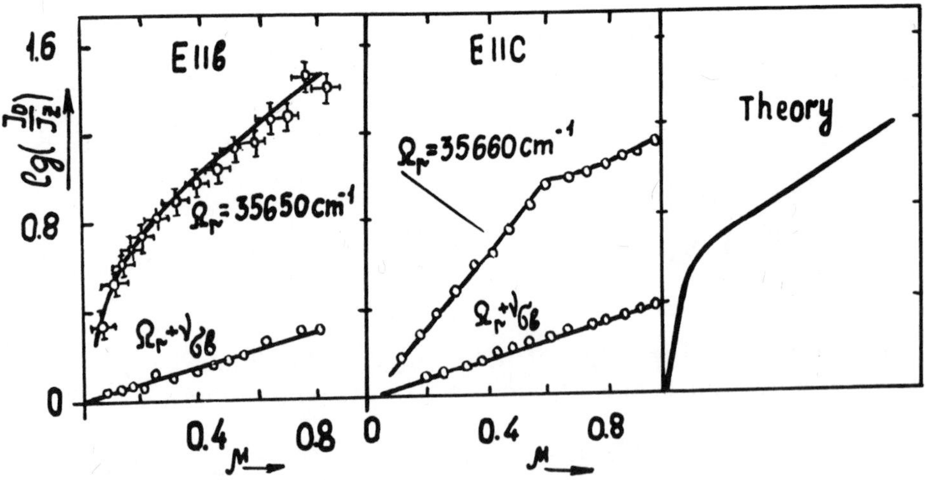

Fig. 4. The dependence of the light transmission coefficient $\ell g(I_0/I_z)$ on the paradichlorbenzol crystal width at 4°K in the region of pure electronic transitions Ω_r and vibronic transitions $\Omega_r + \nu_{\sigma B}$ of small intensity. The experimental data is of Delyukov and Klimusheva.

$$n_{1,2} = n_0 \pm \frac{f_0}{2\Omega_r}\sqrt{\frac{c}{v_0 n_0}} \quad , \quad \kappa = \frac{c\gamma_0}{2v_0\Omega_r} \; ,$$

In this case the energy flux density inside a crystal is determined by an approximate expression

$$S(z) = S(0)e^{-\gamma_0 z/v_0} \cos^2\left(\frac{z f_0}{2\sqrt{c\, n_0 v_0}}\right), \tag{30}$$

The phonon energy flux density decreases, oscillating with a spatial period

$$\Lambda = \frac{2\pi}{f_0}\sqrt{c\, n_0 \gamma_0} \; .$$

These oscillations are due to the periodic transfer of phonon energy into exciton energy and vice versa, because the relaxation parameter γ_0 is small.

Comparing (12) with (30) we see that these expressions would be the same, if the exciton velocity v_0 were the same as the light velocity c/n_0 in a crystal and z were replaced by

ct/n_0 .

B) The case when spatial dispersion plays a minor role

In the resonance region ($\omega = \Omega_r$) the role of spatial dispersion is minor, if the inequality

$$v_0 f_0^2 << c\, n_0\, \gamma_0^2 \tag{31}$$

holds, which is the reciprocal of (26). In this case

$$N_1^2 = \epsilon_0 + i f_0^2 (2\gamma_0 \Omega_r)^{-1} \quad , \quad N_2^2 = \epsilon_0 + i2\, c\, n_0 \gamma_0 (v_0 \Omega_r)^{-1}$$

Then $B_2 = 0$ and the flux density of electromagnetic energy in a crystal varies by the law

$$S(z) = S(0) \exp\left(-2 \frac{\Omega_r}{c} \kappa z\right) \tag{32}$$

where

$$\kappa = \tfrac{1}{2}\left\{\sqrt{\epsilon_0^2 + f_0^4 (2\gamma_0 \Omega_r)^{-2}} - \epsilon_0\right\}^{\frac{1}{2}}$$

Expression (32) at $\Omega_r = \Omega_0$ is exactly the same as a similar expression that may be obtained when there is no spatial dispersion ($v_0 = 0$). The absorption coefficients at vibronic frequencies $\Omega_r + \nu_{\sigma B}$, represented in Fig. 4, correspond to an exciton band with a very low oscillator strength, so that the change in intensity of absorption is determined by the exponential law.

With increasing temperature of a crystal the exciton band width and the exciton velocity v_0 decrease, while the parameter γ_0 increases. Therefore the inequality (26) can change to the inequality (31) for which spatial dispersion is not important. The experimental data of Delyukov and Klimusheva[9,10] confirm the fact that the increase of temperature from 4° to 100°K results in diminishing the role of spatial dispersion (see Fig. 5).

Spatial dispersion is also insignificant far from the resonance when the inequality

$$v(\omega) f^2 \ll c\, n_0 (\omega - \Omega_\omega)^2 \tag{33}$$

holds, where

$$v(\omega) = \frac{\hbar \omega n_0}{m^* c}, \quad \Omega_\omega = \Omega_0 + \frac{\hbar}{2m^*}\left(\frac{\omega n_0}{c}\right)^2 .$$

Then

$$N_1^2 = n_0^2 - f^2[2\omega(\omega - \Omega_\omega)]^{-1} + i\,\frac{\gamma(\omega)f^2}{2\omega(\omega - \Omega_\omega)^2}$$

$$N_2^2 = n_0^2 + \frac{2cn_0}{v\omega}[\omega - \Omega_\omega + i\gamma(\omega)]$$

Therefore the value of B_2 in the superposition (18) is equal to zero and the electromagnetic field strength in a crystal is described by the normal wave

$$E_x(z,t) = \frac{1}{N_1} H_y(z,t) = E_0 \exp\{ -\frac{\omega\kappa}{c} z + i\omega(\frac{nz}{c} - t)\} ,$$

where

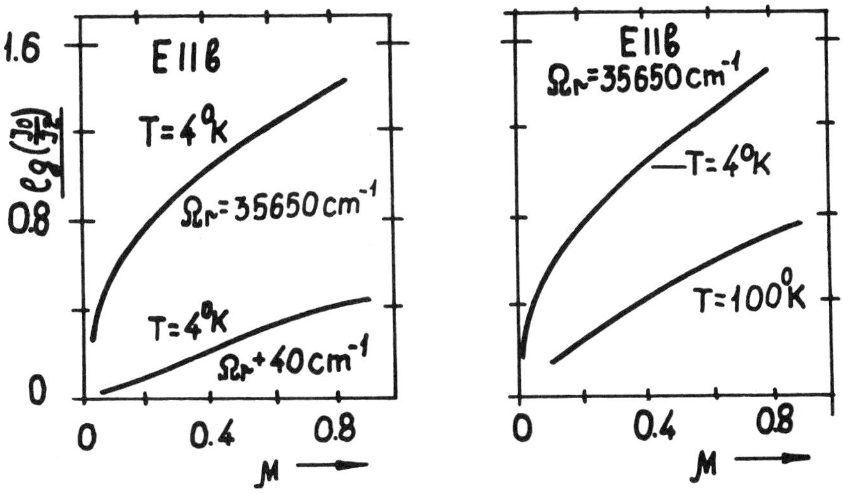

Fig. 5. The dependence of the light transmission coefficient on the crystal width: a) at 4°K for the resonance frequency and for the frequency shifted by 40 cm^{-1}; b) at 4°K and 100°K for the resonance frequency Ω_0. The experimental data is of Delyukov and Klimusheva.

$$n = n_0 - \frac{f^2}{4n_0 \omega (\omega - \Omega_\omega)}, \quad \kappa = \frac{\gamma f^2}{4n_0 \omega (\omega - \Omega_\omega)^2}$$

The diminishing role of spatial dispersion when we go away from the resonance frequency Ω_r by 40 cm^{-1} is illustrated by the experimental data of Delyukov and Klimusheva, given in Fig. 5.

A phenomenological theory of light propagation in a crystal in the frequency range corresponding to a negative effective exciton mass was developed by Eremko.[11] The interaction with quadrupole excitons was discussed by Eremko and the author.[2]

QUANTUM - STATISTICAL THEORY OF LIGHT PROPAGATION IN CRYSTALS

In this section we present some results of a microscopic theory of light propagation in a crystal, developed by Serikov and the author.[4] The theory is based on a quantum-mechanical description of a system of interacting photons, excitons and phonons. It is assumed that the photon frequency corresponds to the exciton absorption region. These photons do not interact directly with phonons.

The true absorption of light in a perfect crystal is caused by the photon energy transfer into the crystal lattice vibration energy. Because of a very great number of degrees of freedom with quasi-continuous frequencies of vibrations and the rapid establishment of thermal equilibrium between phonons, the latter form a dissipative subsystem. The light absorption is realized through an intermediate stage of formation of excitons which interact directly with a dissipative subsystem.

To simplify our notation we use in this section a system of units in which the energy has dimensions of frequency, i.e., $\hbar = 1$. In an optically isotropic crystal with one quasi-continuous band of dipole exciton states the latter are characterized by the Hamiltonian

$$H_{ex} = \sum_{\vec{k}} \Omega_{\vec{k}} A^{+}_{\vec{k}} A_{\vec{k}} \qquad (34)$$

where $A^+_{\vec{k}}$ and $A_{\vec{k}}$ are the creation and the annihilation operators for excitons in states with wave vector $\vec{k}$ and frequency $\Omega_{\vec{k}}$. The sum Σ' is henceforth over all N-3 values of the wave vector in the first Brillouin zone, except for $\vec{k} = 0$. The value $\vec{k} = 0$ characterizes the ground state whose energy is assumed to be the origin of exciton energy.

If $a^+_{\vec{k}}$, $a_{\vec{k}}$ are the creation and annihilation operators for photons of frequency $\omega_{\vec{k}} = c|\vec{k}|n_0^{-1}$, the photon Hamiltonian is

$$H_{ph} = \sum_{\vec{k}}{}' \omega_{\vec{k}} a^+_{\vec{k}} a_{\vec{k}} , \tag{35}$$

n_0 is the refractive index of a crystal due to electronic states disregarded in (34).

The exciton-photon interaction operator is written as

$$H_{int} = \frac{1}{2n_0} \sum_{\vec{k}}{}' f_{\vec{k}} \{a^+_{\vec{k}} A_{\vec{k}} + a_{\vec{k}} A^+_{\vec{k}}\} , \tag{36}$$

where

$$f^2_{\vec{k}} = \frac{4\pi e^2 \Omega_k}{n v \omega_k} F$$

Here v is the unit cell volume of the crystal, and F is the oscillator strength of the quantum transition from the ground state to the exciton state.

The system of excitons and photons interacting with each other and described by the Hamiltonian will be called a dynamic system. The interaction of excitons with a dissipative subsystem of phonons results in the fact that a dynamic system becomes open. If the interaction of excitons with phonons is characterized by phenomenological parameters γ_k the state of an open dynamic system will be described by the density matrix ρ, which satisfies the kinetic equation

$$\frac{\partial\rho}{\partial t} = i[\rho, H] - \sum_{\vec{k}}' \gamma_{\vec{k}} \{[A^+_{\vec{k}} A_{\vec{k}}, \rho]_+ - 2A_{\vec{k}} \rho A^+_{\vec{k}}\} \tag{37}$$

where

$$[a, b]_\pm \equiv ab \pm ba$$

States with definite values of $\vec{k}$ are spatially homogeneous. In a crystal whose volume is taken to be a unit volume, the densities of photons and excitons with wave vectors $\vec{k}$ are defined using the statistical operator $\rho(t)$ by the equations

$$\begin{aligned} W_{ph}(\vec{k}, t) &= Sp\{\rho a^+_{\vec{k}} a_{\vec{k}}\}, \\ W_{ex}(\vec{k}, t) &= Sp\{\rho A^+_{\vec{k}} A_{\vec{k}}\}. \end{aligned} \tag{38}$$

Equation (37) enables us to find (see Ref. 4) equations for the functions (38). Here we give their solutions which correspond to the initial conditions

$$W_{ph}(0) = 1 \;, \quad W_{ex}(0) = 0,$$

at resonance values of k_0 for the two limiting cases: a strong and a weak exciton-photon interaction. We introduce the notation

$$\omega_{k_0} = \Omega_{k_0} = \Omega_0 \;, \quad f_{k_0} = f_0 \;, \quad \gamma_{k_0} = \gamma_0$$

For a strong exciton-photon interaction the inequality

$$f_0^2 >> \gamma_0^2 n_0^2 \tag{39}$$

holds. The change with time of exciton and photon densities is then described by the functions

$$W_{ex}(\Omega_0, t) \approx e^{-\gamma_0 t} \sin^2\left(\frac{f_0 t}{n_0}\right), \tag{40}$$

$$W_{ph}(\Omega_0, t) \approx e^{-\gamma_0 t} \left[\cos\left(\frac{f_0 t}{2n_0}\right) + \frac{n_0 \gamma_0}{f_0} \sin\left(\frac{f_0 t}{2n_0}\right) \right]^2 \qquad (41)$$

For the particular case when $f_0 = 8\gamma_0 n_0$, these functions are represented in Fig. 6 by solid lines.

For a weak exciton-photon interaction the inequality

$$f_0^2 << \gamma_0^2 n_0^2 \qquad (42)$$

holds. Here the change of exciton and photon densities is determined by the functions

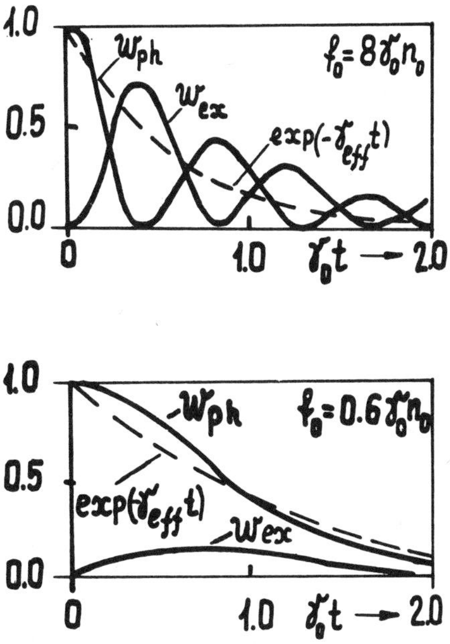

Fig. 6. The change in time of the probability of observing the system in photon $W_{ph}(t)$ and exciton $W_{ex}(t)$ states. The dashed lines represent the effective exponential decrease in photon density.

$$W_{ex}(\Omega_0, t) \approx \frac{f_0^2}{2\gamma_0^2 n_0^2}\{e^{-\frac{f_0 t}{2\gamma_0 n_0}} - 2e^{-\gamma_0 t} + \exp\left(-2\gamma_0 t + \frac{f_0^2 t}{2\gamma_0 n_0}\right)\} \tag{43}$$

$$W_{ph}(\Omega_0, t) \approx e^{-\gamma_0 t}\{\mathrm{csh}(\tfrac{t}{2}\sqrt{\gamma_0^2 - f_0^2 n_0^{-2}}) + \mathrm{snh}(\tfrac{t}{2}\sqrt{\gamma_0^2 - f_0^2 n_0^{-2}})\}^2$$

$$\approx \exp\left(-f_0^2 t\,[2\gamma_0 n_0]^{-1}\right). \tag{44}$$

For the particular case when $f_0 = 0.6\,\gamma_0 n_0$ the function (43), (44) are represented by solid lines in Fig. 6.

When the resonance is absent, the change in time of the functions $W_{ph}(\vec{k}, t)$ and $W_{ex}(\vec{k}, t)$ is defined by very lengthy expressions (see Ref. 4). Therefore, in order to consider the general case it is convenient to introduce an averaged characteristic of the time change of photon density in a state with a definite wave vector $\vec{k}$. For this purpose we replace the exact function $W_{ph}(\vec{k}, t)$ by an exponential function $\exp(-\gamma_{eff}(k)t)$, in which the effective parameter of photon damping is defined from the condition

$$\int_0^\infty W_{ph}(k, t)\,dt = \int_0^\infty \exp(-\gamma_{eff}\,t)\,dt, \quad W_{ph}(\vec{k}, 0) = 1$$

As shown in Ref. 4, we then have

$$\gamma_{eff} \approx \frac{\gamma_0 f_0^2 (2\epsilon_0)^{-1}}{(\omega - \Omega_2)^2 + \gamma_0^2 + f_0^2 (4\epsilon_0)^{-1}} \tag{45}$$

The photons which arise at time $t = 0$ under the influence of an external perturbation and have a well-defined wave vector $\vec{k}$ correspond to a spatially homogeneous electromagnetic field decreasing with time. To investigate spatial motion of photons along the directions z produced by an external source in the plane $z = 0$, it is necessary to represent their states by wave packets with a certain spread of wave vectors about the average value $k = k_z$.

Such a problem was solved by Serikev and the author.[4] By using Eq. (37) we obtained a set of equations for four functions

$$W_\ell(\vec{k} + \vec{\xi}_1, \vec{k} + \vec{\xi}_2; t)$$

in terms of which the functions

$$W_\ell(\vec{k}, \vec{r}, t) = \frac{1}{(2\pi)^3} \sum_{\vec{\xi}_1 \vec{\xi}_2} W_\ell(\vec{k} + \vec{\xi}_1, \vec{k} + \xi_2; t)\, e^{i(\vec{\xi}_1 - \vec{\xi}_2)\vec{r}} \tag{46}$$

are expressed.

From the functions (46) one can obtain (detailed calculations are given in Ref. 4) the functions $W_{ph}(\vec{k}, \vec{r}, t)$ and $W_{ex}(\vec{k}, \vec{r}, t)$ which are proportional to the densities of photons and excitons, respectively, at point $\vec{r}$ and time t. Next, in the space $z \geq 0$ we solved a stationary problem with the boundary conditions

$$W_{ph}(k, 0) = 1, \; W_{ex}(k, 0) = 0, \; (k = k_z, \; k_x = k_y = 0)$$

which corresponds to the passage of a stationary flux of photons through the plane z = 0. Here we shall discuss the calculated results for some limiting cases with photon frequencies corresponding to resonance ($\omega = \Omega_r$).

(A) Relaxation processes are absent ($\gamma_0 = 0$). In the case

$$W_{ex}(k_0, z) = \frac{2c_0}{c_0 + v_0} \sin^2\left(\frac{f_0 z}{2\sqrt{c\, v_0 n_0}}\right),$$

$$W_{ph}(k_0, z) = 1 - \frac{v_0}{c_0} W_{ex}(k_0, z) \tag{47}$$

where $c_0 = c/n_0$; v_0 is the velocity of excitons with frequency Ω_r. From these expressions it follows that the total flux of photons and excitons proportional to

$$v_0 W_{ex}(k_0, z) + c_0 W_{ph}(k_0, z) = c_0$$

does not depend on z (there is no real absorption).

The oscillations of the functions (47) are due to the energy transfer from photons to excitons and vice versa. At $v_0 = c_0$ the energy transfer between photons and excitons is complete as in the case of a time spatially homogeneous problem without damping [see (40), (41)].

(B) Spatial dispersion is absent while the relaxation processes are present. In this case

$$W_{ex}(k_0, z) = \frac{f_0^2}{4\epsilon_0\gamma_0^2 - f_0^2}\left\{\exp\left(-\frac{z f_0^2}{2\gamma_0 c_0 \epsilon_0}\right) - \exp\left(-\frac{2\gamma_0 z}{c_0}\right)\right\},$$

$$W_{ph}(k_0, z) = \frac{f_0^2}{4\epsilon_0\gamma_0^2 - f_0^2}\left\{\frac{4\epsilon_0\gamma_0^2}{f_0^2}\exp\left(-\frac{z f_0^2}{2\gamma_0 c_0 \epsilon_0}\right) - \exp\left(-\frac{2\gamma z}{c_0}\right)\right\} \tag{48}$$

The law for the change of exciton and photon densities is basically exponential at the values $2\gamma_0 z \geq c_0$. In this case the decrease in the photon density with distance is defined by the velocity of the slowest of two processes: (a) the energy transfer from photons to excitons; (b) the energy transfer from excitons to a thermal sink. In the case of a weak exciton-photon interaction (as compared to the interaction of excitons with a thermal sink), i.e., at $f_0^2 << \epsilon_0\gamma_0^2$ the process (a) is slower. In this case

$$W_{ph}(k_0 z) \approx \exp\left(-\frac{z f_0^2}{2\gamma_0 c_0 \epsilon_0}\right), \quad W_{ex}(k_0, z) = \frac{f_0^2}{4\gamma_0^2 \epsilon_0} W_{ph}(k_0, z) \tag{49}$$

Thus, the light absorption coefficient is proportional to the oscillator strength, since $f_0^2 \sim F$.

In the case of a strong exciton-photon interaction, i.e., at $f_0^2 >> \epsilon_0 \gamma_0^2$ the process (b) is slower. Then photons very rapidly exchange energy with excitons and the functions of photon and exciton densities (at $2\gamma_0 z \geq c_0$) are practically coincident

$$W_{ex}(k_0, z) \approx W_{ph}(k_0, z) \approx \exp\left(-\frac{2\gamma_0 z}{c_0}\right) \tag{50}$$

Thus, the light absorption coefficient depends only on the parameter γ_0 that defines the relaxation processes.

(C) Spatial dispersion in the presence of relaxation processes

($v_0 \neq 0$ $\gamma_0 \neq 0$).

The phenomenological investigation of light propagation process has shown that the role of spatial dispersion is essential, if the inequality

$$v_0 f_0^2 >> c n_0 \gamma_0^2$$

is satisfied.

When this equality is satisfied, the microscopic theory (4) leads to the following values

$$W_{ex}(k_0, z) = \frac{c_0}{c_0 + v_0} \left\{ \exp\left(-\frac{2\gamma_0 z}{c_0 + v_0}\right) - e^{-\frac{\gamma_0 z}{v_0}} \cos\left(\frac{z f_0}{\sqrt{e v_0 n_0}}\right) \right.,$$

$$W_{ph}(k_0, z) = \exp\left(-\frac{2\gamma_0 z}{c_0 + v_0}\right) - \frac{v_0}{c_0} W_{ex}(k_0, z), \quad c_0 = c/n_0 \tag{51}$$

It follows from (51) that the photon flux decreases with distance according to the law

$$c_0 W_{ph}(k_0, z) = \frac{c_0}{c_0 + v_0}\left\{ c_0 e^{-\frac{2\gamma_0 z}{c_0 + v_0}} - v_0 e^{-\frac{\gamma_0 z}{v_0}} \cos \frac{z f_0}{\sqrt{e v_0 n_0}} \right\} \tag{52}$$

When investigating the process of light propagation in a crystal in the range of frequencies which does not correspond to resonance, it is convenient to introduce the effective photon absorption coefficient $\kappa_{eff}(k)$ defined by the integral equality

$$\int_0^\infty W_{ph}(k, z)\, dz = \int_0^\infty \exp\,(-\kappa_{eff}(k, z)\, dz$$

from which there follows

$$\kappa_{eff}(k) = \{\int_0^\infty W_{ph}(k, z)\, dz\}^{-1} .$$

For this value in Ref. 4 the following expression

$$\kappa_{eff}(\omega) = \frac{\gamma_0 f_0^2 (2cn_0)^{-1}}{(\omega - \Omega_r)^2 + \gamma_0^2 + f_0^2 (4\epsilon_0)^{-1}} \tag{53}$$

was obtained where the values γ_0 and f_0 are taken at the resonance frequency. Comparing (53) with (44) we see that there is a simple relation

$$\kappa_{eff}(\omega) = \frac{n_0}{c} \gamma_{eff}(\omega) \tag{54}$$

It follows from (53) that the half-width of the absorption curve $\kappa_{eff}(\omega)$ is defined by the expression

$$\Gamma = \sqrt{\gamma_0^2 + (4\epsilon_0)^{-1} f_0^2} \tag{55}$$

The damping γ_0 increase monotonically with increasing temperature. If the inequality $\epsilon_0\gamma_0^2 << f_0^2$ is satisfied at a very low temperature, then with increasing temperature the half-width Γ varies from $f_0(2n_0)^{-1}$ to γ_0 when the inequality $\epsilon_0\gamma_0^2 >> f_0^2$ is satisfied.

The maximal value of the absorption coefficient is defined by the equality

$$\max(\kappa_{eff}) = \frac{\gamma_0 f_0^2 (2cn_0)^{-1}}{\gamma_0^2 + f_0^2 (4\epsilon_0)^{-1}} \tag{56}$$

As the crystal temperature increases this value first increases from the value $2n_0\gamma_0/c$, passes through the maximum $f(2c)^{-1}$, and then decreases by the law $f_0^2(2\gamma_0 n_0 c)^{-1}$.

The integral value of absorption is defined by

$$K_{eff} \equiv \int \kappa_{eff}(\omega)\, d\omega = \frac{\pi f_0^2 \gamma_0 (2n_0c)^{-1}}{\sqrt{\gamma_0^2 + f_0^2 (4\epsilon_0)^{-1}}} \tag{57}$$

Under the conditions of strong coupling (low temperatures) ($f_0^2 >> \gamma_0^2\epsilon_0$) the value $K_{eff} = \pi\gamma_0 fc^{-1}$. As the temperature increases the integral absorption increases and reaches, when the inequality $\gamma_0^2\epsilon_0 >> f_0^2$ is satisfied, the saturation

$$K_{eff} = \pi f_0^2 (2n_0c)^{-1} \tag{58}$$

The equality (58) is usually used for calculating the transition oscillator strength. We see that this is possible only in the case when the inequality $\gamma_0^2\epsilon_0 >> f_0^2$ is satisfied. This inequality is satisfied when the integral absorption K_{eff} reaches a constant value with increasing temperature.

Recently in some papers[9, 10, 12, 13] it was found that the area of the absorption curve of the isolated band increases as the temperature increases. In papers of Delyukev and Klimusheva[9-10] it was found that the area of the curve of the purely electronic absorption band of paradichlorbenzol crystals increases almost by a factor of six as the temperature increases from 4° to 100°. The maximum of the absorption band also changes qualitatively according to (56).

Voigt[12] investigated the absorption in cadmium sulphide in the temperature range from 1.8^{o}K to 180°K in the regions of three exciton bands. It was established that in the band $A_{n=1}(\vec{E} \perp \vec{c})$ which has the smallest excitation energy ($\sim$ 2.55 ev) and oscillator strength $F \approx 0.026$, the integral absorption increases with increasing temperature by a factor of ten, assuming constant values at a temperature exceeding 77°K. In the absorption band $B_{n=1}(\vec{E} \parallel c)$ with a transition energy of 2.57 ev and oscillator strength F = 0.016 the change of integral absorption is less pronounced. The saturation is already observed at temperatures exceeding 4°K. In the band $C_{n=1}$ with an energy of 2.63 ev and oscillator strength 0.008 the temperature dependence of integral absorption is not observed.

Kreingold and Makarov[13] investigated the light absorption by quadrupole excitons in Cu_2O over the temperature range from 1.8^{o}K to 120°K in the frequency region $\Omega_0 \approx 3x10^{15}$ sec^{-1}. A fairly appreciable change of integral absorption with the temperature varying from 1.8^{o}K to 30°K was observed in perfect crystals.

REFERENCES

1. A. S. Davydov, A. A. Eremko, Ukrain. Fis. Journal 28, 1868 (1973).
2. A. S. Davydov, A. A. Eremko, Phys. Stat. Sol. (b)59, 251 (1973).
3. A. S. Davydov, E. M. Myasnikov, Phys. Stat. Sol. (b)63, 325 (1974).
4. A. S. Davydov, A. A. Serikov, Phys. Stat. Sol. (b)56, 351 (1973).
5. V. M. Agranovich, "Theory of Excitons", "Nauka", Moscow, 1968.
6. A. S. Davydov, "Theory of Molecular Excitons", "Nauka", Moscow, 1968.
7. A. S. Davydov, B. M. Myasnikov, Phys. Stat. Sol. 20, 153 (1967); Doklady Akademy Nauk SSSR 173, 1040 (1967).
8. M. S. Brodin, N. A. Davydova, M. I. Strashnikov, Doklady Akademy Nauk Ukrain SSR, series A, N4, 364 (1975).

9. A. A. Deljukov, G. V. Klimusheva, Fisika Tverd. Tela 16, 3255 (1974).
10. A. A. Deljukov, G. V. Klimusheva, Doklady Akademy Nauk Ukrain SSR, series A, N11, 1017 (1974); Ukrain Fisika 19, 341 (1974).
11. A. A. Eremko, Doklady Akademy Nauk Ukrain SSR, series A, N3, 272 (1974).
12. I. Voigt, Phys. Stat. Sol. (b)64, 549 (1974).
13. F. I. Kreingol'd, V. L. Makarov, Pis'ma v GETP 20, 441 (1974).

NEUTRON SCATTERING AND ABSORPTION BY LASER PHOTONS IN CRYSTALS

V. M. Agranovich and I. I. Lalov

Institute of Spectroscopy
Academy of Sciences of the USSR
Podolskii r-n, P.o. Akademgorodok, Moscow, USSR

The cross-section for scattering of neutrons by photons is high in crystals. Since a photon in a crystal (polariton), being a superposition of transverse photon and Coulomb excitations (optical phonon, exciton, etc.), excites during its propagation the motion of the nuclear subsystem, the cross-section for scattering of neutrons by photons turns out to be proportional to the neutron scattering cross-section for nuclei and the phonon strength function at the polariton frequency.

The neutron-photon interaction in vacuum is negligible. However, in condensed media this interaction increases drastically and results in phenomena which can be observed experimentally.

There is also another aspect of the problem discussed below. The point is that while considering neutron-photon interaction in crystals one can notice their close relationship with the Mössbauer effect; in fact, we have here one more analogy to this effect.

To elucidate the main idea of the report we shall first consider light-light scattering.

It is well known that this scattering in vacuum is rather weak. In condensed media, however, a photon, turning into a polariton, "dresses itself", involving in motion both electrons and nuclei. In this case a photon acquires "a coat" consisting of virtual elementary excitations corresponding to electrons and nuclei; it is this very interaction of these "coats" (or "clouds") in condensed media which brings about a considerable increase of light scattering by light.

A similar phenomenon occurs also for the neutron-photon interaction. Since in a photon "coat" virtual phonons are present, the cross-section of neutron scattering by photons in a crystal increases drastically, as its main part turns out to be proportional to the cross-section of neutron scattering by nuclei. The presence of electron excitations in the photon "coat" results in the appearance of a contribution, proportional to the cross-section of magnetic neutron scattering by electrons, in the above mentioned cross-section. In the discussion to follow we shall - according to [1] - take into account the neutron-nuclei interaction only. The role of neutron-electron interaction can be considered in a similar manner.

From purely qualitative considerations it is clear that in crystals neutron scattering by photons with frequency ω should be proportional to the so-called phonon strength function $S(\omega)$, figuring in the polariton theory [3, 2, 4]. This function describes the portion of mechanical energy in a polariton and is equal to the ratio of square atomic amplitudes in a polariton and an optical phonon. For example, for an isotropic medium with dielectric function

$$\epsilon(\omega) = \epsilon_\infty + \frac{(\epsilon_0 - \epsilon_\infty)\Omega_\perp^2}{\Omega_\perp^2 - \omega^2}$$

(ϵ_0 and ϵ_∞ are low and high frequency values of $\epsilon(\omega)$, $\Omega_\perp$ is the optical transverse phonon frequency) the magnitude $S(\omega)$ is given by

$$S(\omega) = \frac{\omega\Omega_\perp\omega_0^2/\epsilon_\infty}{(\omega^2 - \Omega_\perp^2) + \Omega_\perp^2\omega_0^2/\epsilon_\infty} \tag{1}$$

where

$$\omega_0^2 = \Omega_\perp^2(\epsilon_0 - \epsilon_\infty)$$

The notion of a strength function S(ω) and the relationship (1), naturally, can be used also for describing polariton structure in the region of exciton transitions ("exciton strength function", $\Omega_\perp$ being the frequency of the exciton transition) and it is useful because the magnitude S(ω) naturally appears in the theoretical description of inelastic polariton scattering by phonons both in electronic and vibrational spectral ranges. Specifically, the magnitude S(ω) determines the intensity of Raman polariton scattering by phonons at the long wave edge of exciton absorption bands [5]. It appears also in the theory of Raman scattering by polaritons, where it describes the intensity [4], as well as the width of a Raman line [5, 7].

In studies of neutron scattering in crystals the retarded interaction is usually neglected. This interaction is practically insignificant for acoustic and dipole-inactive vibrations; however, for dipole active vibrations of the crystal it gives rise to a substantial reconstruction of spectra in the long wave range.

As a result of this reconstruction in the specified spectral range, instead of optical (Born) phonons and transverse photons there appear new elementary excitations-polaritons; their properties can be studied by means of various optical methods (see, e.g., [8]).

It is just because polaritons are a mixture (superposition) of transverse photons and Coulomb excitations (phonons, excitons, etc.) that the cross-section of neutron scattering by a polariton turns out to be proportional to the cross section of neutron scattering by nuclei. Other things being equal, i.e., at the same energy of quanta and the same flux, this cross section exceeds by many orders the corresponding cross-section of neutron scattering by photons in vacuum. However, due to the low density of states in ordinary conditions a polariton spectral range makes only a small contribution in cross sections of inelastic neutron scattering in crystals, so that it is practically not necessary to take polariton effects into account. The situation changes radically only in the case when sufficiently intensive electromagnetic radiation flux (i.e., polariton flux) from some source is propagating through a crystal. In this case, in a crystal (but not in vacuum) such inelastic neutron scattering becomes possible, when the whole photon (polariton) energy is transmitted to the neutron. If a monochromatic photon source (e.g., a laser) is used and the thermal energy spread of primary neutrons is sufficiently low, then the resulting neutrons acquire a high degree of monochromaticity (at $kT \sim 10^{-3}$ ev and $\hbar\omega \sim 0.1$ ev, the monochromaticity is equal to $2kT/\hbar\omega \sim 10^{-2}$, etc.).

The efficiency of the considered effect is found to be proportional to the Debye-Waller factor, which decreases exponentially with the increase of quantum energy transmitted to the neutron. The occurrence of this factor here does not seem to be a surprise, as it is this very factor (similar to the conditions when the Mössbauer effect is realized), which determines the probability of the considered energy transmission to the neutron being phononless.

However, for estimation of the total cross-section of photon absorption by a neutron at $\hbar\omega$ large enough, and for a low Debye-Waller factor, it is necessary to take into account the processes of creation and absorption of phonons. In this connection we shall obtain a neutron spectrum with due regard for multiphonon processes, and also we shall trace further the analogy with the Mössbauer effect.

Thus, we assume that the retarded interaction is taken into account and that the spectrum of lattice vibrations is of the form shown schematically in Fig. 1 For simplicity, we shall first consider the case of a crystal consisting of identical nuclei. The displacement of a nucleus, situated in the $\vec{n}$-th elementary crystal cell, from an equilibrium position (this displacement arising from lattice vibrations) may be given as

$$\vec{u}_n = \sum_s \vec{\ell}_s (\xi_s e^{i\vec{f}\vec{n}} + \xi_s^* e^{-i\vec{f}\vec{n}}) ,$$

where $\vec{n}$ is a lattice vector; the index $s \equiv \vec{f}, \mu$ is used for designating vibrations of μ type with the wave vector $\vec{f}$ and frequency $\omega_s \equiv \omega_\mu(\vec{f})$; ξ_s and ξ_s^* are the vibration amplitude and its complex conjugated value; ℓ_s is a unit vector in the direction of vibrations. If they do not interact with light, the only nonzero matrix elements of the operator are:

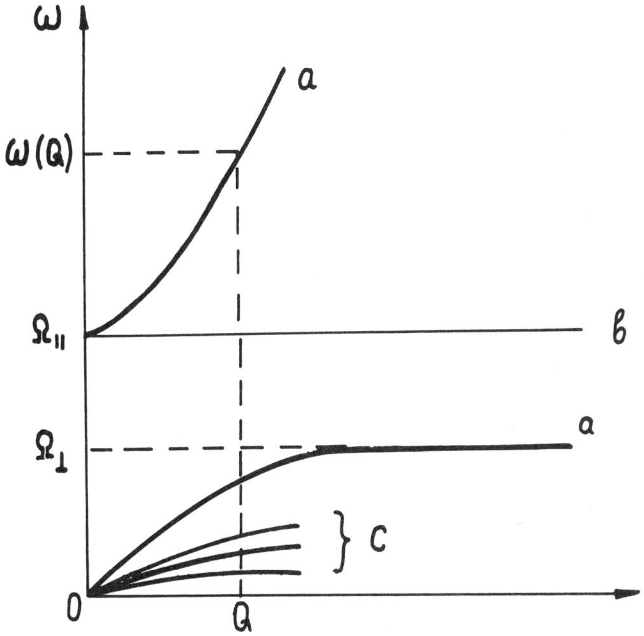

Fig. 1. Vibrational spectrum of a crystal with due regard for radiation. a) polariton branches; b) longitudinal optical vibrations; c) acoustic vibrations.

$$\xi_{n_s, n_s+1} = \sqrt{\frac{\hbar (n_s+1)}{2M_s N\omega_s}} \tag{2a}$$

where M_s is the mass corresponding to the oscillator s; N is the number of nuclei in the crystal; n_s is a quantum number determining the oscillator state. But if the index s corresponds to a polariton of frequency ω, then

$$\xi_{n_s, n_s+1} = \sqrt{\frac{\hbar (n_s+1)}{2M_\perp N\Omega_\perp}}\; S^{\frac{1}{2}}(\omega) \tag{2b}$$

where S(ω) is a phonon strength function [see (1)]. For obtaining cross sections of neutron scattering in crystals, calculations are required which are quite similar to those usually made when retardation is neglected (see e.g., [9]).

As a result, we find that the differential cross section per atom is determined by

$$d\sigma = \frac{\sigma_o}{8\pi^2} d\theta' dE \frac{p'}{p} \int_{-\infty}^{\infty} e^{i\mu(E'-E)+G(\mu)} d\mu \tag{3}$$

where E, E' and $\vec{p}, \vec{p}'$ are neutron energies and momenta before and after scattering; $d\theta'$ is an element of solid angle in which the neutron momentum $\vec{p}'$ is found after scattering; σ_o is the total cross section of neutron scattering by a bound nucleus;

$$G(\mu) = \sum_s [(n_s+1)e^{i\mu\hbar\omega_s} + n_s e^{-i\mu\hbar\omega_s} - 2n_s - 1]q_s^2 \tag{4}$$

where

$$q_s^2 = \frac{(\vec{p}-\vec{p}', \vec{\ell}_s)^2}{2M_s N\hbar\omega_s} \tag{5}$$

If the index s corresponds to lattice oscillators which do not interact with light, then

$$q_s^2 = \frac{(\vec{p}-\vec{p}', \vec{\ell}_s)}{2M_\perp N\hbar\Omega_\perp} S(\omega) \tag{6}$$

for a polariton mode.

In (3) not only multiphonon processes are considered, but multiphoton ones as well. Here we are interested only in one-photon (one-polariton) processes. Hence, we shall expand (3) into a power series of quantum numbers of polariton oscillators n_p and take into account only the term which is linear in n_p. Assuming $n_p \gg 1$ and designating the total number of polaritons in a crystal by N_p,

$$N_p = \frac{VQ^2 d\vec{Q}}{(2\pi)^3} n_p,$$

(where V is the crystal volume, $\vec{Q}$ is the wave vector of the polariton produced by an external source which lies within the range $d\vec{Q} \equiv dQd\theta$; $d\theta$ is an element of solid angle), we obtain the required cross-section, with due regard also for (6), in the form

$$d\sigma(E') = \frac{\sigma_o}{8\pi^2} d\theta' dE' \frac{p'}{p} \frac{(\vec{p}-\vec{p}_1', \vec{\ell}_p)^2}{2M\hbar\Omega_\perp} \frac{N_p}{N} S(\omega) I(E'), \tag{7}$$

where

$$I(E') = \int_{-\infty}^{\infty} e^{i\mu(E' - E - \hbar\omega) + g(\mu)} d\mu \tag{8}$$

In (8) the function $g(\mu)$ is of the form (4), however, regardless of polariton contributions. The integral I(E') in (7), which determines to a great extent the relationship of the cross-section σ to the final neutron energy E', is similar to that used in the theory of neutron scattering in crystals [9] and also in the theory of Mössbauer effect or its optical analogue [10]. In this connection the previous results can be used here.

It was shown in [9, 10] that it is in this very region of scattered neutron energies E', where $E' \simeq E + \hbar\omega$, that the range of large μ values makes the main contribution into the above-mentioned integral. In this region of energies E' the value I(E') is determined by

$$I(E') \cong e^{-2w} \delta(E' - E - \hbar\omega) , \tag{9}$$

where $2w = \sum_s q_s^2(1 + 2n_s)$

hence, Eq. (7) becomes

$$d\sigma = \frac{\sigma_o}{4\pi} d\theta' dE' \frac{p'}{p} \frac{(\vec{p} - \vec{p}_1', \vec{\ell}_p)^2}{2M_\perp \hbar\Omega_\perp} \frac{N_p}{N} S(\omega) e^{-2w} \delta(E' - E - \hbar\omega) \tag{10}$$

Equation (10) has already been used in [1] for evaluating the cross-section of neutron absorption of the total photon energy and now we shall not consider this relationship any more. It should be noted also that for $\hbar\omega >> E$ we have also $p' >> p$, so that according to (5)

$$q_s^2 \simeq \frac{m\omega \cos^2 \varphi_s}{M_s \omega_s N}$$

where φ_s is the angle between vectors $\vec{p}$ and $\vec{l}_s$, and m is the neutron mass. When using the Debye approximation and $T << \theta$, according to (9), we obtain

$$2w \simeq \frac{3\hbar\omega}{4k\theta}\left(\frac{m}{M}\right)$$

where k is Boltzmann's constant, M is the nucleus mass and θ is the Debye temperature. This means that the Debye-Waller factor (i.e., the value e^{-2w}) decreases exponentially with the increase of ω, so that phononless processes can make an appreciable contribution to the value of the total cross-section of neutron scattering by photons only in the case when the energy of photons is sufficiently low. Specifically, at small values of $\hbar\omega$ and low crystal temperature ($kT << \hbar\omega$), when $w \lesssim 1$, only one-phonon processes, alongside with a phononless peak, are of importance. In this case the spectrum of scattered neutrons may be schematically represented by the distribution shown in Fig. 2a. If $w >> 1$, processes involving the creation of many phonons make the main contribution to the cross-section of neutron scattering.

The energy thus transmitted to the lattice equals $\epsilon = E + \hbar\omega - E'$. If $\epsilon >> k\theta$ then small values of μ (approximation of small collision time; see in [9, 10]) play the main role in the integral $I(E')$ (see [8]). Expanding the value $q(\mu)$ into a power series μ and restricting ourselves to the consideration of only linear and quadratic terms in μ, we find that in the discussed energy range E' (we assume $p' >> p$, as before) the required cross-section is*

*It can be shown that a similar result (a transition to the cross-section of scattering by free nuclei is obtained for lattices, composed of different nuclei (see, e.g., [11]), the only difference being that for such lattices in Eq. (11), the value σ_0 equals the total cross-section of neutron scattering by a bound nucleus ($\sigma_0 = \sigma_0^{coh} + \sigma_0^{incoh}$).

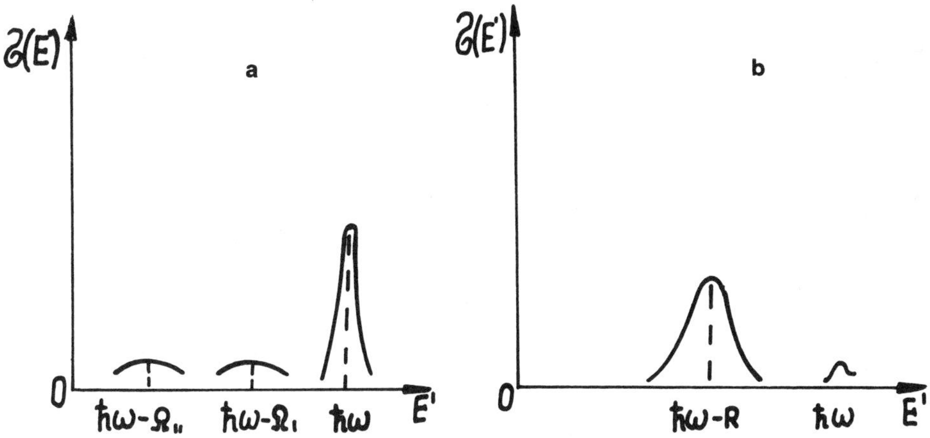

Fig. 2. Spectrum of neutrons scattered by photons in a crystal: a) the value w << 1; multiphonon processes are negligible; b) the value w >> 1; the intensity of the process is completely determined by a multiphonon wing.

$$\frac{d\sigma}{dE'} = \frac{\sigma_o}{2\sqrt{\pi}} \frac{p'}{p} \frac{E'}{\hbar\Omega_\perp} \frac{m}{M} \frac{N_p}{N} S(\omega) \frac{1}{\sqrt{R\epsilon_o}} e^{-(E'-\hbar\omega+R)^2/4R\epsilon_o} \quad (11)$$

where

$$R = \frac{|p-p'|^2}{2M} \simeq \frac{m}{M} E'$$

is the recoil energy from neutron scattering by a free nucleus, while the value ϵ_o (in Debye approximation) is determined by

$$\epsilon_o = 3kT\left(\frac{T}{\theta}\right)^3 \int_0^{\theta/T} \left(\frac{1}{e^t - 1} + \frac{1}{2}\right) t^3 dt , \quad (12)$$

so that for $T \gg \theta$ we have $\epsilon_o = kT$, and for $T \ll \theta$ we obtain

$\epsilon_0 \simeq \frac{3}{8} k\theta$. Thus, the multiphonon wing is of the Gaussian type (see Fig. 2b), its centre being at $\overline{E'} = \hbar\omega/(1+m/M)$. Specifically, in hydrogenous media $\overline{E'} = \hbar\omega/2$, so that in this case, on the average, about a half of the quantum energy $\hbar\omega$ in the wing region is transmitted to the lattice.

Assuming $\hbar\omega \simeq 1$ ev >> kT, one may see that the integrated cross-section, corresponding to the wing contribution (11), is given by

$$\sigma_{tot} = \int_0^{\hbar\omega+E} \frac{d\sigma}{dE'} dE' \simeq \frac{\sigma_o}{3} \frac{m}{M} \frac{N_p}{N} \frac{\omega}{\Omega_\perp} S(\omega) \left[\frac{\hbar\omega}{\epsilon_o(1+\frac{m}{M})}\right]^{\frac{1}{2}} \tag{13}$$

where $\sigma_a = \sigma_o(1+m/M)^{-2}$ is the cross-section of neutron scattering by a free nucleus (the temperatures of the neutrons and the lattice are assumed to coincide). The wing halfwidth (11) is approximately equal to 2Δ, where Δ is the value of the dispersion. It follows from calculations that for the differential cross-section (11) we have

$$\Delta = \sqrt{\epsilon_o \frac{m}{2M} \frac{\hbar\omega}{(1+\frac{m}{M})^3}} \tag{14}$$

The general view of the scattered neutron spectrum is shown in Fig. 2.

Now we shall estimate numerically the foregoing characteristics of neutron scattering by photons in crystals. In so doing, for definiteness, we shall take media with hydrogen nuclei (m = M). Since for such nuclei the value of the cross-section σ_a is rather high and, besides, they are mostly intensively involved in motion while a polariton is propagating, only neutron scattering by hydrogen nuclei will be considered. If we take into account that

$$\Omega_\perp^2 \frac{\epsilon_o - \epsilon_\infty}{\epsilon_o} = \Omega_\parallel^2 - \Omega_\perp^2$$

where $\Omega_\parallel$ and $\Omega_\perp$ are the frequencies of longitudinal and transverse optical phonons, then we obtain for the integral cross-section of scattering related to one hydrogen atom:

$$\sigma_{tot} \simeq \frac{\sigma_a}{3}\frac{n_p}{n}\frac{\Omega_{\parallel}^2 - \Omega_{\perp}^2}{\omega^2}\left(\frac{\hbar\omega}{2kT}\right)^{\frac{1}{2}} \tag{15}$$

where n_p and n are the polariton and nucleus concentrations, respectively, while

$$\Delta = \tfrac{1}{4}\sqrt{\epsilon_o(T)\hbar\omega}\ . \tag{16}$$

At $T \simeq \epsilon_o/k = 100^oK$ and $\hbar\omega \simeq 1$ ev the wing halfwidth Γ equals approximately the value $2\Delta \simeq 0.05$ ev, while the value of the integral macroscopic cross-section $n\sigma_{tot}$, e.g., for a LiH crystal ($\Omega_{\parallel} = 1130\ cm^{-1}$, $\Omega_{\perp} = 590\ cm^{-1}$) is found to be

$$\Sigma_{tot} = n\sigma_{tot} \simeq n_p\ 10^{-24}\ cm^2\ ,$$

so that for $n_p = 10^{16}\ cm^{-3}$ we have $\Sigma_{tot} \simeq 10^{-8}\ cm^{-1}$, the appropriate partial lifetime of a neutron being $\tau_p \simeq 10^{-3}$ sec in the medium ($\tau_p = (\Sigma_{tot}\ v)^{-1}$) where $v \simeq \sqrt{3kT/m}$ is the average neutron velocity. If τ_o is the lifetime of a neutron in the medium, when photons are absent, then under steady-state conditions the efficiency of scattering by photons is equal to

$$\nu = \frac{1}{\tau_p}\left(\frac{1}{\tau_p} + \frac{1}{\tau}\right)^{-1} = \frac{\tau}{\tau + \tau_p}\ .$$

The lifetime of a neutron in the medium with hydrogen is usually $\tau \approx 2\text{x}10^{-4}$ sec; thus, in the case considered we have $\nu \simeq \tau/\tau_p \simeq 10^{-7}$.

In connection with these evaluations it should be noted that the monochromaticity of radiation is not so significant for studying a neutron scattered by photons in the wing region, as it was in the case for a phononless peak. Consequently, for high values of σ_{tot} to be obtained, instead of lasers it is possible to use powerful nonmonochromatic radiation sources, that are used, for example, for pumping of lasers (if $n_p \simeq 10^{18}\ cm^{-3}$, $\nu \simeq 10^{-5}$ and so on). It should be borne in mind that due to the linear relationship between the cross-section of the process in question (see, e.g., Eq. (13) and (15)) and photon concentration, focusing of radiation becomes unnecessary; this fact may be of importance in connection with the problem of the strength of the exposed crystals.

REFERENCES

1. V. M. Agranovich, I. I. Lalov, Phys. Letters 53A, 169 (1975); Zh. Eksp. Teor. Fiz. 69, 647 (1975).
2. M. Born, Kun Huang, Dynamic theory of crystal lattices (Oxford, 1954).
3. V. M. Agranovich, Theory of excitons. (Izd. "Nauka", Moscow, 1968).
4. E. Burstein, Comments on Solid State Physics 1, 202 (1969); E. Burstein, D. L. Mills, Comments on Solid State Physics 2, 93, 111 (1969); 3, 12 (1970).
5. V. M. Agranovich, Yu. V. Konobeev, Fiz. tverd. tela. 3, 360 (1961).
6. H. J. Benson, D. L. Mills, Phys. Rev. B1, 4835 (1970).
7. V. M. Agranovich, V. L. Ginzburg, Zh. Eksp. Teor. Fiz. 61, 1243 (1971).
8. Polaritons, Proceedings of the First Taormina Conference (1972, Italy). Edited by E. Burstein and F. De Martini.
9. A. Ahieser, I. Pomerantchuk, Some questions of the nuclear theory. CITTL, Moscow, 1950.
10. A. Maradudin, Solid State Physics 18, 273 (1966); 19, 1 (1966).
11. I. I. Gurevich, L. V. Tarasov, Physics of the low energy neutrons. Izd. "Nauka", Moscow, 1965.

SPATIAL DISPERSION AND DAMPING OF PLASMON-PHONON MODES IN SEMICONDUCTORS

V. I. Zemski, E. L. Ivchenko, D. N. Mirlin, and I. I. Reshina

A. F. Ioffe Physico-Technical Institute
Academy of Sciences of the USSR
Leningrad, USSR

It is well known that, if the plasma frequency in a semiconductor is comparable to that of the LO-phonon, the interaction of these excitations leads to the formation of coupled plasmon-phonon modes. The eigenfrequencies of the modes are given by zeros of the longitudinal dielectric function. The dispersion curves for plasmon-phonon modes ω_+ and ω_- are shown schematically in Fig. 1.

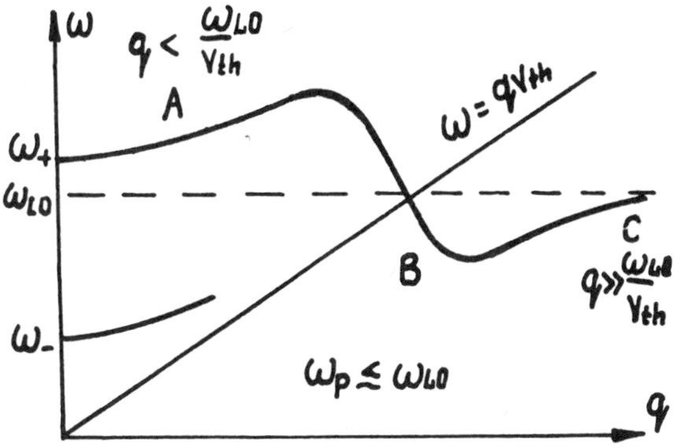

Fig. 1. The dispersion curves for plasmon-phonon modes. ω is the frequency, q is the wave-vector, v is the electron thermal velocity. The curves represent the case when the plasma frequency ω is less than ω_{LO}.

The long wavelength plasmon-phonon modes have been investigated in detail [1-3]. The data were obtained by means of an infrared reflectivity technique and in Raman scattering experiments with the incident radiation energy well below the band gap of the crystal. The positions of lines $\omega_\pm$ in the Raman scattering spectra were in good agreement with the theoretical estimate. The linewidths were consistent with the data obtained from mobility measurements.

The dispersion of plasmon-phonon modes, like plasmon dispersion, is governed by $(q/q_D)^2$ where q_D is a screening momentum. Moreover, for nonzero q, besides the collision damping it is necessary to take into account the Landau damping of the modes.

For III-V semiconductors the LO-phonon and plasma frequencies are comparable in magnitude for an electron concentration of about 10^{17} cm^{-3}. The corresponding value for q_D is of the order of 10^6 cm^{-1} at room temperature . Therefore, for most crystals the dispersion of plasmon-phonon modes is appreciable in Raman scattering experiments only with the incident light frequency within the region of fundamental absorption. So, such experiments need to be conducted in the back-scattering configuration.

The use of Raman scattering in determining the wave-vector dependence of the coupling between LO phonons and plasmons was first demonstrated by Burstein and co-workers [4, 5]

In the present report we review the results of our detailed investigation of the dispersion and damping of plasmon-phonon modes by means of a Raman back-scattering technique. The measurements were performed using a He-Ne laser for excitation (λ = 6328 Å). The samples were n-type GaAs and InAs. The experimental conditions allowed us to study the region A in the case of GaAs and the region B in InAs (see Fig. 1).

Figure 2 shows Raman scattering spectra obtained at room temperature from three GaAs samples with different electron concentrations. The scattering configurations are given in a well-known notation. The lines labelled as TO and LO are due to a transverse phonon and an unshifted longitudinal phonon, the latter owing to the scattering within a surface depletion layer. Besides these lines one may see two broad lines ω_+ and ω_- corresponding to plasmon-phonon modes. With increasing electron concentra-

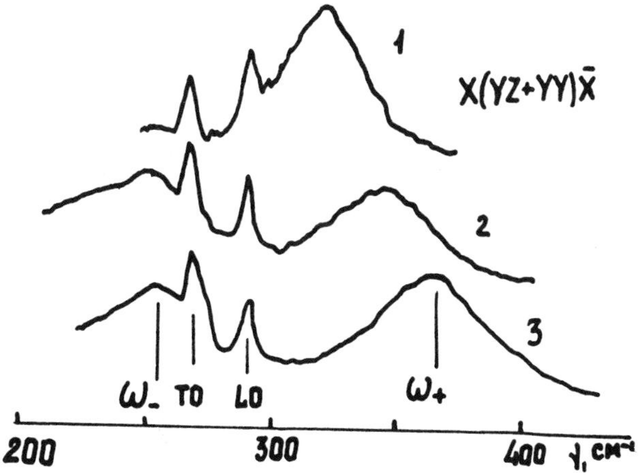

Fig. 2. Raman scattering spectra for GaAs samples with different electron concentrations: 1 - 3.4×10^{17} cm^{-3}, 2 - 5×10^{17} cm^{-3}, 3 - 6.7×10^{17} cm^{-3}.

tion, as expected, the lines $\omega_\pm$ shift towards higher frequencies and the linewidths increase. The polarization dependences for the lines $\omega_\pm$ were in accordance with the selection rules.

The Raman scattering cross-section by plasmon-phonon modes was calculated by McWhorter and Mooradian neglecting the optical phonon damping [1]. The expression for the scattering cross-section can be written for T symmetry crystals as

$$\frac{d^2\omega}{d\Omega d\omega} = -(n_\omega+1)\,\mathrm{Im}\,\frac{1}{\epsilon(\omega,q)}\,[A(\epsilon_\infty+\epsilon_{ph})^2(\vec{e}_1\vec{e}_2)^2 + B(\Sigma|e_{ijk}|e_{1i}e_{2j}e_{0k})^2(1+\frac{c\omega_{t0}^2}{\omega_{t0}^2-\omega^2})] \,, \qquad (1)$$

$$\epsilon(\omega,q) = \epsilon_\infty + \epsilon_{ph} + \epsilon_{e\ell}$$

Here ϵ_∞ is the background dielectric constant, ϵ_{ph} and $\epsilon_{e\ell}$ are phonon and electron contributions to the dielectric function;

i, j, k are the principal symmetry axes of the crystal, $\vec{e}_1$ and $\vec{e}_2$ are the polarization unit vectors of the incident and scattered light, $\vec{e}_0$ is the unit vector parallel to the scattering wave-vector, and e_{ijk} is the unit antisymmetrical tensor. Parameters A and B are independent of ω and q. The first term in the formula is due to charge-density fluctuations. The second one is connected to the electro-optic and deformation-potential mechanisms of scattering. For simplicity these terms will be hereafter referred to as electron and phonon contributions, respectively. We extended the formula for Raman efficiency taking into account the phonon damping rate. In the general formula the expressions for electron and phonon contributions are rather complicated, but their polarization dependences remain exactly the same.

In order to separate the phonon and electron contributions we measured Raman spectra from a (100) surface of GaAs. The results for the mode ω_+ are shown in Fig. 3. In the diagonal configuration only charge-density fluctuations contribute to the Raman intensity. On the contrary, in the nondiagonal configuration only the phonon contribution is present. It should be noted

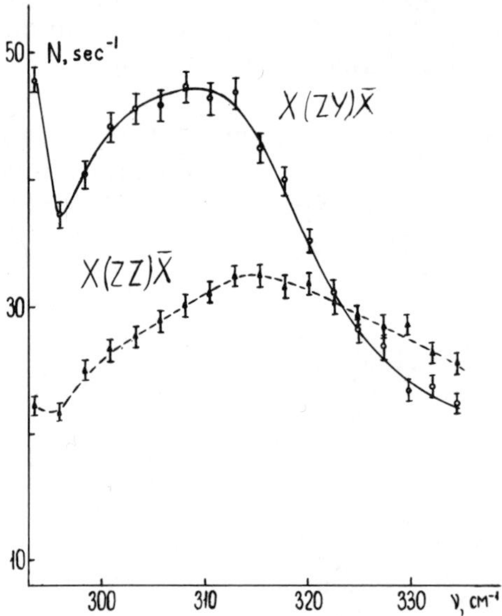

Fig. 3. Raman scattering spectra from a (100) surface of GaAs with N = 3.4×10^{17} cm^{-3}.

that the lines in Fig. 3 are shifted relative to each other. This may be understood taking account the additional frequency dependent factors in the expression (1). The calculated difference between the maxima of the lines is in agreement with the measured value.

In addition to Raman spectra we have measured, for the same samples, infrared reflectance spectra and obtained from these data the frequencies of plasmon-phonon modes in the long wavelength limit. In Fig. 4 we compare reflectance and Raman spectra. The frequencies in the long wavelength limit are shown by arrows.

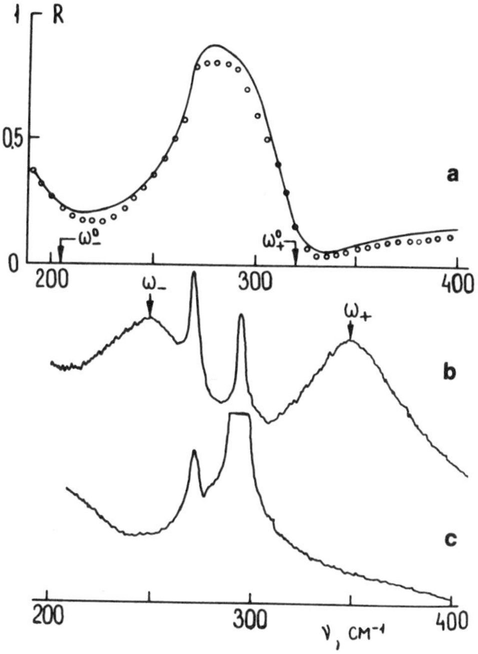

Fig. 4. (a) Infrared reflectivity spectrum for a GaAs sample (N = 5×10^{17} cm^{-3}). The dotted curve shows the measured spectrum and the solid curve shows the calculated one. The frequencies $\omega_{\pm}$ (0) in the long wavelength limit are denoted by arrows. (b) Raman scattering spectrum in the same sample obtained using a He-Ne laser excitation. (c) Raman scattering spectrum for an argon laser excitation.

We see that the dispersion effects in our experiments are appreciable. The lower curve in Fig. 4 represents the Raman scattering spectrum obtained with an argon laser excitation. The line used was 4881 Å. In this case the scattering wave vector is appreciably higher than that for a He-Ne laser excitation. Accordingly the Landau damping is much greater. So, in this case the lines $\omega_{\pm}$ are too broad to be observable. This is one of the reasons why in a number of works plasmon-phonon modes have not been observed in Raman back-scattering spectra.

In Table 1 we compare for three GaAs samples the experimental and computed frequencies ω_+ and ω_-. The agreement between the measured and calculated values is good. In Table 2 the measured values of $\omega_{\pm}(0)$ and $\omega_{\pm}(q)$ for $q = 7.5x10^5$ cm^{-1} are presented.

Table 1. Measured and Calculated Frequencies of Plasmon-Phonon Modes for $q = 7.5x10^5$ cm^{-1}

Sample	ω_+ (cm^{-1})		ω_- (cm^{-1})	
	calc.	meas.	calc.	meas.
GaAs-8	325	326		
GaAs-13	350	350	248	250
GaAs-14	373	370	254	259

Table 2. Sample Parameters and Plasmon-Phonon Frequencies in cm^{-1}

Sample	N(cm^{-3})	$\omega_+(0)$	$\omega_+(q)$	$\omega_-(0)$	$\omega_-(q)$
GaAs-8	$3.4x10^{17}$	308	321	173	
GaAs-13	$5.0x10^{17}$	321	351	204	250
GaAs-14	$6.7x10^{17}$	337	370	222	259

The linewidth in Raman spectra is greater than the damping rate of long wavelength plasmon-phonon modes. Additional broadening of the lines is especially appreciable with increasing temperature. Figure 5 shows the temperature dependence of Raman spectra for the mode ω_+. The ratio qv/ω_{LO} changes from 0.6 to 0.75 within the temperature range 295-470°K and the linewidth increases approximately twice. In Table 3 the linewidths of the mode ω_+ are compared to those calculated. It should be emphasized that the linewidth increase is predominantly associated with the Landau damping increase. For the experimental

Table 3. Damping of the Plasmon-Phonon Mode ω_+ in Sample N 8 ($N = 3.4 \times 10^{17}$ cm^{-3})

T (K)	γ_+ calc.	γ_+ meas.
295	28	34
345	45	45
390	65	54
430	80	60
470	84	70

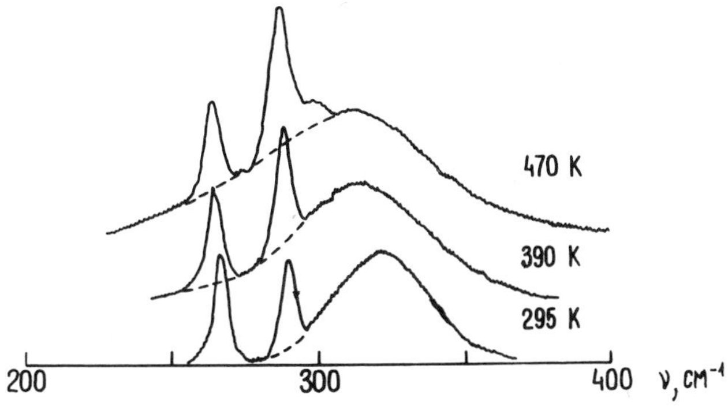

Fig. 5. Temperature dependence of Raman spectra in GaAs.

conditions in InAs the ratio qv/ω was greater than unity. In this region plasma collective excitations are absent. However, due to free carrier screening, the LO-phonon frequency changes. For the scattering wave-vector $q \simeq 8x10^5$ cm^{-1} the Raman line was expected to shift towards lower frequencies with the electron concentration increase. Figure 6 shows such a shift of the line in InAs. The electron concentration increase from $7x10^{14}$ to $3x10^{16}$ cm^{-3} was achieved by a change of temperature. A reasonable fit with experiment is obtained assuming a greater value of electron concentration near the sample surface compared to the bulk electron concentration.

So in the present work we have observed and investigated the effects of spatial dispersion of plasmon-phonon modes: the shift and broadening of the lines $\omega_\pm$ in Raman spectra.

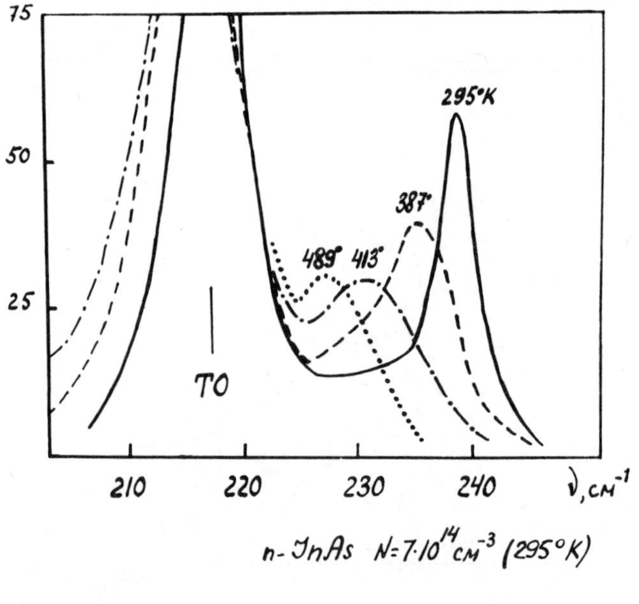

Fig. 6. Temperature dependence of Raman spectra in InAs (N = $7x10^{14}$ cm^{-3} at T = 295°K).

REFERENCES

1. A. Mooradian, A. L. McWhorter, Proc. Intern. Conf. Light Scattering Spectra Solids, p. 297, Springer, N.Y., 1969.
2. M. V. Klein, B. N. Ganguly, P. J. Colwell, Phys. Rev. B6, 2380 (1972).
3. P. M. Platzman, P. A. Wolf, Waves and Interactions in Solid State Plasmas (Sol. St. Phys.-Suppl. 13), N.Y., 1973.
4. A. Pinczuk, L. Brillson, E. Burstein, E. Anastassakis, Proc. of the 2nd Int. Conf. Light Scattering in Solids (Ed. by M. Balkanski), Paris, 1971, p. 115.
5. S. Buchner, E. Burstein, Phys. Rev. Lett., 33, 908 (1974).

THE SCATTERING OF LIGHT FROM SURFACE POLARITONS: LINE INTENSITIES AND LINE SHAPES

D. L. Mills†

Dept. of Physics, University of California
Irvine, California 92664 USA

and

Y. J. Chen and E. Burstein*

Dept. of Physics and Laboratory for LRSM
University of Pennsylvania
Philadelphia, Pennsylvania 19174 USA

In this paper, we present a summary of our recent work on the theory of light scattering from surface polaritons. The emphasis is placed on contact with the results of Ushioda et al, who have observed surface polaritons in the Raman spectrum of light forward-scattered from a GaAs film on a sapphire substrate, with the laser beam incident through the substrate. The theory accounts very well for the relative intensity of the surface polariton peaks in the near forward direction, the large forward/backward asymmetry in the cross section, and the shapes of the structures observed in the Raman spectrum.

†Supported by Grant No. AFOSR 76-2887 of the Air Force Office of Scientific Research, Office of Aerospace Research, USAF.

*Supported in part by the National Science Foundation Materials Research Laboratory through Grant No. GH-33633 and by the Army Research Office-Durham under Contract DAHC-0181.

Since the time of Sommerfeld,[1] it has been well known that under a variety of conditions,[2] surface polaritons may propagate along the interface between a semi-infinite dielectric, or along the interface between two dielectrics. A surface polariton is an electromagnetic wave which propagates parallel to the interface, with electromagnetic fields that fall to zero exponentially as one moves away from the interface into the material on either side.

For the particular case of a semi-infinite dielectric with isotropic, frequency dependent dielectric constant $\epsilon(\omega)$, the dispersion relation for a surface polariton which propagates along the interface between the dielectric and vacuum assumes the form[2]

$$\frac{c^2 k_\parallel^2}{\omega^2} = \frac{\epsilon(\omega)}{1+\epsilon(\omega)} . \tag{1}$$

The surface polariton may propagate only in frequency regions where $\epsilon(\omega) < 0$ (for the moment we assume $\epsilon(\omega)$ is real), and note from Eq. (1) that necessarily $c k_\parallel > \omega$. This condition must be satisfied for the electromagnetic field in the vacuum to decay to zero exponentially as one moves into the vacuum away from the interface. In Fig. 1, we sketch the dispersion relation for the surface polariton for two simple cases, a semi-infinite free electron metal where the dielectric constant is negative below the electron plasma frequency, and a semi-conductor or insulator with a single infrared active TO phonon. In the latter instance, the dielectric constant is negative between the TO phonon frequency ω_{TO} and the LO phonon frequency ω_{LO}.

One may couple an incident electromagnetic wave to surface polaritons through use of a grating or prism coupling device.[2] By such methods, one may study the dispersion relation of the modes,[3] or one may launch them, allow them to propagate along the substrate a macroscopic distance and detect them either with a grating or prism coupler.[4]

Since surface polaritons are elementary excitations of the semi-infinite dielectric, in principle one should be able to observe them through the technique of Raman scattering of light from the crystal. This method has the virtue that the modes may be studied on the free surface, unperturbed either by the presence of a prism or a grating. This paper summarizes our theoretical investigation

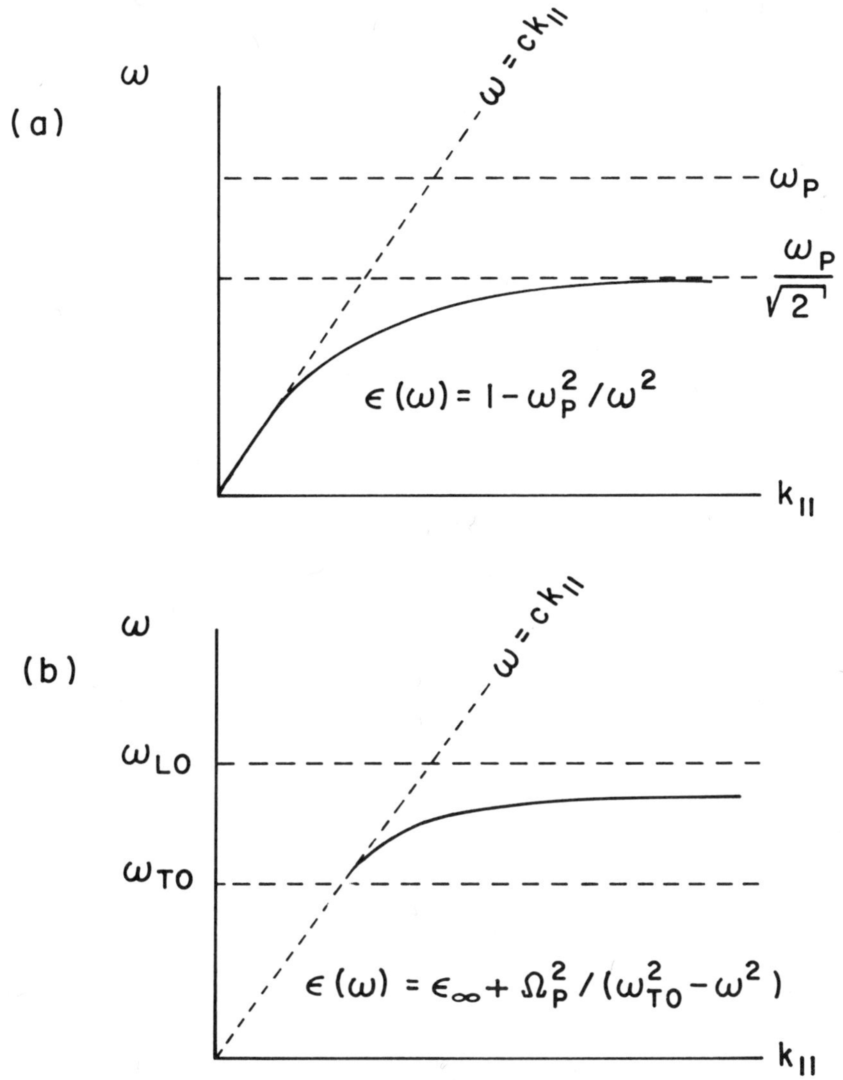

Fig. 1. The dispersion relation for surface polaritons for two simple cases (a) the interface between vacuum and a semi-infinite nearly free electron metal with dielectric constant $\epsilon(\omega) = 1 - \omega_p^2/\omega^2$, and (b) the interface between vacuum and an insulator with dielectric constant $\epsilon(\omega) = \epsilon_\infty + \Omega_p^2/(\omega_{TO}^2 - \omega^2)$. In case (a), $\epsilon(\omega)$ is negative below the electron plasma frequency ω_p and in case (b), $\epsilon(\omega)$ is negative between ω_{TO} and the LO phonon frequency ω_{LO}.

of the Raman scattering of light from surface polaritons on semiconductor surfaces.

We begin with some introductory remarks of a qualitative nature.

It is clear that such a light scattering study must be carried out under circumstances where the signal produced by scattering from surface polaritons is not swamped by the signal from the volume excitations (volume TO polaritons or volume LO phonons). The scattering volume should thus be confined to a region near the surface with depth the order of the penetration depth of the electromagnetic fields associated with the surface polariton. This requires the experiment to be carried out either in a thin film, or in backscattering from a substrate opaque to the incident radiation.

An elementary argument suggests that under the conditions described in the preceding paragraph, the scattering efficiency associated with scattering from the surface polariton should be comparable to that from the volume excitations. To see this, we reason as follows, for scattering from a surface polariton on a semiconductor surface.

Consider a single volume LO phonon quantum. If this mode is spread over a volume V, then the average of the fluctuating electric field associated with the mode is given by

$$\langle E^2 \rangle \approx \frac{4\pi \hbar\omega_{LO}}{V} \quad . \tag{2}$$

If one scatters an incident photon from the mode, and the photon penetrates into the material a depth δ, the scattering efficiency is proportional to

$$S_{LO} \propto \int_0^{\delta} dz \, \langle E^2 \rangle = \frac{4\pi \hbar\omega_{LO}\delta}{V} \quad . \tag{3}$$

This is for one mode, and if the penetration depth of the radiation is δ, one knows the LO phonon line contains contributions from all modes with wave vectors k_z normal to the surface which lie in the range $-\delta^{-1} < k_z < \delta^{-1}$.[5] The total number of such modes is, if the sample has thickness L,

$$\Delta n = \frac{L}{2\pi} \int_{-\frac{1}{\delta}}^{+\frac{1}{\delta}} dk_z = \frac{L}{\pi\delta} \tag{4}$$

so the scattering efficiency from LO phonons is proportional to

$$S_{LO}^{(TOT)} = S_{LO}\, \Delta n \propto \frac{4\hbar\omega_{LO}}{A} , \tag{5}$$

where A is the surface area of the crystal

We make a similar estimate for scattering from surface polaritons. If $\langle E^2 \rangle_0$ is the average of E^2 at the crystal surface for a single surface polariton quantum, and α the decay constant of the electromagnetic field as one moves into the dielectric from the surface, one may estimate $\langle E^2 \rangle_0$ from the relation

$$\int_0^\infty dz \, \langle E^2 \rangle_0 \, e^{-2\alpha z} \approx \frac{4\pi \hbar\omega_s}{A} , \tag{6}$$

or

$$\langle E^2 \rangle_0 \cong \frac{8\pi \hbar\omega_s}{A} \alpha . \tag{7}$$

If the penetration depth δ of the radiation is small compared to α^{-1}, then the square of the field $\langle E^2 \rangle$ associated with the surface polariton is constant over the volume exposed to the radiation, and the scattering efficiency scales as

$$S_{sp} \propto \int_0^\delta dz \, \langle E^2 \rangle = \frac{2\hbar\omega_s}{A} (4\pi\alpha\delta) . \tag{8}$$

This estimate is admittedly very crude, but since $\hbar\omega_s$ and $\hbar\omega_{LO}$ lie close to each other, when $\alpha\delta$ is of order unity, we expect the scattering efficiency from surface polaritons to be comparable to that from the volume LO phonon and TO polariton modes.

We are aware of a number of attempts by experimentalists to observe Raman scattering from surface polaritons under conditions where $\alpha\delta \sim 1$, in backscattering from the surface of semi-infinite semiconductors opaque to the incident laser radiation. Such experiments have been carried out in our laboratory at the University of Pennsylvania, for example. While one may observe scattering from the volume LO phonon and volume TO polaritons under these conditions, there is no evidence of structure in the Raman spectrum of scattering from the surface polaritons, contrary to the expectations based on the argument presented above.

The first (and to this date the only) experimental Raman scattering investigations of surface polaritons have been carried out by Ushioda and his collaborators at Irvine.[6,7] These studies employ a GaAs film about 2500 Å thick placed on a sapphire substrate transparent to the incident laser radiation. The experiments are forward scattering studies. The incident laser beam passes through the sapphire substrate to strike the GaAs film from below, and the radiation is scattered in the near forward direction in the vacuum above the film. In the spectra obtained by the Ushioda group, the intensity of the surface polariton features in the forward scattering spectrum are indeed comparable in strength to the lines produced by the volume excitations, as our estimate suggests. (In the structure explored by Ushioda and coworkers, there are two branches to the surface polariton spectrum.[8] Both the upper branch[6] and the lower branch[7] have been studied experimentally.)

We have undertaken a theoretical investigation of the Raman scattering from surface polaritons in the three media geometry employed in these experiments in order to understand the physical origin of the very large forward/backward asymmetry in the scattering efficiency for scattering from surface polaritons, and to understand the relative intensities and the shapes of the surface and volume excitations in the film. The remainder of this paper presents a brief summary of this work, which is described in detail elsewhere.[9,10] Before we begin the description of our work, we note that Nkoma and Loudon[11] have presented an analysis of the backscattering of light from surface polaritons on the surface of a semi-infinite dielectric, and Nkoma[12] has discussed forward scattering through a film placed on a substrate.

We have approached this problem by two distinct methods. The first treats the surface polariton as a long-lived well-defined excitation of the vacuum-film-substrate system, and calculates the scattering efficiency from quantum mechanical perturbation theory.[9] It provides simple and workable expressions for the scattering efficiency that provide insight into the physics of the scattering process. The calculation provides expressions for the intensity of the surface polariton line in excellent accord with the data. This method provides no information on the shape of the Raman spectrum, however. As a consequence, we have developed a Green's function description of the scattering process which may be used to study the lineshapes as well as the line intensities.[10] We shall provide here a brief description of these two investigations, and a summary of the results we have obtained.

In the calculations which follow, we presume that the laser radiation couples to the surface polaritons associated with the structure through the electro-optic tensor $b_{\alpha\beta\gamma}$ and the atomic displacement contribution $a_{\alpha\beta\gamma}$ to the Raman tensor of the film.

We first turn our attention to the description of surface polariton line intensities through use of quantum mechanical perturbation theory. First consider the backscattering of light from surface polaritons on the surface of a semi-infinite dielectric. The scattering intensity is proportional to the quantity[5] (for the Stokes side of the spectrum)

$$S_B(\Omega) = \sum_{\beta} \left| \sum_{\gamma\sigma\rho} g_{\beta\gamma} \Gamma_{\sigma\zeta} \left[a_{\gamma\sigma\rho} + b_{\gamma\sigma\rho} \frac{(\omega_{TO}^2 - \Omega^2)\bar{m}}{e_T^*} \right] \left(d_{\parallel} + i \frac{k}{\alpha_1} d_{\perp} \right) \right|^2$$

$$\times \frac{\alpha^* \Omega (1 + \bar{n})}{(\Delta k_z')^2 + (\Delta k_z'' + \alpha_1)^2} . \tag{9}$$

Here Ω is the frequency of the surface polariton, $\bar{m}$ is the reduced mass of the ions in the unit cells, and e_T^* the transverse effective charge. The factor $[a_{\gamma\sigma\rho} + b_{\gamma\sigma\rho}(\omega_{TO}^2 - \Omega^2)\bar{m}/e_T^*]$ is familiar from the theory of light scattering from volume polaritons. The function $\Gamma_{\sigma\zeta}$ is the amplitude of the Cartesian component σ of the incident field in the medium, assuming outside the medium the field is polarized along the ζ direction. The function $g_{\beta\gamma}$ is a transfer function which relates the Cartesian component γ of the scattered field in the medium to the β component outside. In addition, $d_{\parallel}$ and $d_{\perp}$ measure the amplitude of the atomic displacement $\vec{u}$ of the mode in the directions parallel and perpendicular to the surface, with α_1 the attenuation constant of the field in the medium, and $\alpha^* \cong \alpha_1$. The quantity $\Delta k_z''$ are the real and imaginary parts of $\Delta k_z^{(B)}$, where

$$\Delta k_z^{(B)} = k_z^{(I)} + k_z^{(S)} \tag{10}$$

is the sum of the (complex) z component of the wave vector of the incident and scattered radiation in the film. The denominator in Eq. (9) has its origin in the fact that the scattering efficiency is proportional to the integral

$$I = \int_0^{\infty} dz\, e^{i(\Delta k_z^{(B)} + i\alpha_1)z} \quad . \tag{11}$$

As stated above, Eq. (9) applies to backscattering of surface polaritons from a semi-infinite material. It is straightforward to extend the result to the discussion of forward or backward scattering from the three medium configuration employed in the experiments of Ushioda et al. We have presented the generalized expression elsewhere.[9]

The main qualitative difference between the expression for the forward and backscattering intensity is the form of Δk_z that enters. A backscattering event produces a contribution to the scattering efficiency similar to that in Eq. (9), with $\Delta k_z'$ and $\Delta k_z''$ the real and imaginary parts of $\Delta k_z^{(B)}$ in Eq. (9).

However, for a forward scattering event, the quantities $\Delta k_z'$ and $\Delta k_z''$ that enter the expression for the scattering efficiency are the real and imaginary parts of

$$\Delta k_z^{(F)} = k_z^{(I)} - k_z^{(S)} \quad . \tag{12}$$

It is the difference between $\Delta k_z^{(F)}$ and $\Delta k^{(B)}$ that accounts for the very large forward/backward asymmetry referred to earlier in this paper. The quantity $\Delta k_z^{(B)}$ is typically larger in absolute magnitude than $\Delta k_z^{(F)}$ by an order of magnitude, and when one follows this through to an expression for the cross section, one finds the backward scattering intensity smaller than the forward scattering intensity by roughly two orders of magnitude. We present detailed calculations shortly.

We now see the error in the handwaving argument presented in the introductory remarks of this paper. There we estimated the relative strength of the surface polariton/volume LO phonon cross section by comparing the magnitude of the modulation of the dielectric constant by the fluctuations associated with each mode. The scattering efficiency indeed is proportional to this quantity, but in the description of scattering from surface polaritons, it is crucial to do the calculation with sufficient care to properly include the matrix element which couples the three modes which enter the scattering process (incident photon, scattered photon and surface polariton).

We have extracted numerical values for the surface polariton line intensities from the forward scattering data of Ushioda et al, and compared these values with the theory. In addition we have calculated the intensities for backscattering from the structure. These results, which also have been presented elsewhere,[9] are summarized in Table 1. The calculation gives a good account of the data, and provides the explanation of the very large forward/backward asymmetry responsible for frustrating earlier attempts to observe this phenomenon.

The calculation outlined above provides an account of the integrated intensity of the surface polariton features in the Raman spectrum, for both forward and backscattering configurations, as we have seen. It also provides insight into the factors which control the very large forward/backward asymmetry. However, it provides no information about lineshapes. The experiments are performed under conditions where the surface polariton lines overlap the features in the spectrum produced by scattering from the volume TO polaritons and volume LO phonon. Indeed, the upper branch of the surface polariton dispersion relation as the vacuum-GaAs-sapphire substrate configuration lies very close to the frequency of the GaAs LO phonon. The two lines overlap to the extent that they are not resolved in the forward scattering spectrum, and an elaborate method of data analysis based on the large forward/backward asymmetry is required to obtain information about the upper branch.[7] At small scattering angles, the

Table 1. Calculated and experimental values of the scattering efficiencies for scattering from surface polaritons

(All of the surface polariton intensities listed below are normalized to the scattering efficiency for scattering from volume TO phonons. The angle θ is the scattering angle inside the film.)

	Lower Branch				
θ	1°	1.75°	3.5°	6°	8°
Frequency (cm^{-1})	273.5	276.1	279.6	281.4	281.9
Forward Scatt. Intensity (Theory)	0.24	0.27	0.29	0.26	0.21
Forward Scatt. Intensity (Expt)	0.18	0.19	0.27	0.32	0.26
Back Scatt. Intensity (Theory)	0.002	0.002	0.003	0.005	0.006
	Upper Branch				
θ	1°	1.75°	3.5°	6°	8°
Frequency (cm^{-1})	292.9	292.6	292.0	291.6	291.5
Forward Scatt. Intensity (Th.)	0.21	0.22	0.24	0.24	0.23
Back Scatt. Intensity (Th.)	0.003	0.003	0.004	0.006	0.008

lower branch of the surface polariton merges with the line produced by scattering from volume TO polaritons. Under the circumstances just described, one requires a description of the scattering process which takes full account of damping effects to analyze the data. We turn next to a summary of a Green's function approach we have developed for this purpose. The calculations are described in detail elsewhere,[10] so we only sketch the method and summarize the results here.

The method we have used is an extension to the three layer geometry of the Green's function approach that we applied earlier to the analysis of the backscattering of light from opaque materials.[5] We presume that excitation of a surface polariton or volume mode in the film leads to time and spatially varying fluctuations $\delta\epsilon_{\mu\nu}(\vec{x}, t)$ in the dielectric tensor of the film. To first order in $\delta\epsilon_{\mu\nu}(\vec{x}, t)$, the scattered electromagnetic field may be written

$$E_{\alpha}^{(s)}(\vec{x}, t) = \left(\frac{\omega_0}{c}\right)^2 \sum_{\beta\gamma} \int \frac{d^3x'dt'}{4\pi} D_{\alpha\beta}(\vec{x}\vec{x}';t-t')\delta\epsilon_{\beta\gamma}(\vec{x}', t') E_{\gamma}^{(o)}(\vec{x}'t') \quad (13)$$

where $D_{\alpha\beta}(\vec{x}, \vec{x}';t-t')$ is a set of Green's functions for the Maxwell equations that describe the three layer structure, in the absence of fluctuations and $E_{\gamma}^{(o)}(\vec{x}', t)$ the incident laser field. The integration over $\vec{x}'$ in Eq. (13) is confined to the film within which the incident radiation couples to the excitations. If we write

$$D_{\mu\nu}(\vec{x}\vec{x}';t-t') = \int \frac{d\omega}{2\pi} D_{\mu\nu}(\vec{x}\vec{x}';\omega)e^{-i\omega(t-t')} \quad (14)$$

then $D_{\mu\nu}(\vec{x}\vec{x}';\omega)$ satisfies the set of equations

$$\sum_{\lambda} \left\{\frac{\omega^2}{c^2} \epsilon(z, \omega)\delta_{\mu\lambda} - \frac{\partial^2}{\partial x_{\mu} \partial x_{\lambda}} + \delta_{\mu\lambda} \nabla^2 \right\} D_{\lambda\nu}(\vec{x}\vec{x}';\omega) \quad (15)$$

$$= 4\pi \delta_{\mu\nu} \delta(\vec{x} - \vec{x}') .$$

In Eq. (15), $\epsilon(z, \omega)$ is the dielectric constant of the three layer structure, i.e. $\epsilon(z, \omega)$ assumes the value unity in the vacuum, the (complex) value $\epsilon_1(\omega)$ in the film and the (complex) value $\epsilon_2(\omega)$ in the substrate. Explicit expressions for $D_{\mu\nu}(\vec{x}, \vec{x}';\omega)$ have been constructed for an investigation of a different physical phenomenon.[14]

We may relate $\delta\epsilon_{\beta\gamma}(\vec{x}, t)$ to the electric field $\vec{\mathcal{E}}(\vec{x}, t)$ and the atomic displacement $\vec{u}(\vec{x}, t)$ associated with the surface polariton:

$$\delta\epsilon_{\beta\gamma}(\vec{x}t) = \sum_{\delta} b_{\beta\gamma\delta} \mathcal{E}_{\delta}(\vec{x}, t) + \sum_{\delta} a_{\beta\gamma\delta} u_{\delta}(\vec{x}, t) \quad . \quad (16)$$

If we Fourier transform out the time variable in Eq. (16), and relate $\vec{u}(\vec{x}, t)$ to $\vec{\mathcal{E}}(\vec{x}, t)$ in the standard manner, for the Fourier transform we have

$$\delta\epsilon_{\beta\gamma}(\vec{x}, \Omega) = \sum_{\delta} \tilde{b}_{\beta\gamma\delta}(\Omega)\, \mathcal{E}_{\delta}(\vec{x}, \Omega) \tag{17}$$

where

$$\tilde{b}_{\beta\gamma\delta}(\Omega) = b_{\beta\gamma\delta} + \frac{\epsilon_1(\Omega) - \epsilon_1^{\infty}}{4\pi n e_T^*}\, \alpha_{\beta\gamma\delta} . \tag{18}$$

It is a straightforward though tedious matter to form expressions for the intensity of the Raman scattered radiation from the expressions displayed above. The Raman scattering efficiencies may be related to the correlation function that describes electric field fluctuations within the film:

$$\int dt\, e^{i\Omega t} \langle \mathcal{E}_{\delta'}(\vec{x}t)\, \mathcal{E}_{\delta}(\vec{x}'\, o) \rangle \equiv [1 + n(\Omega)]\, D_{\delta'\delta}(\vec{x}\vec{x}';\Omega) . \tag{19}$$

The task that remains is then to construct the spectral density function that appears on the right hand side of Eq. (19). As Abrikosov, Gor'kov and Dzyaloshinskii have pointed out,[15] these spectral densities may be constructed from the same classical Green's functions $D_{\lambda\nu}(\vec{x}\vec{x}';\omega)$ defined in Eq. (15) that entered our description of the scattering process. The prescription is

$$D_{\delta'\delta}(\vec{x}, \vec{x}';\Omega) = \frac{\Omega^2}{i c^2} [D_{\delta'\delta}(\vec{x}, \vec{x}';\Omega - i\eta) - D_{\delta'\delta}(\vec{x}, \vec{x}';\Omega - i\eta)] . \tag{20}$$

Thus, once we have the form of the electromagnetic Green's functions in hand, then the procedure outlined above allows us to construct expressions for the Raman efficiency per unit solid angle, per unit frequency, $d^2S/d\Omega d\Omega(\hat{k}_s)$. The complete final expression can be applied to either an analysis of the forward scattering spectrum, or the backscattering spectrum.

We have carried out numerical calculations of the shape of both the forward scattering spectrum (laser incident on GaAs film from the substrate side, scattered field measured in vacuum) and for the back scattering configuration (laser incident on the GaAs film from the vacuum, and scattered field measured in the vacuum). We conclude with a presentation of the results of these numerical calculations.

One can decompose the Raman spectrum into contributions from fluctuations normal to the surface, and fluctuations parallel to it. The calculations reported here involve the use of a single adjustable parameter, which controls the admixture of fluctuations parallel to or normal to the surface. This parameter was adjusted to reproduce the LO/TO ratio observed for the film at the largest scattering angle.

In Fig. 2 we show our calculations for the spectrum observed in the near forward direction, for light scattered through the film, for parameters chosen appropriate to the work of Ushioda and co-workers. The calculated spectra look remarkably similar to the experimental spectra in all significant details. One may see the feature produced by the lower surface polariton branch located prominently between the volume LO phonon and TO phonon peaks, at the largest scattering angle. As the scattering angle decreases, one sees this peak move toward and then merge with the volume TO polariton peak.

The effect of the upper branch of the surface polariton spectrum is also in evidence, if one examines the figure carefully. At the largest scattering angles, the peak near the volume LO phonon line is displaced to a frequency a distinct amount below the volume LO phonon frequency. This peak is actually a superposition of that from the upper surface polariton branch, and the volume LO phonon. The two modes have a frequency separation a bit smaller than their intrinsic width, so one sees in the spectrum a single feature shifted a bit down in frequency below the volume LO phonon frequency. As the scattering angle decreases, the frequency of the upper branch mode shifts upward (see Table 1) toward the volume LO phonon frequency, and the structure shifts to higher frequencies. We see from this that the high frequency peak contains a substantial contribution from scattering from the upper branch of the surface polariton dispersion relation at all scattering angles, in agreement with the expectation from Table 1, and the analysis of Prieur and Ushioda.[7]

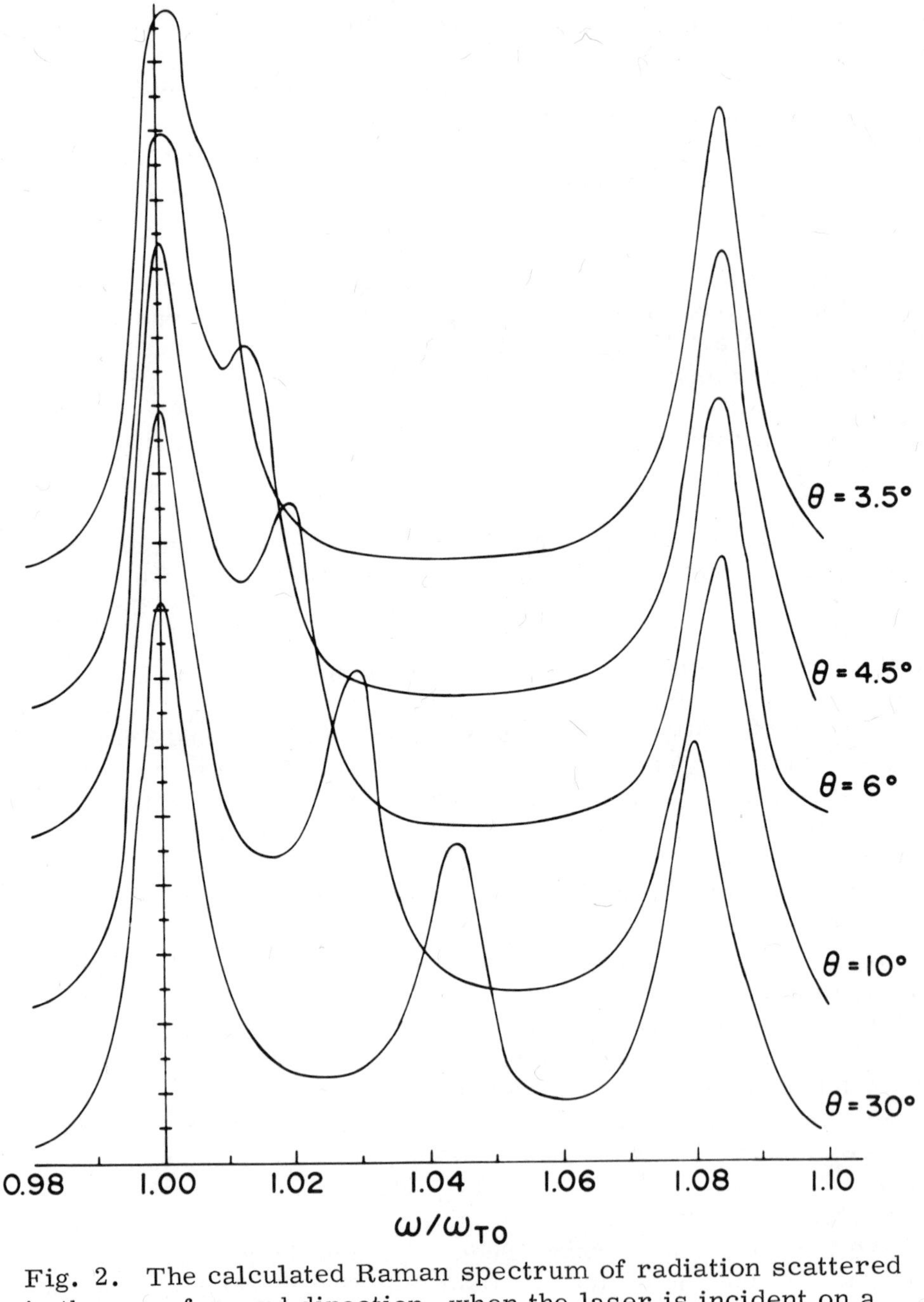

Fig. 2. The calculated Raman spectrum of radiation scattered in the near forward direction, when the laser is incident on a 2500 Å GaAs film through a sapphire substrate. The scattering angle displayed in the figure is the scattering angle <u>outside</u> the film.

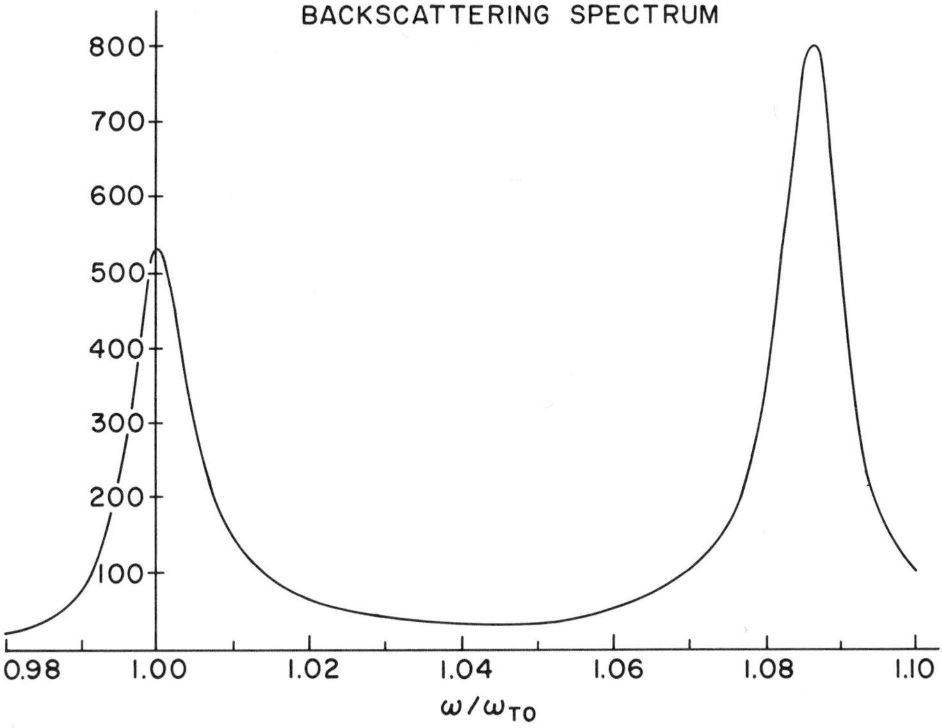

Fig. 3. The calculated Raman spectrum of radiation backscattered into the vacuum near the normal, for the case of radiation normally incident on a 2500 Å GaAs film from the vacuum. The film is presumed on a sapphire substrate.

In Fig. 3, for the same parameters we show the calculated spectrum of light backscattered from the film, into the vacuum, when the laser beam strikes the sample at normal incidence. Here one sees no hint of structure from the surface polaritons. There is only a single feature centered about the volume TO polariton frequency, and the structure near the volume LO frequency is accurately centered at ω_{LO}.

To summarize briefly, our theoretical investigation has provided an excellent account of the experimental Raman studies of surface polaritons to date. Particularly with the Green's function approach, one may readily calculate the shape of the Raman spectrum for any film/substrate geometry of interest.

REFERENCES

1. An excellent brief discussion of these waves from a historical point of view may be found in G. Joos, Theoretical Physics (Hafner, New York, 1934).
2. For a general discussion of the properties of these waves, see section X of D. L. Mills and E. Burstein, Reports on Progress in Physics 37, 817 (1974).
3. For a very beautiful experimental study, see V. V. Bryxin, D. N. Mirlin and I. I. Reshina, Solid State Communications 11, 695 (1972).
4. J. Shoenwald, E. Burstein and J. Elson, Solid State Communications 12, 185 (1973).
5. D. L. Mills, A. A. Maradudin and E. Burstein, Annals of Physics 56, 504 (1970).
6. D. J. Evans, S. Ushioda and J. D. McMullen, Phys. Rev. Letters 31, 369 (1973).
7. J. Y. Prieur and S. Ushioda, Phys. Rev. Letters 34, 1012 (1975).
8. D. L. Mills and A. A. Maradudin, Phys. Rev. Letters 31, 372 (1973).
9. Y. J. Chen, E. Burstein and D. L. Mills, Phys. Rev. Letters 34, 1516 (1975).
10. D. L. Mills, Y. J. Chen, and E. Burstein, preprint entitled "Raman Scattering of Light by Polaritons in Thin Films; Surface Polaritons and Size Effects", to be published.
11. J. Nkoma and R. Loudon, J. Phys. C. 8, 1950 (1975).
12. J. S. Nkoma, Proceedings of the Third International Conference on Light Scattering in Solids (Campinas, Brazil, 1975), to be published.
13. The derivation of this result is straightforward through use of the method developed in reference 5.
14. D. L. Mills and A. A. Maradudin, preprint entitled "Surface Roughness and the Optical Properties of a Semi-Infinite Material: The Effect of a Dielectric Overlayer", Phys. Rev. B (to be published).
15. A. A. Abrikosov, L. P. Gor'kov and I. E. Dzyaloshinskii, Methods of Quantum Field Theory in Statistical Physics (Prentice Hall, Inc., Englewood Cliffs, New Jersey).

COMMENT

(by V. M. Agranovich)

In connection with the problems touched on in this report, I would like to point out some new possibilities for studying optical properties of solids offered by Raman scattering by surface polaritons.

a. Nowadays Raman scattering by bulk polaritons is one of the main techniques of finding the dielectric permeability tensor in IR spectrum region. But in media with inversion centers this method does not "work", as the tensor of nonlinear susceptibility $\chi_{ij\ell}$ goes to zero.

The situation is different, however, if we use the "method of broken symmetry" [1] and turn to Raman scattering by surface polaritons under conditions when the surface-active medium being considered, with an inversion center along its surface, borders upon a transparent nonlinear medium (i. e. a medium with $\chi_{ij\ell} \neq 0$).

Since the electromagnetic field in a surface polariton differs from zero on both sides of the interface plane and, consequently, also in the region where $\chi_{ij\ell} \neq 0$, the intensity of Raman scattering turns out to differ from zero and is strong enough to be observed experimentally. It is of interest that in the case considered the surface lines are not lost against the background of the lines from bulk polaritons of the surface-active medium, as these lines under the conditions of the broken symmetry method are not excited (see also [4]).

b. It follows from the continuity of the normal component of the induction vector on the media interface that normal components of the electric field strength vector in a surface polariton always differ in sign (for simplicity we take here isotropic media). As shown in [3] (see also [4]) this circumstance leads to a compensation effect, i.e. to the appearance (for a certain scattering angle) of a deep and, in some cases, rather narrow gap in the Raman scattering intensity by a surface polariton. As the position of this gap is determined by a ratio of nonlinear polarizabilities of the media in contact, the observation of the compensation effect allows (when the value of χ for one medium is known) to find the value of χ for the other medium (without measuring absolute values of the cross section of the process).

REFERENCES

1. V. M. Agranovich, ZhETF Pis. Red. 19, 18 (1974).
2. V. M. Agranovich, Uspekhi Fiz. Nauk. 115, 199 (1975).
3. V. M. Agranovich, Optics Communications 11, 389 (1974).
4. V. M. Agranovich, T. A. Leskova, Fiz. Tverd. Tela 17, 1367 (1975).

COMMENT

(by V. L. Ginzburg)

I would like to point out a somewhat new kind of scattering-transitional scattering (see V. L. Ginzburg, V. N. Tsytovich, JETP 65, 1818, 1973; Radiophysica 18, 173, 1975). In the simplest case, when applied to a three-dimensional problem, the following is meant: A charge is assumed to be fixed in a medium with dielectric permeability $\epsilon(\omega)$. Then, when a permeability wave, with

$$\delta\epsilon(\omega_o, \vec{k}_o) = a \cos(\vec{k}_o \vec{r} - \omega_o t)$$

falls upon the charge, there occurs an additional polarization of the medium around the charge (i.e., polarization

$$\vec{P} = \frac{\delta\epsilon}{4\pi}\vec{E},$$

where $\vec{E}$ is the charge field). As a result, the charge environment turns into a source of electromagnetic waves. Permeability waves may be represented by acoustic waves, plasma waves, etc. Concerning scattering processes on the surface, I would like to note that the aforesaid effect should occur on the surface as well. Let us assume, for example, that a surface acoustic wave is propagating, while on the surface or nearby there are fixed ions (charged centres) present. Then, an acoustic wave at these centres will transform (scatter) into electromagnetic waves, all waves capable of propagating in the system considered and being excited for local oscillations of the medium polarization. Naturally, in the simplest case of a surface bordering upon vacuum, only usual electromagnetic waves will propagate in vacuum. But in the case of a crystal we speak about two normal waves and - with due regard to the spatial dispersion - probably of a still greater number of waves.

LIGHT SCATTERING IN CRYSTALS WITH SURFACE CORRECTIONS

Melvin Lax

City College of CUNY*, New York 10031 and
Bell Laboratories, Murray Hill, New Jersey 07974

and

Donald F. Nelson

Bell Laboratories, Murray Hill, New Jersey 07974

A summary is given of techniques and results for calculating light scattering in anisotropic crystals valid for arbitrary directions of the incident beam, the scattered beam and the crystal surface normal relative to the crystal axes. A dyadic Green's function that distinguishes ray and propagation directions leads to a scattering efficiency <u>inside</u> the crystal that involves the Gaussian curvature of the surface of wave normals. After taking account of the solid angle expansion, and the source volume demagnification at the crystal surface, a scattering formula is given suitable for comparison with experiments done <u>outside</u> the crystal. Application is made to Brillouin and Raman scattering.

INTRODUCTION

The availability of laser sources has led to a great upsurge in light scattering experiments in recent years. Many experiments have been performed on crystals of low symmetry and large opti-

*Supported in part by AROD and ONR.

cal anisotropy while a few experiments[1, 2] have used light scattering to measure quantitatively the nonlinear polarization causing the scattering. Such numerical measurements require a theory which (1) accounts correctly for the noncollinearity of the ray vector and the wavevector that results from the optical anisotropy and (2) relates the nonlinear polarization to measuring instrument parameters which are outside the crystal. Previous theories[3] have often mishandled the first requirement and have always ignored the second.

The purpose of our study of the theory of light scattering, of which we present a summary here, has been to remedy these deficiencies of previous theories. To remedy the first deficiency resulting from optical anisotropy we have based our theory on a Green's function solution of the electric field wave equation.[4, 5] This technique is appropriate because the scattering volume is typically very small compared to the size of the crystal under study when laser sources are used. Because even the crystal surface is in the far field, an asymptotic evaluation of the Green's function inside the crystal is adequate. Our first - and incorrect - evaluation[4] used a stationary phase integration over two variables followed by a residue integration over the third variable. The appealing but nevertheless incorrect result was what would have been calculated for an isotropic medium with the subsequent replacement of the isotropic refractive index by the one appropriate to the correct ray direction in the crystal. This replacement procedure has been used by others.[6] Our second - and correct - evaluation[5, 7] follows that of Kogelnik[8] and Kogelnik and Motz[9] who used it on a magnetoionic media problem. Their work was based on a stationary phase technique of Lighthill[10] in which the residue integration was performed first and the stationary phase method was applied second. The correct asymptotically evaluated Green's function in an anisotropic crystal[5, 7] is presented in Sec. II. This procedure takes proper account of the noncollinearity of the ray and propagation directions in an anisotropic crystal. In Sec. III the Green's function solution is used to find an expression for the scattered power inside the crystal. The expression differs in several ways from the best previous treatment of Brillouin scattering in anisotropic media by Motulevich.[11] One difference is the appearance of the Gaussian curvature K of the surface $\omega(k) = \omega$ in our formula for the scattered power inside the crystal. Motulevich's treatment was not based on a Green's function approach but rather on Ginzburg's[12] Hamiltonian approach using an anisotropic Coulomb gauge.

In Secs. IV and V we present the remedy to the second deficiency of previous theories, that is, to relating the scattered power inside the crystal to quantities characterizing the detector outside the crystal. In Sec. IV the solid angle inside the crystal is related to the expanded solid angle outside the crystal. The formulas presented are completely general in applying to any orientation of the crystal axes, the scattered ray, and the surface normal. In Sec. V the length of the scattering volume along the incident beam is related to the corresponding demagnified length as seen outside the crystal by a field stop of the detection optics. The demagnification formula is also completely general in applying to any orientation of the crystal axes, the scattered ray, and the surface normal. The width of the scattering volume (the unscattered beam) when laser sources are used is typically much less than the width of the field stop of the detection optics and so does not enter the formulas explicitly. The solid angle expansion and source volume demagnification expressions are then combined with the scattered power formula of Sec. III to produce in Sec. VII a scattered power formula applying to measurements made outside the crystal.

The scattered power formulas of Secs. III and VII are formulated in terms of an arbitrary mechanism of light scattering. In Sec. VIII the power formula is specialized to two important mechanisms, Brillouin scattering[1, 13] (from acoustic phonons) and Raman scattering[14] (from optic phonons). The final formulas are completely general yet compact and convenient.

II. INSIDE GREEN'S FUNCTION

The wave equation for the electric field $\vec{E}(\vec{r}) \exp(-i\omega t)$ in an anisotropic dielectric generated by the nonlinear polarization $\vec{P}^{NL}(\vec{r}') \exp(-i\omega t)$ at the single frequency ω is

$$\nabla \times (\nabla \times \vec{E}) - \frac{\omega^2}{c^2} \overleftrightarrow{\kappa}(\omega) \cdot \vec{E} = \frac{\omega^2}{c^2} \frac{\vec{P}^{NL}}{\epsilon_0} . \tag{2.1}$$

It has the Green's function solution[4]

$$\vec{E}(\vec{r},\omega) = \int \overleftrightarrow{G}(\vec{r}-\vec{r}\,') \cdot \vec{P}^{NL}(\vec{r}\,')d\vec{r}\,'e^{-i\omega t}/\epsilon_o \tag{2.2}$$

where

$$\overleftrightarrow{G}(\vec{R}) = \int \frac{\exp(i\vec{k}\cdot\vec{R})}{\overleftrightarrow{\alpha}(\vec{k},\omega)} \frac{d\vec{k}}{(2\pi)^3} , \tag{2.3}$$

$$\overleftrightarrow{\alpha}(\vec{k},\omega) \equiv (c/\omega)^2 [k^2\overleftrightarrow{1} - \vec{k}\vec{k}] - \overleftrightarrow{\kappa}(\omega) , \tag{2.4}$$

and $\overleftrightarrow{\kappa}(\omega)$ is the frequency dependent dielectric tensor.

We have previously given[7] an asymptotic evaluation of Eq. (2.3) valid for $kR >> 1$. With a slight change of notation, the asymptotic dyadic Green's function can be written

$$\overleftrightarrow{G}(\vec{R}) = [\frac{\omega}{c}]^2 \sum_{\Phi=1,2} \frac{\hat{e}^{\Phi}\hat{e}^{\Phi}}{\cos^2\delta^{\Phi}} g^{\Phi}(\vec{R}) \tag{2.5}$$

where $\hat{e}^{\Phi}$ is the unit electric field vector associated with a given mode Φ (e.g., extraordinary) whose ray direction $\hat{t}$ is parallel to the direction of observation

$$\hat{t} = \vec{R}/R , \tag{2.6}$$

and δ^{Φ} is the angle between the ray vector $\hat{t}$ and the wavevector $\vec{k}^{\Phi}$ associated with the above ray direction. The scalar Green's function becomes

$$g^{\Phi}(\vec{R}) = f^{\Phi} \frac{\exp[i\vec{k}^{\Phi}\cdot\vec{R}]}{4\pi R} \tag{2.7}$$

where

$$f^{\Phi} \equiv \frac{\cos\delta^{\Phi}}{k^{\Phi}\sqrt{K^{\Phi}}} \tag{2.8}$$

and K^{Φ} is the Gaussian curvature of the $\omega(\vec{k})$ surface at $\vec{k}^{\Phi}$, Eq. (2.7) has been shown[7] to agree precisely, in the uniaxial case, with the Green's function obtained without asymptotic approximations. This Green's function disagrees by the factor f^{Φ} with the intuitive notion that for a given direction of observation one may use the Green's function for an isotropic medium with the index of refraction appropriate to direction $\vec{k}^{\Phi}$.

III. SCATTERED POWER INSIDE THE CRYSTAL

If Eqs. (2.2), (2.5), (2.7) and (2.8) are used to calculate the electric field and if the corresponding magnetic field, $\vec{H} = (\nabla \times \vec{E})/(i\omega\mu_0)$, is also found, then the Poynting vector for a given mode Φ is found to be

$$\vec{S}^{\Phi} = \frac{\mu_0 \omega^4}{32\pi^2 c} \frac{n^{\Phi} |f^{\Phi}|^2 |c^{\Phi}|^2 \hat{t}}{(\cos \delta^{\Phi})^3} \tag{3.1}$$

where n^{Φ} is the index of refraction appropriate to wavevector $\vec{k}^{\Phi}$ and

$$c^{\Phi} \equiv \hat{e}^{\Phi} \cdot \int_{V^S} \vec{P}^{NL}(\vec{r}) \exp(-i\vec{k}^{\Phi} \cdot \vec{r}) d\vec{r} . \tag{3.2}$$

The ratio of the scattered power inside the crystal,

$$P^{scat}_{ins} = |\vec{S}^{\Phi}| r^2 d\Omega^r , \tag{3.3}$$

(where $d\Omega^r$ is a solid angle of rays in $\vec{r}$ space) to the incident power,

$$P^{inc}_{ins} = A|\vec{S}^{\theta}| = A\tfrac{1}{2}cn^{\theta} |\vec{E}^{\theta}|^2 \cos \delta^{\theta} , \tag{3.4}$$

may be expressed as

$$\frac{P_{ins}^{scat}}{P_{ins}^{inc}/A} = RV^S d\Omega_{in}^r \ . \tag{3.5}$$

Here V^S is the scattering volume accepted by the detector, A, the cross-sectional area of the incident beam inside the crystal and R, the scattering efficiency (the scattered power per unit incident power, per unit solid angle, per unit path length), is given by

$$R = [\frac{\omega}{c}]^4 \frac{n^\Phi}{8\pi^2 n^\theta \cos\delta^\Phi \cos\delta^\theta} \frac{J}{(k^\Phi)^2 K^\Phi} \ . \tag{3.6}$$

The nonlinear phenomena that give rise to the scattering are included in the quantity

$$J = \frac{\dfrac{|\hat{e}^\Phi \cdot \int \vec{P}^{NL}(\vec{r})\, e^{-i\vec{k}^\Phi \cdot \vec{r}} d\vec{r}\,|^2}{V^S}}{2\epsilon_0^2 |\vec{E}^\theta|^2 V^S} \ . \tag{3.7}$$

One striking way that the expression, Eq. (3.6), for the scattering efficiency differs from previous expressions is by its dependence on the Gaussian curvature K^Φ of the Φ branch of the $\omega(\vec{k}) = \omega$ surface.

IV. SOLID ANGLE EXPANSION

The ratio of the solid angles subtended by the rays of a beam inside and outside a crystal can be factored as

$$\frac{d\Omega_{in}^r}{d\Omega_{out}} = \frac{d\Omega_{in}^r}{d\Omega_{in}^k} \frac{d\Omega_{in}^k}{d\Omega_{out}} \ . \tag{4.1}$$

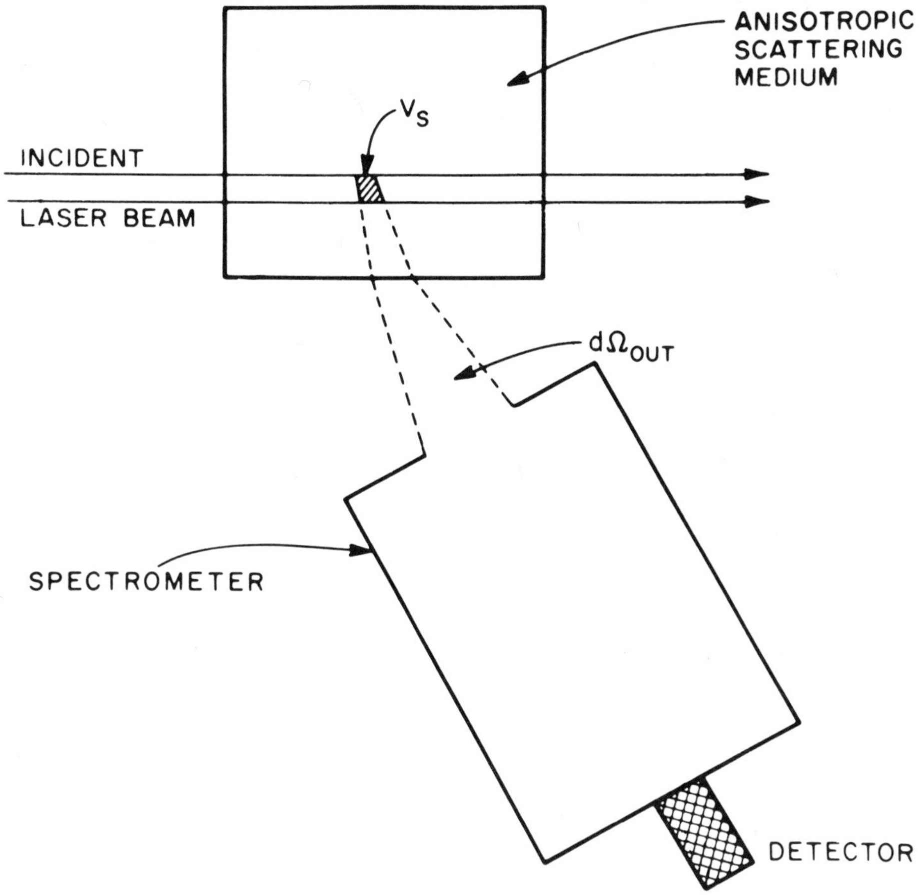

Fig. 1. A typical experimental setup for a Raman scattering experiment which displays the expansion of the solid angle on emerging from the crystal. For a narrow laser beam, the source volume $V^S = A\ell_S$ from which scattered light is accepted is limited by the input beam area A and a length ℓ_S determined by the field stop.

The first factor describes the ratio of the solid angles in ray vector space and wavevector space and is independent of the existence of a crystal surface. The second factor, describing the change in wavevector solid angles, is completely determined by Snell's law.

Since $d\Omega_{in}^{k}$ is readily computed in terms of the area of a patch dA^{k} of the $\omega(\vec{k})$ surface, see Fig. 2, and $d\Omega_{in}^{r}$ is related to the same area by Gauss' theorema egregium, see Fig. 3, we obtain our previously quoted result[7]

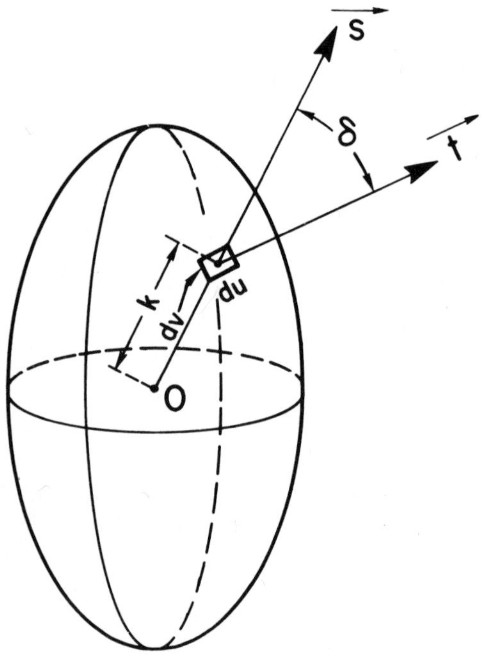

Fig. 2. The solid angle $d\Omega^{k}$ associated with a patch of area dA^{k} = dudv on the surface $\omega(\vec{k}) = \omega$ of free-wave $\vec{k}$ vectors is given by $d\Omega^{k} = dA^{k} \cos\delta/k^{2}$, where $dA^{k}\cos\delta$ is the component of the area dA^{k} normal to $\vec{k}$, since dA^{k} is normal to $\vec{t}$ or $\nabla\omega(\vec{k})$.

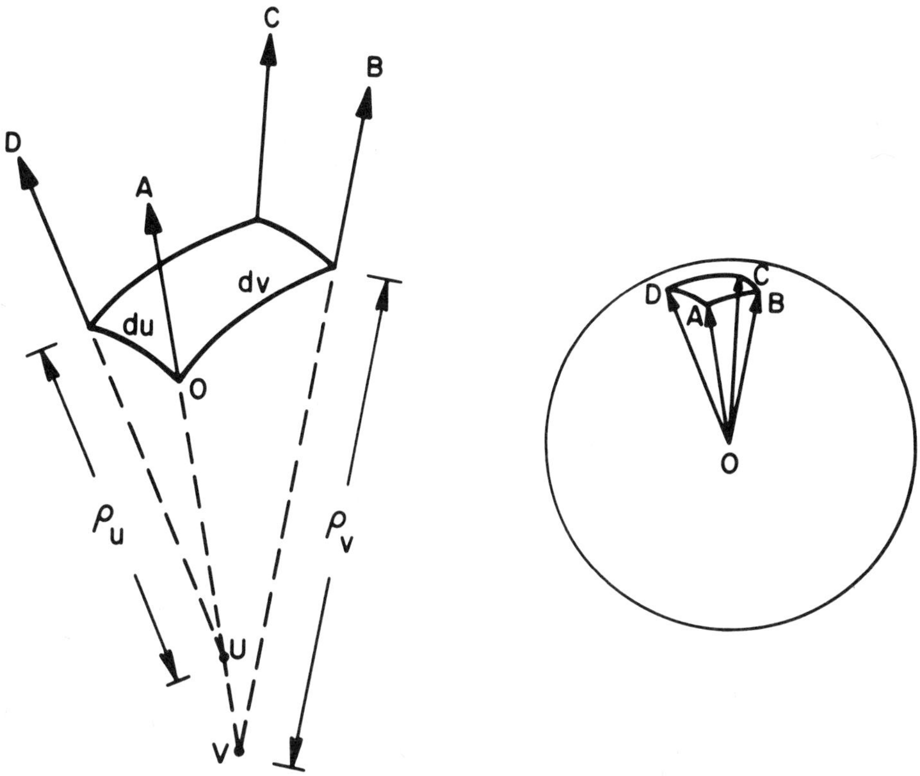

Fig. 3. Gauss's bump theorem relates the area dA = dudv of an element of surface to the area $d\Omega^r$ on the surface of a unit sphere subtended by the unit normals to dudv that have been shifted in a parallel manner until their origins coincide at O, the sphere center. The solid angle $d\Omega^r = d\theta d\Phi$ where $d\theta = du/\rho_u$ is the angle AUD and $d\Phi = dv/\rho_v$ is the angle AVB where ρ_u and ρ_v are the principal radii of curvature. Thus $d\Omega^r = KdA^k$ where $K = (\rho_u\rho_v)^{-1}$ is the Gaussian curvature and dA^k = dudv is the area of the patch on the $\omega(\vec{k})$ surface.

$$d\Omega^r_{in}/d\Omega^k_{in} = K^\Phi(k^\Phi)^2/\cos\delta^\Phi . \tag{4.2}$$

Across the surface of a crystal (nominally in the 3 or z direction) Snell's law guarantees the continuity of the transverse components of the propagation vector and, hence, of dk_1dk_2. Since dk_1dk_2 is simply related to the patch area dA^k, see Fig. 4, and hence to the solid angle $d\Omega^k$, see Fig. 2, Snell's law leads to the relation[7]

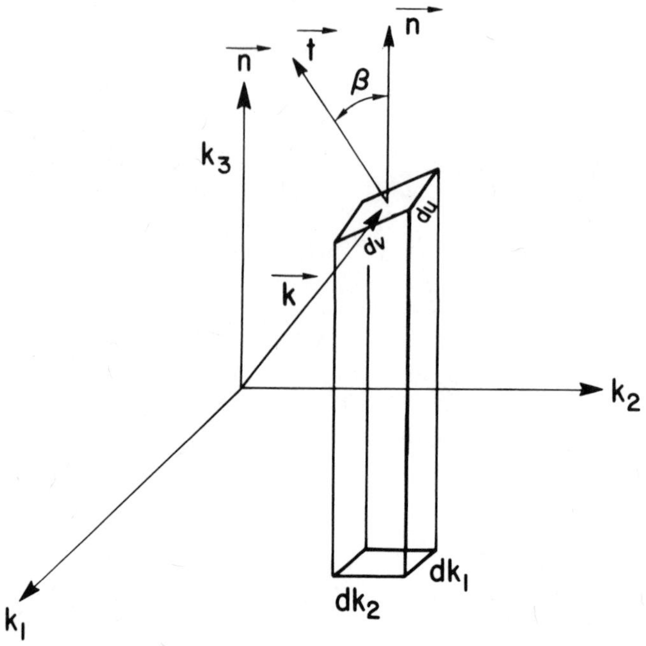

Fig. 4. The patch $dA^k = dudv$ on the surface $\omega(\vec{k}) = \omega$ of Fig. 3 is plotted in a k_1, k_2, k_3 coordinate system in which the three-direction is along the direction $\vec{n}$ of a unit normal to the surface of the crystal. The projection of dA^k onto the 1-2 plane yields $dk_1\, dk_2 = dA^k \cos\beta$ because β is the angle between the normal $\vec{t}$ to the patch (the ray direction) and the normal $\vec{n}$ to the surface.

$$\frac{d\Omega^k_{in}}{d\Omega_{out}} = \frac{\cos\delta^{\Phi}\cos\alpha}{(n^{\Phi})^2\cos\beta} \tag{4.3}$$

where β is the angle of arrival of the ray inside the crystal to the surface normal, δ^{Φ} is the angle between ray and wavevectors, as before, and α is the angle between the departure ray outside the crystal and the surface normal.

The product of Eqs. (4.2) and (4.3) yields the desired solid angle expansion

$$d\Omega_{in}^{r}/d\Omega_{out} = (\omega/c)^{2} K^{\Phi} \cos\alpha/\cos\beta \,. \tag{4.4}$$

A slight rearrangement of this equation suggests that

$$d\Omega^{r} \cos\beta/K \tag{4.5}$$

is an invariant for the passage of a beam from one material to another, a result we have recently proved quite generally.[15] Equation (4.4) is a special case of this invariance in which the second medium is a vacuum with $K_{vac} = (c/\omega)^2$.

V. SOURCE VOLUME DEMAGNIFICATION

When the laser beam, scattered ray, surface normal, and departure ray outside the crystal are all in one plane, it is possible to derive the length ℓ_S along the laser beam in the crystal from which radiation is admitted by a detector field stop of length ℓ_D. As seen from Fig. 5, these lengths are related by

$$\frac{\ell_S \sin\theta_S}{\cos\beta} = \frac{\ell}{\cos\beta} = \frac{\ell_D}{\cos\alpha} \tag{5.1}$$

where θ_S is the scattering angle.

When the rays mentioned above are not all coplanar, Eq. (5.1) must be replaced by

$$\frac{\ell_S}{\ell_D} = \frac{N\cos\beta}{\sin\theta_S \cos\alpha} \tag{5.2}$$

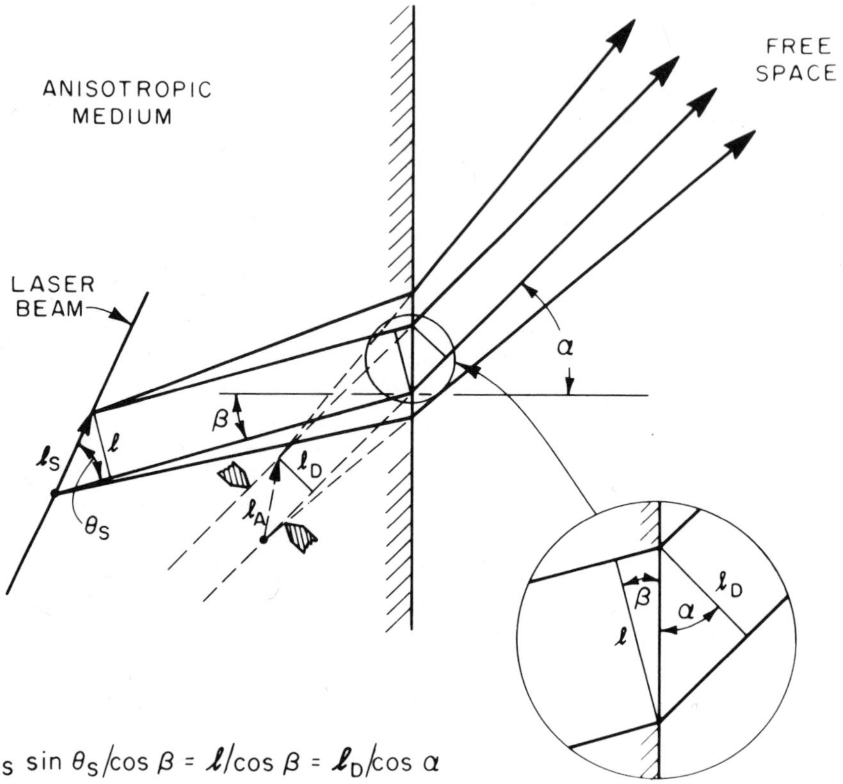

Fig. 5. Demagnification corrections when the arrival ray, departure ray, and surface normal are all in one plane. The detector field stop (see Fig. 1) is represented in image space by the knife edges, which accept a dimension ℓ_D perpendicular to the beam. The portion ℓ_S of the laser beam accepted is determined by the geometrical conditions shown, independent of the orientation of the virtual image ℓ_A: $\ell_S \sin \theta_S / \cos \beta = \ell / \cos \beta = \ell_D / \cos \alpha$.

where N is the noncoplanarity correction,[16]

$$N = 1/(\cos\Phi\cos\Phi' - \cos\beta\,\sin\Phi\,\sin\Phi'). \tag{5.3}$$

Here Φ' is the angle of tilt between the plane defined by the unscattered laser beam and the normal to the input surface and the plane defined by the scattered ray and exit surface normal; Φ is the angle of tilt between the latter plane and the departure plane defined by the normal to the exit surface and the scattered ray after it has left the crystal.

VI. SCATTERED POWER OUTSIDE THE CRYSTAL

To convert Eq. (3.5) to a power scattering formula outside the crystal, we use

$$P_{ins}^{scat} = P_{out}^{scat}/T^{exit} \tag{6.1}$$

$$P_{ins}^{inc} = P_{out}^{inc}\,T^{ent} \tag{6.2}$$

where T^{ent} and T^{exit} are entrance and exit transmission factors. If we note that $V^S/A = \ell_S$, Eqs. (4.4) and (5.2) can be combined to yield

$$\ell_S d\Omega_{in}^{r} = [\frac{\omega}{c}]^2 \frac{N\ell_D d\Omega^D K^\Phi}{\sin\theta_S} \tag{6.3}$$

where we have written $d\Omega^D$ for $d\Omega_{out}$ to remind us that it is the detector solid angle. A detailed derivation of this result has been given elsewhere.[7,14,15,16]

If Eqs. (6.1), (6.2) and (6.3) are combined with Eq. (3.5), we obtain

$$\frac{P_{out}^{scat}}{P_{out}^{inc}} = \frac{R\,T^{ent}\,T^{exit}\,\ell_D d\Omega^D N K^\Phi}{\sin\theta_S}[\frac{\omega}{c}]^2 . \tag{6.4}$$

If Eq. (3.6) is used for the scattering efficiency, we obtain

$$\frac{P_{out}^{scat}}{P_{out}^{inc}} = [\frac{\omega}{c}]^4 \frac{N\ell_D \, d\Omega^D}{\sin\theta_S} \frac{T^{ent} T^{exit} J}{8\pi^2 n^\Phi n^\theta \cos\delta^\Phi \cos\delta^\theta} . \tag{6.5}$$

Equation (6.5) incorporates all the geometric optics of the crystal surface. The mechanism of the scattering is contained in J, whose evaluation will be discussed in the next section.

VII. APPLICATION TO LIGHT SCATTERING

To apply our final scattering equation, Eq. (6.5), to Brillouin, Raman, or some other scattering mechanism, it is necessary to evaluate J, of Eq. (3.7), which involves the nonlinear optical properties of the scattering medium. Since the volume V^S is large compared to all relevant wavelengths, it is permissible to take the limit as V^S approaches infinity. The Weiner-Khinchin theorem[17] applied to spatial variables rather than the time then permits Eq. (3.7) to be rewritten as an autocorrelation,

$$J = \frac{e_i^\Phi e_a^\Phi \int d\vec{r} \exp(-i\vec{k}^\Phi \cdot \vec{r}) \langle P_i^{NL}(0)^* P_a^{NL}(\vec{r}) \rangle}{2\epsilon_0^2 |\vec{E}^\theta(\vec{r})|^2} \tag{7.1}$$

where the limits are infinite.

With the understanding

$$\vec{E}^\theta(\vec{r},t) = \tfrac{1}{2}[\vec{E}^\theta(\vec{r})e^{-i\omega_L t} + \vec{E}^\theta(\vec{r})^* e^{-i\omega_L t}], \tag{7.2}$$

$$\vec{P}^{NL}(\vec{r},t) = \tfrac{1}{2}[\vec{P}^{NL}(\vec{r})e^{-i\omega_B t} + \vec{P}^{NL}(\vec{r})^* e^{-i\omega_B t}], \tag{7.3}$$

where ω_L is the input or laser frequency and ω_B is the Brillouin (or Raman) scattered frequency, we can write

$$P_a^{NL}(\vec{r}, t) = \epsilon_o \chi_{ab}(\vec{r}, t) E_b^{\theta}(\vec{r}, t) \tag{7.4}$$

where the "susceptibility" χ_{ab} is space and time dependent because it is induced by the phonon field in Brillouin or Raman scattering. Equation (7.1) can then be simplified to

$$J = \tfrac{1}{2} \int \exp[-i(\vec{k}^{\theta} - \vec{k}^{\theta}) \cdot \vec{r}] \langle N^{\Phi\theta}(0,0) N^{\Phi\theta}(\vec{r}, 0) \rangle d\vec{r} \tag{7.5}$$

where

$$N^{\Phi\theta}(\vec{r}, t) = e_a^{\Phi} \chi_{ab}(\vec{r}, t) e_b^{\theta} \tag{7.6}$$

and we have used the plane-wave character $\exp(i\vec{k}^{\theta} \cdot \vec{r})$ of the unscattered wave $\vec{E}^{\theta}(\vec{r})$.

If one is concerned with line shape, Eq. (7.5) can be decomposed by

$$J = \int_0^{\infty} J(\omega)\, d\omega/2\pi \tag{7.7}$$

where

$$J(\omega) = \tfrac{1}{2} \int\int \exp[i(\omega - \omega_L)t] \exp[-i(\vec{k}^{\Phi} - \vec{k}^{\theta}) \cdot \vec{r}] \langle N^{\Phi\theta}(0,0)\, N^{\Phi\theta}(\vec{r}, t) \rangle d\vec{r}\, dt\,, \tag{7.8}$$

and only the positive frequency components of $\exp(-i\omega_L t) N^{\Phi\theta}(\vec{r}, t)$ are included so that $J(\omega)$ vanishes for $\omega < 0$.

Brillouin Scattering

The first application of our modified Green's function, Eq. (2.7), with surface corrections was made in connection with a detailed study of Brillouin scattering in calcite. Because the formulas in this paper give a clear factorization of our Brillouin scattering formula[1] into intrinsic and geometric components, we indicate here the evaluation of J based on Eq. (7.5). To be consistent with our definition[1] of the photoelastic susceptibility as a relation between the positive frequency components of $\vec{P}^{NL}$, $\vec{E}^{\theta}$ and $u_{c,d}$ (the displacement gradient), we must write

$$\chi_{ab} = 2\chi_{abcd}\, u_{c,d} \, . \tag{7.9}$$

The averages, $\langle u_{i,j}(0,0) u_{c,d}(\vec{r},t)\rangle$, can be evaluated by using the expansion of u_c in terms of normal coordinates,[18]

$$\vec{u}(\vec{r},t) = \sum_{\vec{q}} \left[\frac{\hbar}{2\rho\Omega\omega(\vec{q})}\right]^{\frac{1}{2}} \vec{b}\left[e^{i\vec{q}\cdot\vec{r}} a(\vec{q},t) + e^{-i\vec{q}\cdot\vec{r}} a^{\dagger}(\vec{q},t)\right], \tag{7.10}$$

where $\vec{q} \equiv \vec{k}^A$ is the acoustic phonon propagation vector, ρ is the crystal density and Ω is its volume so that $\rho\Omega = MN$ = mass per unit cell × number of cells. The unit displacement vector $\vec{b}$ is characteristic of the type of mode (e.g., transverse), and the sum over $\vec{q}$ also implies a sum over types of modes. The mode amplitudes in the quantum mechanical case obey

$$\langle a^{\dagger}(\vec{q},t)\, a(\vec{q}\,',t)\rangle = \bar{n}\,\delta(\vec{q},\vec{q}\,') \, , \tag{7.11}$$

$$\langle a(\vec{q},t)\, a^{\dagger}(\vec{q}\,',t)\rangle = (\bar{n}+1)\delta(\vec{q},\vec{q}\,') \tag{7.12}$$

where $\bar{n}$ is the actual phonon excitation number that reduces to

$$\bar{n} = \frac{1}{\exp[\hbar\omega(\vec{q})/kT] - 1} \tag{7.13}$$

in the thermal equilibrium case. Average of the type, Eq. (7.12), contribute to Stokes scattering, whereas those of type, Eq. (7.11), contribute to anti-Stokes scattering. Because we usually have $\hbar\omega << kT$, both Stokes and anti-Stokes scattering have an intensity proportional $\bar{n} \approx kT/\hbar\omega$. If we combine Eqs. (7.5) and (7.9)-(7.13), we obtain

$$J = \frac{kT}{\rho v_A^2} G, \quad G \equiv |e_i^{\Phi} e_j^{\theta} \chi_{ijkl} b_k a_l|^2 \tag{7.14}$$

where $a_l \equiv q_l / |\vec{q}| = k_l^A / |\vec{k}^A|$ is a unit vector in the direction of phonon propagation and

$$v_A \equiv \omega(\vec{k}^A)/k^A \tag{7.15}$$

is the sound velocity.

Equation (6.5) for the Stokes (or anti-Stokes) power scattering by a single type of acoustic mode can be rewritten as

$$\frac{P_{out}^{scat}}{P_{out}^{inc}} = [\frac{\omega}{c}]^4 \frac{N\ell_D \, d\Omega^D}{\sin\theta_S} \frac{T^{ent} \, T^{exit} kTG}{8\pi^2 n^{\Phi} n^{\theta} \cos\delta^{\Phi} \cos\delta^{\theta} \rho v_A^2}, \tag{7.16}$$

where G of Eq. (7.14) can also be expressed in terms of the Pockels tensor, p_{ijkl}, by means of

$$G = (\tfrac{1}{4})(n^{\Phi} n^{\theta})^4 (\cos\delta^{\Phi} \cos\delta^{\theta})^2 F, \tag{7.17}$$

$$F \equiv |d_i^{\Phi} d_j^{\theta} p_{ijkl} b_k a_l|^2, \tag{7.18}$$

Here $\hat{d}^{\theta}$ and $\hat{d}^{\Phi}$ are unit electric displacement vectors of the input and scattered beams, and ω is the Brillouin scattered frequency ω^B. Equation (7.16) was used in the analysis of our experimental results[1, 19] except for the noncoplanarity factor, N, which was not needed then since all experiments were conducted in a symmetry plane.

Raman Scattering

A detailed analysis of Raman scattering by polaritons using the fluctuation-dissipation approach of Barker and Loudon[6] has been given.[14] To evaluate Eq. (7.8) for $J(\omega)$, needed to obtain the line shape in Raman scattering, we express the nonlinear susceptibility defined in Eq. (7.4) in the usual way,

$$\chi_{ij}(\vec{r},t) = 2\sum_{\mu} A^{\mu}_{ij}\, w^{\mu}(\vec{r},t) + 2B_{ijk}\, E_k(\vec{r},t), \tag{7.19}$$

as an ionic contribution associated with the displacement w^{μ} of mode μ plus an electronic contribution proportional to the electric field $E(\vec{r},t)$. The factors of 2 assure conformity with definitions in our previous work.[14] Because field and displacement are correlated in polariton motion, it was simpler to evaluate the correlations with the help of the fluctuation-dissipation theorem.[14,6] If T_{BA} represents the response of $\langle B\rangle$ to a unit force at frequency ω applied to A, Eq. (7.8) yields[20]

$$J(\omega+\omega^{L}) = 4\hbar\, n(\omega)\, e_i^{\Phi}\, e_j^{\theta}\, e_a^{\Phi}\, e_b^{\theta}\, \mathrm{Im}[\,J_{ijab}\,] \tag{7.20}$$

where

$$J_{ijab} = \sum_{\mu\nu} (A^{\mu}_{ij})^{*} A^{\nu}_{ab}\, T_{\nu\mu^{\dagger}} + B^{*}_{ijk} B_{abc}\, T_{ck^{\dagger}} + \sum_{\mu} (A^{\mu}_{ij})^{*} B_{abc}\, T_{c\mu^{\dagger}} + \sum_{\nu} B^{*}_{ijk} A^{\nu}_{ab}\, T_{\nu k^{\dagger}} \tag{7.21}$$

$$= \Big[\sum_{\mu} (A^{\mu}_{ij})^{*} q^{\mu}_{k} B^{\mu}(\omega) + B^{*}_{ijk}\Big]\, T_{ck^{\dagger}}\, \Big[\sum_{\nu} A^{\nu}_{ab}\, q^{\nu}_{c}\, B^{\nu}(\omega) + B_{abc}\Big] + \sum_{\mu} (A^{\mu}_{ij})^{*} A^{\mu}_{ab}\, B^{\mu}(\omega) + \sum_{\mu} (A^{\mu}_{ij})^{*} B^{\mu}(\omega).$$

The second form, Eq. (7.22), has used the equations of motion to express all response functions in terms of the field-field response, i.e., the response of the electric field to a unit applied external polarization, as

$$T_{ck^{\dagger}} = E_c / P_k^{ext} = (\overleftrightarrow{\alpha}^{-1})_{ck} / \epsilon_o$$

$$= \sum_{\Phi = 1, 2} \frac{e_c^{\Phi} e_k^{\Phi}}{\epsilon_o \cos^2 \delta^{\Phi} [(ck/\omega)^2 - (n^{\Phi})^2]} - \frac{S_c S_k}{\epsilon_o \vec{S} \cdot \overleftrightarrow{k}(\omega) \cdot \vec{S}} . \tag{7.23}$$

The first two terms in Eq. (7.23) represent the transverse modes while the third term represents the longitudinal mode.

The mode w^{μ} will generally obey an equation of the form[20]

$$m^{\mu} [(\omega^{\mu})^2 - \omega^2 - i\omega \gamma^{\mu}(\omega)] w^{\mu} = q_c^{\mu} E_c \tag{7.24}$$

where m^{μ} is the effective mass, ω^{μ} the effective (angular) frequency, γ^{μ} (ω) the effective frequency dependent damping constant and q_c^{μ} is the c component of the effective charge. The response of w^{μ} to a unit applied force needed in Eq. (7.22), is given by

$$B^{\mu}(\omega) = \{ m^{\mu} [(\omega^{\mu})^2 - \omega^2 - i\omega \gamma^{\mu}(\omega)] \}^{-1} . \tag{7.25}$$

We may combine Eqs. (7.20) - (7.25) to obtain $J(\omega + \omega^L)$. If $J(\omega + \omega^L)$ is inserted in place of J in Eq. (6.5), the result, in view of Eq. (7.7), is the ratio of the Raman scattered power per unit frequency interval to the incident power, both computed outside the crystal.

REFERENCES

1. D. F. Nelson, P. D. Lazay and M. Lax, Phys. Rev. B6, 3109 (1972).
2. G. Hauret, J. P. Chapelle and L. Taurel, Phys. Stat. Sol. 11, 255 (1972).
3. See, for example, O. Keller, Phys. Rev. B11, 5095 (1975); see also ref. 1 for a discussion and comparison of previous theories.
4. M. Lax and D. F. Nelson, Phys. Rev. B4, 3694 (1971).

5. M. Lax and D. F. Nelson, "Crystal Electrodynamics" in Atomic Structure and Properties of Solids, E. Burstein, editor (Academic Press, New York, 1972) pp. 48-118.
6. See, for example, A. S. Barker and R. Loudon, Rev. Mod. Phys. 44, 18 (1972). This paper also provides an excellent review of Raman scattering calculations.
7. M. Lax and D. F. Nelson, in Coherence and Quantum Optics, L. Mandel and E. Wolf, editors (Plenum Press, New York, 1973) p. 415.
8. H. Kogelnik, J. of Research, Nat'l Bureau of Standards, Div. of Radio Propagation 64D, 515 (1960).
9. H. Kogelnik and H. Motz, Symposium on Electromagnetic Theory and Antennas, Copenhagen, 1962 (Pergamon, New York, 1963) p. 477.
10. M. J. Lighthill, Phil. Trans. Roy. Soc. (London) A252, 397 (1960).
11. G. P. Motulevich, Trudy Fiz. Inst. P. N. Lebedev 5, 9-62 (1950); see I. L. Fabelinskii, Molecular Scattering of Light, (Plenum Press, New York, 1968) pp 139.
12. V. L. Ginzburg Zh. Eksp. i Teor. Fiz 10, 601 (1940).
13. M. Lax and D. F. Nelson, Comment on Keller's Theory of Brillouin Scattering, Phys. Rev. B, submitted.
14. M. Lax and D. F. Nelson, in Polaritons, edited by E. Burstein and F. De Martini (Pergamon, New York, 1974) pp 27-40.
15. M. Lax and D. F. Nelson, J. Opt. Soc. of Amer. 65, 68 (1975).
16. M. Lax and D. F. Nelson, "Imaging Through a Surface of An Anisotropic Medium with Application to Light Scattering", J. Opt. Soc. of Amer., submitted.
17. See for example, M. Lax, "Fluctuations and Coherence Phenomena in Classical and Quantum Physics" in Statistical Physics, Phase Transitions and Superfluidity, edited by M. Chretien, E. P. Gross and S. Deser, (Gordon and Breach Science Publishers, New York, 1968), Vol. II, pp 270-478, especially Eqs. (4A3) and (4A4).
18. M. Lax, Symmetry Principles in Solid State and Molecular Physics, (John Wiley, New York, 1974), Eqs. (11.2.13), (11.2.29).
19. D. F. Nelson and P. D. Lazay, Phys. Rev. Letters 25, 1187 (1970); 25, 1638 (1970).
20. M. Lax, J. Phys. Chem. Solids, 25, 487 (1964).

Section VI

Resonance Scattering and Scattering from Local Excitations

ON THE THEORY OF RESONANT SECONDARY RADIATION: SCATTERING, LUMINESCENCE, AND HOT LUMINESCENCE

K. K. Rebane, I. Y. Tehver and V. V. Hizhnyakov

Institute of Physics
Estonian Academy of Sciences
Tartu, USSR

The aim of the present report is to examine the changes in the theory of light scattering when an excitation frequency is in resonance with an absorption band. To be more precise, it is not only the scattering that is in question, but the whole flux of radiation from the excited sample, i.e., resonant secondary radiation (RSR). Our examination will concern rapidly relaxing impurity centers in crystals, i.e., the centers whose vibrational relaxation time Γ_ℓ^{-1} is much smaller than the radiative decay time $\gamma^{-1}(\Gamma_\ell >> \gamma)$. Usually the impurity centers are like this. The main results, referred to below, were obtained in [1-4] (see also reviews [5-7]).

First of all let us make some general remarks.

It has been established long ago that when approaching resonance, the characteristics of secondary radiation undergo qualitative changes. In off-resonance, secondary radiation represents the Rayleigh and Raman scattering, while in resonance we have luminescence, the radiation whose spectral and time characteristics make it essentially different from scattering.

However, luminescence does not represent the whole RSR, although its total intensity may constitute up to 99.999% of the whole RSR intensity. This is understandable just from the fact that the elastic Rayleigh scattering is always present both in off-resonance as well as in resonance. Its cross section can be estimated using the optical theorem

$$I_o \sim |\mathrm{Im}\Phi(\omega_o)|^2 = \frac{\omega_o^2}{16\pi^2 C^2}\, \kappa^2(\omega_o) \sim \frac{\gamma}{\Gamma_t}\, \kappa(\omega_o)\,. \qquad (1)$$

Here $\Phi(\omega_o)$ is the amplitude of the elastic forward scattering, $\kappa(\omega_o) \sim \gamma C^2/\Gamma_t \omega_o^2$ is the absorption cross section, Γ_t is the width of the resonant band (line) of the absorption, ω_o is the excitation frequency. Hence, in resonance, the Rayleigh scattering makes up as much as γ/Γ_t of the whole RSR. As it follows from physical considerations, besides the elastic Rayleigh scattering there exists the inelastic Raman scattering (in resonance as well as in off-resonance). One can show [1] that its cross section also makes up as much as γ/Γ_t of the total cross section of RSR.

In RSR, besides the luminescence and scattering mentioned above, there is one more kind of radiation, that is, the luminescence from partially relaxed vibrational states. The existence of such radiation and its difference from the other components of RSR can best be seen from the example of the centers with a local mode, whose frequency changes considerably with the electronic transition (Fig. 1). If, at excitation, the mode occurs at a high level of the number r_o', then in the relaxation process each level $r' \leq r_o'$ is occupied for some time. As a result, in the RSR spectrum the lines caused by the transitions from all these levels, will be present. The spectral position of the latter differs from that of the Raman scattering line as well as of the low temperature luminescence lines. This luminescence from partially relaxed vibrational state of local modes was experimentally detected for the impurity molecule NO_2^- in alkali halides [8, 9] and it was called hot luminescence (HL).

Thus, RSR may consist of different components: luminescence, scattering and hot luminescence. In addition, the whole picture of RSR may be complicated by the interference effects of its components [3, 4].

The following questions arise:

a) What quantum electrodynamic formulas ought to serve as a basis for the theory which would describe all the components of RSR?

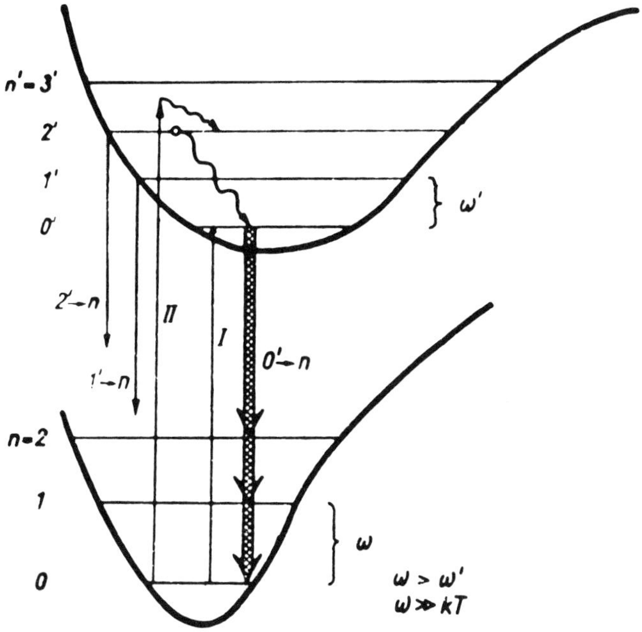

Fig. 1. Local mode level diagram in the ground and excited electronic states. At the excitation in the phonon wing of the quasiline 0 → 2′, besides the series of the conventional luminescence lines (the transitions 0′ → 1), there appear the series of weaker HL lines, corresponding to the transitions 1′ → n and 2′ → n. The spectral position of these lines differs from that of the Raman lines.

b) How should the experimental data on RSR be interpreted and when (and in which way) should the components of RSR be distinguished? Here is where we face the well-known classical problem of the classification of RSR.

Below, based on RSR theory [1-4], the answers to these questions will be briefly reviewed and some concrete results referred to.

The processes from which the components of secondary radiation both in resonance and off-resonance originate are evidently the two-photon ones. Hence, they can be described by not lower than a second order formula of perturbation theory for the interaction of light with matter. The corresponding probability per unit time is of the well-known form

$$I = 2\pi \sum_f \left| \sum_m \frac{\langle i|H'|m\rangle\langle m|H'|f\rangle}{\epsilon_i - \epsilon_m + i\gamma^m} \right|^2 \delta(\epsilon_i - \epsilon_f) =$$

$$= 2\pi \sum_f \sum_m \sum_{m'} p_{if}^{(m)} (p_{if}^{(m')})^* \delta(\epsilon_i - \epsilon_f). \tag{2}$$

Here H' is the interaction operator of the system with the electromagnetic field, $|i>$, $|f>$ and $|m>$ are the initial, final and intermediate states of the matter and the electromagnetic field, ϵ_i, ϵ_f and ϵ_m the corresponding energies,

$$p_{if}^{(m)} = \langle i|H'|m\rangle\langle m|H'|f\rangle \, (\epsilon_i - \epsilon_m + i\gamma_m)^{-1}$$

is the contribution of the m-th intermediate level to the polarizability of the transition $|i\rangle \rightarrow |f\rangle$. The constant of the radiative decay γ^m of the intermediate state is essential only in resonance.

It is often considered that formula (1) describes only the scattering of light and hot luminescence. However, it is not so: the radiation determined by this formula contains also luminescence, as was shown in [1]. The question of the separation of the luminescence will be discussed below. Note that in formula (2) luminescence corresponds to the diagonal part of the twofold sum over the intermediate states

$$\sum_m \sum_{m'} (\sim \delta_{mm'})$$

on the condition that the spectrum is quasicontinuous. In other words, luminescence corresponds to the Van Hove diagonal singularity.

Form (2) of the formula for two-photon processes is not convenient. Another, invariant form of the formula, representing the δ-function and the two resolvents as integrals proves to be more convenient. In this case RSR is described by the following threefold integral:

$$I(\omega_0, \Omega) = B\int_0^\infty d\tau \int_{-\infty}^{\infty} dt \int_{-\tau}^{\tau} ds\, e^{i\omega_0(t-2s)-i\Omega t-\gamma\tau} A(\tau, t, s), \tag{3}$$

where Ω is the frequency of the emitted light, B = const, and the three-time correlation function is determined by

$$A(\tau, t, s) = \langle H'(t-2s)H'(\tau+t-s)H'(\tau-s)H'\rangle, \tag{4}$$

In the adiabatic approximation for the electronic and vibrational states

$$A(\tau, t, s) = \sum_{\alpha\alpha'} \sum_{\beta\beta'} i_\alpha i_{\alpha'} n_\beta n_{\beta'} A_{\alpha\beta\beta'\alpha'} (\tau, t, s), \tag{5}$$

$$\begin{aligned} A_{\alpha\beta\beta'\alpha'}(\tau, t, s) = \langle M_\alpha e^{i(\tau+s)H_1} M_\beta e^{-tH_0} M_{\beta'} e^{i(s-\tau)H_1} \\ \times\ M_{\alpha'} e^{i(t-2s)H_0}\rangle \end{aligned} \tag{6}$$

where $\vec{i}$ and $\vec{r}$ are the unit vectors of the polarization of the exciting and emitted light, respectively, H_0 and H_1 are the vibrational Hamiltonians of the ground and intermediate electronic states, M_α is the electronic matrix element, and

$$\langle \ldots \rangle_i \equiv S_p(\ldots e^{-H_i/kT}) / S_p(e^{-H_i/kT}) .$$

In case the electronic levels are degenerate, or there are some other closely located levels up and down, H_0, H_1 and M_α will be matrices (see [7, 10, 11]).

Besides calculational convenience, formula (3) has the following notable advantages [12]. First, the same three-time correlation function determines not only the independent, but also the time dependent, RSR spectrum. Second, what is especially important is that the variables τ, t, s have the meaning of relaxation times. In particular, τ is the lifetime of the system in the intermediate (excited) electronic state and it describes the change in time of the diagonal elements of the vibrational density matrix in the intermediate state; i.e., the correlator $A(\tau, t, s)$ regarded as a function of τ describes the longitudinal (energetic) relaxation of this state. The variables t and s describe the time dependence of the nondiagonal matrix elements in the intermediate (excited) and final (ground) states; i.e., the correlator $A(\tau, t, s)$ as a function of t and s describes the transverse phase relaxation of the system in the intermediate and final electronic states.

Based on the physical meaning of τ, t and s as relaxation times, it is easy to solve the problem of the classification of RSR. In this case it is essential to distinguish the longitudinal and transverse relaxation. The components of RSR can be determined as follows:

1) The conventional impurity luminescence is the radiation from the thermal equilibrium vibrational state, i.e., from the state where transverse as well as longitudinal vibrational relaxations are finished.

2) HL is the radiation from a state where the phase memory of the excitation has failed already; the vibrational distribution, however, has not yet reached equilibrium. In other words, HL

occurs after the phase relaxation but before the energetic vibrational relaxation.

3) Resonance scattering is the radiation from a state which has a phase memory of excitation, i.e., before the energetic as well as phase vibrational relaxations in the intermediate state are finished.

It follows from these determinations that there corresponds to luminescence a region of large $\tau \sim \gamma^{-1}$ (of the order of radiative lifetime; it is usually far longer than the characteristic time of the longitudinal and transverse relaxation: $\gamma^{-1} >> \Gamma_{\ell}^{-1} \gtrsim \Gamma_t^{-1}$). To hot luminescence there corresponds the region of medium $t = \Gamma_{\ell}^{-1}$, and to scattering the region of small $\tau \lesssim \Gamma_t^{-1}$. In this case the actual values of t and s for conventional and hot luminescence are $|t| \sim |s| \sim \Gamma_t^{-1}$, whereas for scattering they may be greater, since the violation of coherence for the Raman scattering is due only to the dispersion and damping of the generated phonons; in the Rayleigh scattering the coherence is not violated. Therefore, to the Rayleigh scattering there corresponds $t \to \infty$ (or $s \to \infty$).

The examination of concrete models lends support to the correctness of our classification. Particularly, restricting oneself to the case of nondegenerate ground and excited electronic states, then at $\tau >> \Gamma_{\ell}^{-1}$ one can use the asymptotics

$$A(\infty, t, s) = \langle e^{-itH_0} M_\alpha e^{itH_1} M_{\alpha'} \rangle_0 \langle e^{i(t-2s)H_0} M_\beta e^{-i(t-2s)H_1} M_{\beta'} \rangle_1 . \tag{7}$$

A substitution of expression (7) into (3) gives

$$I_o(\omega_0, \Omega) = I_L(\Omega) \kappa(\omega_0), \tag{8}$$

the luminescence spectrum $I_L(\Omega)$ (normalized to unity) multiplied by the absorption cross section of an excitation photon. So, the region of large $\tau >> \Gamma_{\ell}^{-1}$ actually corresponds to luminescence.

It is also easy to find the asymptotic behavior of $A(\tau, t, s)$ for $t \to \infty$,

$$A(\tau, \infty, s) = \langle e^{-i(\tau+s)H_0} M_\alpha e^{i(\tau+s)H_1} M_\beta \rangle_0 \langle e^{i(s-\tau)H_0} M_{\alpha'} e^{i(\tau-s)H_1} M_{\beta'} \rangle_0 . \tag{9}$$

The integration of (9) gives the Rayleigh line, the intensity of which is in accordance with the optical theorem.

Note the following:

1. Intermediate time regions describe interference effects. These effects are especially essential in the cases when longitudinal and transverse times are equal. In this case distinguishing HL and Raman scattering is not always possible [3].

2. Above, a monochromatic excitation was assumed. In more general cases, the correlation function of the substance should be multiplied by the correlation function of the excitation field [12]. Therefore, distinguishing the components of RSR depends, generally speaking, on the excitation conditions. In particular, at "white" (instantaneous) excitation RSR contains only ordinary and hot luminescence.

Let us dwell briefly on some concrete results. Consider a model which takes into account only the shifts of the equilibrium positions of the nuclei at the electronic transition. For this model the correlation function $A(\tau, t, s)$ can be calculated exactly [1]. As a result, the Raman cross section for scattering of arbitrary order is given by the function of one-phonon transitions

$$I(\omega) = (2\pi \langle (V - \langle V \rangle_0)^2 \rangle_0)^{-1} \int dt e^{-i\omega t} \langle (V(t) - V) V \rangle_0$$

(where $V = H_1 - H_0$) and the function

$$\Phi_0(\omega) = i\pi\kappa(\omega) + \int dx \frac{\kappa(x)}{\omega - x} \qquad [12]$$

$$I_m(\omega_0, \Omega) = \frac{\omega_0^2}{16\pi^4 C^2 m!} \int dx_1 \int dx_2 \ldots \int dx_m \, I(x_1) I(x_2) \ldots I(x_m)$$

$$\times \left| \sum_{p=0}^{m} (-1)^p \binom{m}{p} \Phi(\Omega + x_1 + x_2 + \ldots x_m) \right|^2 \delta(\Omega + x_1 + x_2 + \ldots x_m - \omega_0). \qquad (10)$$

These formulae can be considered as a generalization of the optical theorem for the inelastic scattering in this model.

In the case of broad absorption spectra for not very large $|\omega_0 - \Omega|$ in (10), the functions Φ can be expanded into a series and the first terms, different from zero, can be accounted. As a result still simpler formulae are obtained, where the total cross section of scattering of a fixed order is determined only by the absorption spectrum [2] (Fig. 2).

$$I_m = \text{const} \, |\Phi_m(z)|^2 = \frac{\text{const}}{m!} \left| \frac{d^m}{dz^m} \Phi_0(z) \right|^2 \qquad (11)$$

($z = (\omega_0 - \mu_1)\sqrt{2\bar{\mu}_2}$, where the μ_1 and $\bar{\mu}_2$ are first and second moments of absorption spectrum). In this case hot luminescence represents a structureless background of the structural Raman spectrum. The shape of the HL spectrum at $\omega_0 - \Omega >> \bar{\omega}$ ($\bar{\omega}$ is the mean phonon frequency) is very simple [13] (Fig. 3).

$$I_{HL} = \frac{\text{const}}{\sqrt{\omega_0 - \Omega}} . \qquad (12)$$

Fitchen and Buchenauer [14] have compared formulas (11) for the total cross sections of the Raman spectra of different orders

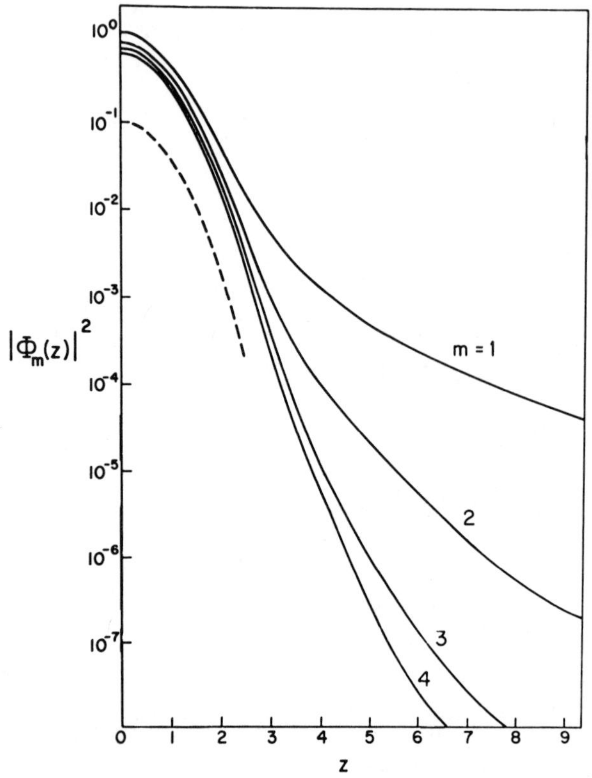

Fig. 2. m-th order Raman cross section as a function of the dimensionless excitation frequency z. Dashed line denotes the absorption spectrum.

with their experimental data on the resonant Raman scattering by F-centers in alkali halides. The theory was found to agree well with the experiment.

In [3, 4] the RSR of the centers with local modes characterized by quasilinear absorption spectra was also calculated. Here the anharmonic decay of the local mode and large frequency change (as compared with the anharmonic decay constant) of the mode at electronic transition had to be taken into account. Here are some of the results:

1) The Raman and hot luminescence lines can be essentially asymmetric due to interference. Moreover, under certain excitation conditions it is impossible to distinguish some Raman lines

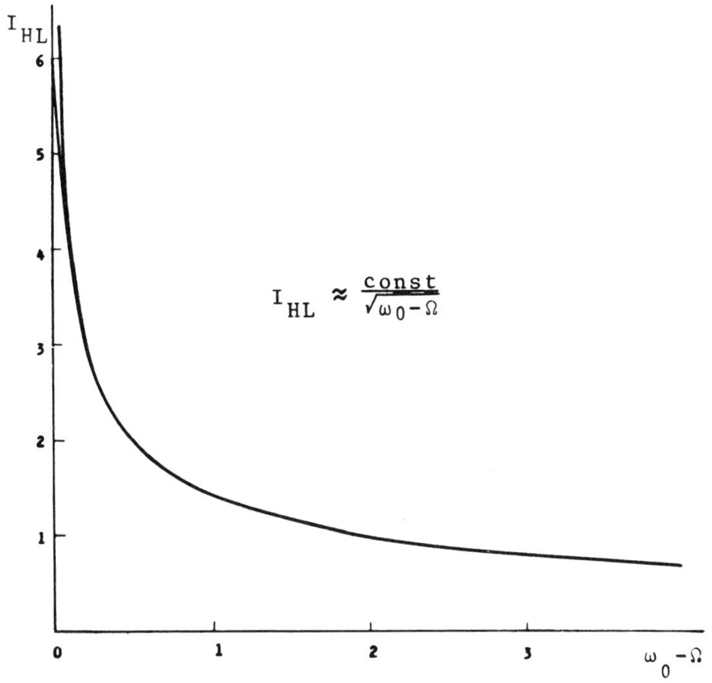

Fig. 3. Calculated HL spectrum in the case of large Stokes losses. Upper curve-semiclassical calculation; lower curve-quantum calculation.

having close frequencies from HL lines; anyhow, there is no sense in doing this.

2) The hot luminescence line width can essentially differ from the widths $\Gamma_{ww} = \Gamma(n + n')$ determined by the Weisskopf-Wigner formula (n and n′ are the numbers of the vibrational levels in ground and excited electronic states, between which the transition occurs, Γ is the constant on anharmonic decay of the local mode). For example, at "white" excitation the shape of the HL lines is determined by

$$I_{nn'}(\omega_0, \Omega) = \frac{1}{\Gamma_n} \sum_{p=0}^{n} \frac{F_p(n, n')\Gamma(n' - n + 2p)}{(\Omega - \Omega_0 + n\omega - n'\omega')^2 + \Gamma^2(n' - n + 2p)^2} \quad (13)$$

where Ω_0 is the frequency of the pure-electronic transition, ω and ω' are the frequencies of the local mode in the ground and excited electronic states, $F_p(n,n') \geq 0$ (for the expression for $F_p(n,n')$ see [4]), i.e., the HL line widths for $F_p(n,n')$ are always smaller than the Weisskopf-Wigner width. It is no wonder that the Weisskopf-Wigner formula can be violated here, because their formula was obtained for a two-level system. In our case, the presence of lower vibrational levels leads to processes with the virtual generation and destruction of phonons. The number of these virtual phonons (in (13) it equals p) can be different; consequently, the resulting line is built up from lines of different widths.

A detailed experimental study of RSR of centers with local vibrations was carried out in [8, 9, 15]. The experiments confirm the theory. Note that their experimental results on hot luminescence enabled them to determine directly the characteristic relaxation times of the excited vibrational levels of the NO_2^- molecule. It is worthwhile to note that these data can be used as time sampling of ultrafast processes on the basis of spectroscopic methods, i.e., without pico-second techniques. The data obtained by Saari on the shape of rotational structure of some hot luminescence lines of NO_2^- measured at 1.9 K are given in Fig. 4.

In conclusion, let us mention a few problems which have been solved within the framework of the RSR theory. A theory of RSR with regard to the degeneracy of electronic levels was developed [10]; the effect of non-totally symmetric vibrations on the spectral and polarization characteristics of RSR was studied. The time characteristics of RSR were studied under different excitation conditions [12]. A theory of RSR with electronic energy transfer between centers was developed [13]. The theory enables one to take into account the hot transfer. Finally note that the results obtained can be used more widely than just for impurity centers. Particularly, they can be used for solutions and some of them also for pure crystals [12].

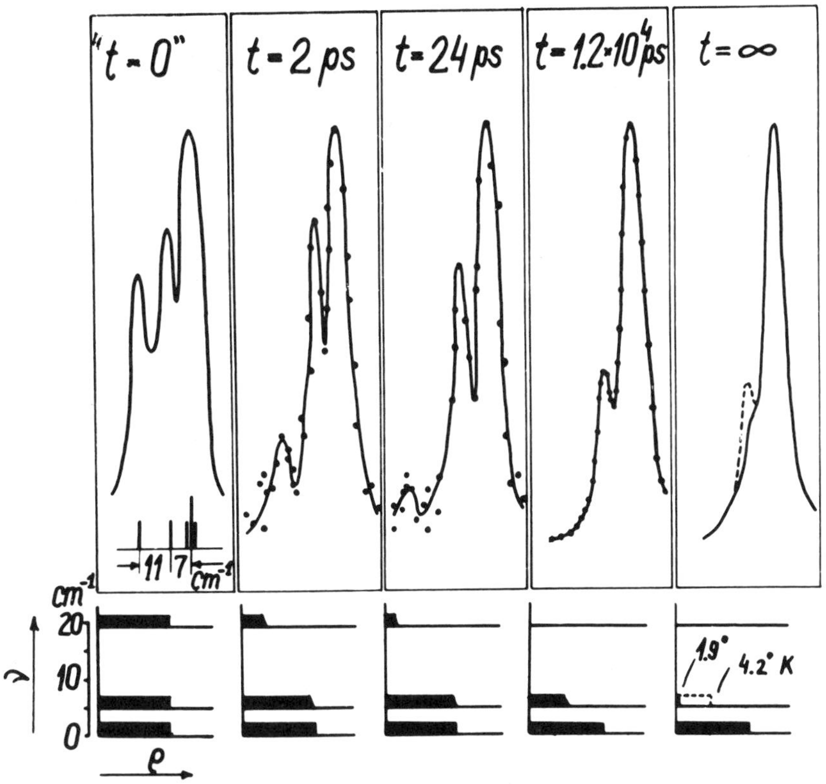

Fig. 4. Rotational structure of the RSR quasilines of NO_2^- in KCl at 1.9 K for different lifetimes of the photon emitting vibrational levels. At the bottom is the energetic scale and the populations of the first three rotational levels. The first and last curves are the calculated ones: "t = 0" corresponds to the equal populations of the rotational levels at the initial moment, "t = ∞" to the thermal equilibrium populations. The other three curves are experimental: "t = 2ps" and "t = 24 ps" correspond to HL lines, and "t = 1.2 x 10^4 ps" - to conventional luminescence lines. The lifetimes of the vibrational levels 2 ps and 24 ps have been found in [15].

REFERENCES

1. I. Tehver, V. Hizhnyakov, Izv. AN ESSR, ser. phys.-mat. i techn. nauk, 15, 9 (1966).
2. V. Hizhnyakov, I. Tehver, phys. stat. sol. 21, 755 (1967).
3. V. Hizhnyakov, K. Rebane, I. Tehver, Proc. Intern. Conf. Light Scattering Spectra of Solids, New York 1968, Springer-Verlag, Inc., 1969 (p. 513).
4. V. Hizhnyakov, I. Tehver, phys. stat. sol. 39, 67 (1970).
5. K. Rebane, V. Hizhnyakov, I. Tehver, Izv. AN SSSR, ser. phys.-mat. i techn. nauk, 16, 207 (1967).
6. K. Rebane "Vtorichnoye svecheniye primesnogo centra kristalla", Konspekt lekcii, prochitannoi v letnei shkole po luminescencii, Baikal, Bukhta Peschanaya, 1969, Rotaprint AN ESSR, Tartu, 1970.
7. K. K. Rebane, V. V. Hizhnyakov "Vtorichnoye svecheniye primesnogo centra - luminescenciya, goryachaya luminescenciya i masseyaniyc", Preprint FAI-28, Tartu, 1973.
8. K. Rebane, P. Saari, Izv. AN ESSR, phys.-mat. 17, 241 (1968).
9. P. Saari, K. Rebane, Solid State Comm. 7, 887 (1969).
10. V. V. Hizhnyakov, I. Yu. Teher, v. sb. "Phyzika primesnykh centrov v kristallakh", Tallinn, 1972 (str. 607).
11. V. Hizhnyakov, I. Tehver, Proc. Second Intern. Conf. on Light Scattering in Solids, Paris 1971, Flammarion Sciences, 1971 (p. 57).
12. V. V. Hizhnyakov "Teoriya rezonansnogo vtorichnogo svecheniya primesnykh centrov kristallov", Avtoreferat dissertacii na soiskaniye uchenoy stepeni doktora phys.-mat, nauk, Tartu 1972.
13. V. V. Hizhnyakov, I. Yu. Tehver "Radiationless Transfer of Electronic Excitation during Vibrational Relaxation", Preprint FI-31, Tartu 1974; I. Yu. Tekver, V. V. Hizhnyakov, Zh. Exper. Teoret. Phys., 69, (1975).
14. D. B. Fitchen, C. J. Buchenauer, in "Physics of Impurity Centres in Crystals", Tallinn 1972 (p. 277).
15. K. K. Rebane, R. A. Avarmaa, L. A. Rebane, P. M. Saari, Proc. Second Intern. Conf. on Light Scattering in Solids, Paris 1971, Flammarion Sciences, 1971 (p. 72).

COMMENT

(by Y. R. Shen)

In my opinion, there is clear physical distinction between resonant Raman scattering (RRS) and hot luminescence (HL) if we believe there is clear distinction between transverse and longitudinal resonant excitations. It can be shown explicitly for a three-level system that RRS is connected to transverse excitation with its decay governed by transverse relaxation, while HL is connected to longitudinal excitation (excess population in the intermediate state) with its decay governed by longitudinal relaxation. RRS is a direct two-photon process while HL is a two-step process. In magnetic resonances and quantum electronics, similar problems exist and are well-known. For example, consider the case of optical mixing of three beams at ω_ℓ, ω_s, and ω_ℓ', in a medium (with $\omega_\ell - \omega_s \approx \omega_0$ where ω_0 is the resonant frequency of an excitation) to generate an antiStokes field at $\omega_a = \omega_{\ell'} + (\omega_\ell - \omega_s)$. The one-step process gives the so-called coherent antiStokes scattering (discussed by Professor S. A. Akhmanov in this symposium) while the two-step process (creation of longitudinal excitation at ω_0 followed by scattering of the probe beam at $\omega_{\ell'}$ off the longitudinal excitation) gives the incoherent spontaneous antiStokes scattering. These two processes can be distinguished by their angular distributions. They have recently been used to measure transverse and longitudinal relaxation times of the material excitations in the picosecond range.

COMMENT

(by V. V. Hizhnyakov)

In connection with discussing the problem of the separation of resonant Raman scattering and hot luminescence, the following should be emphasized: We do not consider that resonant secondary emission can always be divided into luminescence, hot luminescence and scattering. On the contrary, as it follows from our theory, such a separation is impossible in the general case. There exist cases, however (mentioned in the report), where it can be done in good approximation. We have studied theoretically a number of such models. As it was noted in the report, the conclusions of the theory are in agreement with the experiments on hot luminescence made at Tartu.

RESONANT RAMAN SCATTERING NEAR EXCITONIC TRANSITIONS

Y. R. Shen

Department of Physics, University of California, and
Inorganic Materials Research Division,
Lawrence Berkeley Laboratory, Berkeley, CA 94720 USA

We use the cases of Cu_2O and MnF_2 to illustrate the principles and usefulness of the resonant Raman technique. We show that the resonant Raman results can give not only a better understanding of the phonon (magnon) structure but also more details about the excitonic structure and the coupling between excitons (or electrons and holes) and phonons (magnons).

INTRODUCTION

The advent of tunable dye lasers has led the field of Raman spectroscopy into a new era. With the incoming laser frequency variable around particular electronic excitations, Raman scattering in solids can now yield not only better understanding of phonon structure, but also detailed information about electronic structure and electron-phonon interaction for the specific electronic excitations. Over the past three years, resonant Raman scattering (RRS) work has been carried out in a number of solids. In each case, the results are interesting and exciting. We do not have space here to discuss all these interesting cases. Instead, I have decided to choose RRS in Cu_2O as an example to illustrate the RRS technique. Cu_2O is the well-known semiconductor which led to the first junction diodes. It is also the most frequently quoted crystal for exhibiting clear, well-defined series of excitonic transitions. The late Professor E. F. Gross had made great contribution to the understanding of the excitonic properties in Cu_2O. Here, we shall show that by RRS, we can now learn even more about the excitonic properties, such as the exciton mass and exciton lifetime. In addition, we can also obtain a thorough understanding of the phonon spectra of Cu_2O and detailed information about phonon dispersion and exciton-phonon interaction in Cu_2O.[2-4]

Resonant Raman study can of course be extended to other elementary excitations besides phonons. As an example, we shall also discuss briefly here the case of RRS by magnons in MnF_2.[5] In some respects, this case is a close analog of RRS by phonons in Cu_2O.

BACKGROUND INFORMATION ON Cu_2O

The crystal structure of Cu_2O is shown in Fig. 1. It is a cubic lattice with 6 atoms per unit cell and an O_h^4 space group. From group theoretical analysis, we expect to find the following zone-center optical phonons: 6 infrared active ones, $\Gamma_{15}^{-(1)}(3)$ and $\Gamma_{15}^{-(2)}(3)$, 3 Raman active ones, $\Gamma_{25}^{+}(3)$, and 6 silent ones,

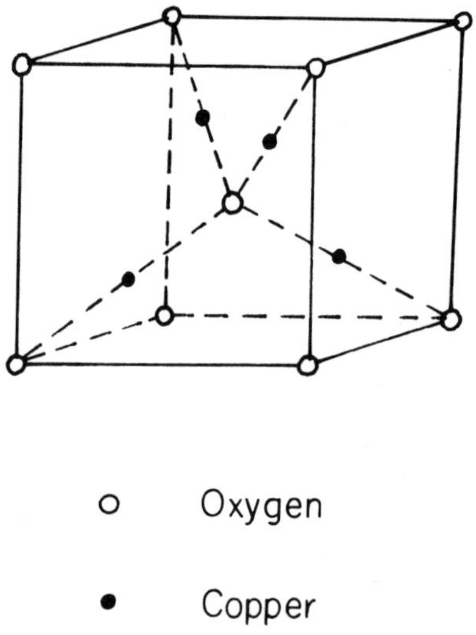

Fig. 1. Crystal structure of Cu_2O.

Table 1. Phonon Energies of Cu_2O Obtained by Various Techniques

Symmetry Assignment of Phonons		Infrared Absorption (Room Temp.)	Optical Absorption	Raman Scattering (Liquid Helium Temp.)	Luminescence
Γ_{25}^-	(Silent)		88	86	87
Γ_{12}^-	(Silent)		110	109	110
$\Gamma_{15}^{-(1)}$(TO)	(Infrared)	146.3		153	152
$\Gamma_{15}^{-(1)}$(LO)	(Infrared)	149.3			152
Γ_2^-	(Silent)			350	350
Γ_{25}^+	(Raman)			515	515
$\Gamma_{15}^{-(2)}$(TO)	(Infrared)	609		640	633
$\Gamma_{15}^{-(2)}$(LO)	(Infrared)	638		660	662

$\Gamma_2^-(1)$, $\Gamma_{12}^-(2)$, and $\Gamma_{25}^-(3)$. The frequencies of these modes have been obtained by various methods, including RRS, as shown in Table 1. All these modes have definite parities because of inversion symmetry of the crystal.

An off-resonant Raman spectrum of Cu_2O is shown in Fig. 2.[6] This spectrum was taken with a Kr ion laser at 6471 Å, which is about 1050 cm^{-1} below the first exciton absorption line or 2200 cm^{-1} below the band gap. Intuitively, one would expect the spectrum to be dominated by the only Raman-active mode $\Gamma_{25}^+(3)$ at 515 cm^{-1}. However, the spectrum of Fig. 2 actually consists of a large number of sharp lines; the expected Raman-active mode, on the other hand, does not even show up. This indicates that the usual off-resonant selection rules are not operative here and multiphonon modes must have dominated the spectrum. As we shall see later, such a Raman spectrum can be understood only with the help of RRS.

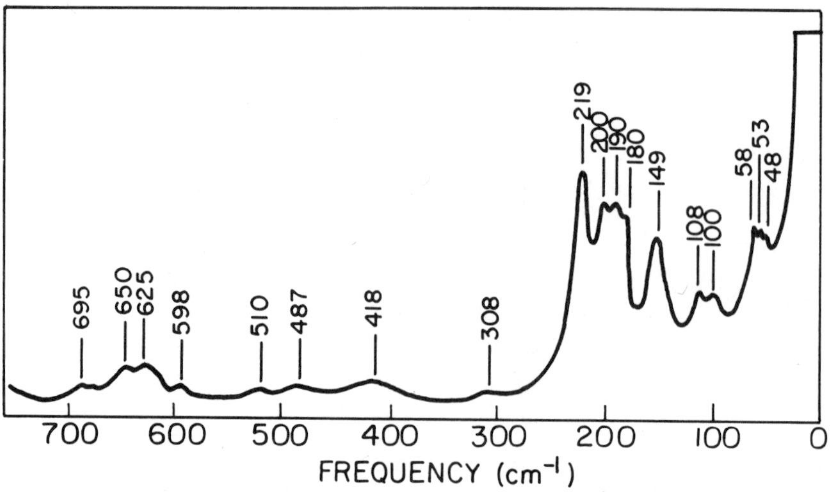

Fig. 2. Off-resonance Raman spectrum of Cu_2O obtained with a Kr ion laser at 6471 Å (after Ref. 6).

In RRS measurements, we tune the incoming laser frequency around optical absorption lines or bands. The absorption spectrum of Cu_2O near the absorption edge is shown in Fig. 3a. It consists of four series of sharp lines superimposed on a continuum. The four series of sharp lines are the well-known exciton series formed by electrons of the two lowest conduction bands and holes of the two top valence bands as shown in Fig. 3b.[7] The first series is known as the yellow exciton series. By the parity selection rule, electric-dipole excitation of the 1s yellow exciton is forbidden; the electric-quadrupole excitation is responsible for the observed 1s line. However, electric-dipole excitation of the forbidden 1s exciton becomes allowed with the assistance of an odd-parity phonon. This phonon-assisted excitation of the 1s yellow exciton leads to the absorption continuum in Fig. 3a. The phonon which contributes most to the continuum is the Γ_{12}^- mode. This explains why the absorption continuum begins at 110 cm^{-1} above the 1s yellow exciton line at $\omega_{1s}(o)$ and has a frequency dependence $[\omega - \omega_{1s}(o) - \omega(\Gamma_{12}^-)]^{\frac{1}{2}}$ near the absorption edge. Careful studies have shown that the absorption edge is actually described by a linear superposition of $[\omega - \omega_{1s}(o) - \omega(\Gamma)]^{\frac{1}{2}}$ functions with several different $\omega(\Gamma)$ resulting from participation of different Γ phonons. Here, we shall discuss only the RRS results obtained with the exciting laser frequency tuned over the yellow exciton lines and the continuum near the absorption edge.

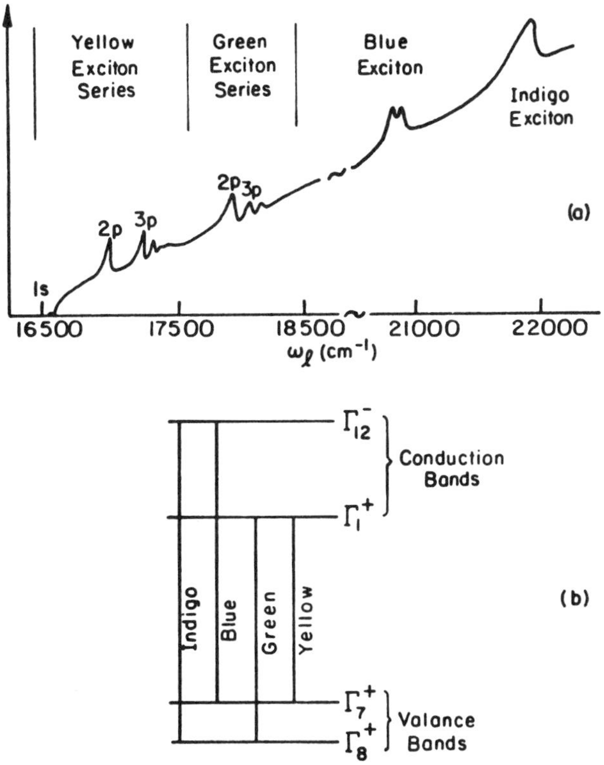

Fig. 3(a) Sketch of absorption spectrum of Cu_2O near the absorption edge. (b) Sketch of the two top valence and two bottom conduction bands near the center of the Brillouin zone.

EXPERIMENTAL ARRANGEMENT AND SIMPLE THEORETICAL BACKGROUND FOR RRS

We used a CW dye laser pumped by an argon ion laser as the exciting source. The dye laser could be tuned from 5250 Å to 6500 Å with a half-width $\lesssim 0.1$ Å and a typical output power > 30 mW. A double monochromator and a photon counting system were used in the detection of scattered radiation. The sample was kept in superfluid He bath at $\sim 2^{o}$K. The Raman spectrum was normalized against the intensity of a Raman line of calcite measured under similar condition.

Strictly speaking, Stokes emission in RRS has contributions from both resonant fluorescence and resonant scattering, and the theoretical expression is in general quite complicated.[8] However, in simple cases where lifetime broadening dominates the linewidths, one finds that the total resonant Stokes emission cross-section σ_R for a Raman mode is given by a simple relation[8, 9]

$$\sigma_R = \text{(Absorption Cross-section)}$$
$$\times \text{(Quantum Efficiency of Stokes Emission)}. \quad (1)$$

In the following section, we shall always use Eq. (1) to discuss our results.

ONE-PHONON RRS NEAR THE 1s YELLOW EXCITON IN Cu_2O

We want to find out how RRS varies when the exciting laser frequency ω_ℓ scans over the absorption spectrum in Fig. 3a. We first consider the case where ω_ℓ is near the 1s yellow exciton line. The experimental results of Compaan and Cummins[3] are presented in Figs. 4 and 5. The Raman spectra in Figs. 4a and

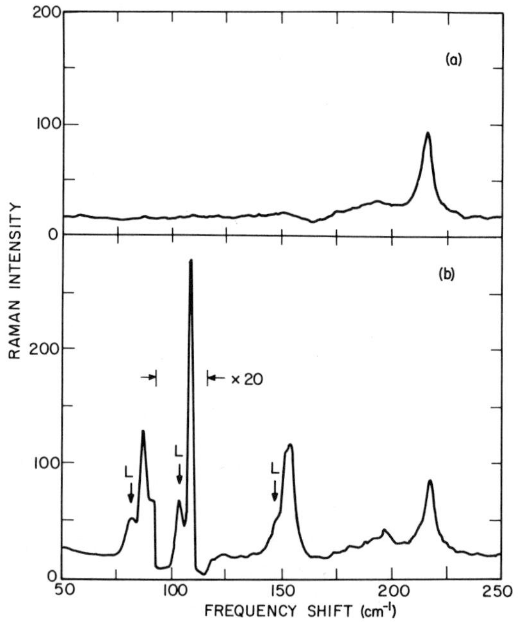

Fig. 4. Raman spectra of Cu_2O at low temperature (a) Laser frequency 10 cm^{-1} above the 1s yellow exciton frequency. (b) Laser frequency in resonance with the 1s yellow exciton. The features labelled L are due to phonon-assisted luminescence (after Ref. 3).

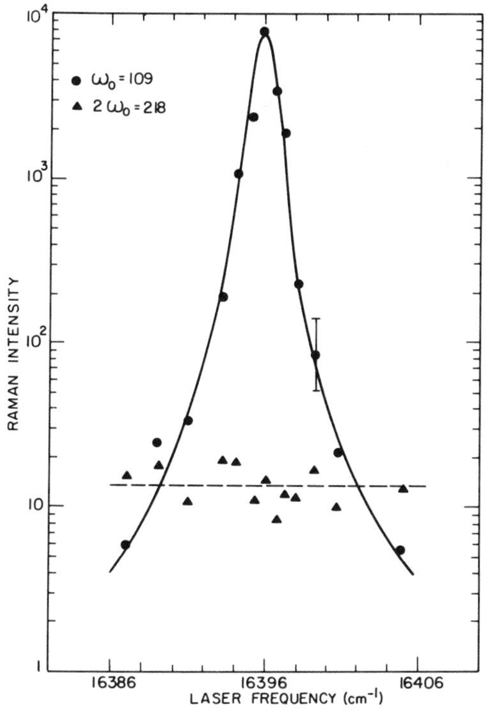

Fig. 5. Resonance enhancement of the 109-cm^{-1} Γ_{12}^{-} Raman line in comparison with the nonresonant behavior of the 218-cm^{-1} $2\Gamma_{12}^{-}$ Raman line (after Ref. 3).

4b were obtained respectively with ω_ℓ on and off resonance with the 1s yellow exciton line. The on-resonance spectrum has the appearance of a set of new lines which correspond to the forbidden one-phonon modes. The RRS enhancement of the Γ_{12}^{-} modes is shown in Fig. 5.

The results can be easily understood. The RRS process can be considered as excitation of the 1s yellow exciton followed by emission of a Stokes photon and a phonon. The exciton is excited by electric-quadrupole transition. Since the exciton has the same parity as the ground state of the system, the phonon-assisted recombination of the exciton must involve an odd-parity, and hence Raman-forbidden, zone-center phonon mode. The results in Fig. 5 agree well with the prediction of Eq. (1). Let $\alpha(\omega_\ell)$ be the absorption cross-section of the electric-quadrupole excitation, $\gamma_{rad}(\Gamma)$ the Γ-phonon-assisted radiative decay rate, and γ_{tot} the total decay rate of the exciton. Then, Eq. (1) becomes

$$\sigma_R(\omega_\ell, \Gamma) = \alpha(\omega_\ell)\gamma_{rad}(\Gamma)/\gamma_{tot} . \tag{2}$$

Since $\gamma_{rad}(\Gamma)/\gamma_{tot}$ is expected to be independent of ω_ℓ near the exciton line, we then find $\sigma_R(\omega_\ell) \propto \alpha(\omega_\ell)$.

Equation (2) shows

$$\sigma_R(\omega_\ell, \Gamma)/\sigma_R(\omega_\ell, \Gamma') = \gamma_{rad}(\Gamma)/\gamma_{rad}(\Gamma').$$

The radiative decay rate $\gamma_{rad}(\Gamma)$ is directly proportional to the square of the matrix element for phonon-assisted exciton recombination. Therefore, the ratio of $\sigma_R(\omega_\ell, \Gamma)$ for different phonons can yield information about the relative coupling strength between the 1s yellow exciton and the phonons.

One-phonon RRS can also occur in the following way. The zone center 1s exciton is first excited by simultaneous absorption of a ω_ℓ photon and emission of an odd-parity phonon; the exciton then decays by emission of a Stokes photon at ω_s via electric-quadrupole transition. It is easy to show, using Eq. (1), that in this case σ_R is proportional to $\alpha(\omega_s)$.

TWO-PHONON RRS NEAR THE ABSORPTION EDGE IN Cu_2O

The 1s exciton with a finite momentum ($K > \omega_\ell n/c$ away from the zone center) can also be excited via phonon-assisted transition. Neglecting the optical wave vector we find that the exciton frequency is given by

$$\begin{aligned} \omega_{1s}(\vec{K}) &= \omega_{1s}(o) + \hbar K^2/2M \\ &= \omega_\ell - \omega(\Gamma(\vec{K})) \end{aligned} \tag{3}$$

where M is the exciton effective mass and $\omega(\Gamma(\vec{K}))$ is the frequency of the Γ phonon with a wave vector $\vec{K}$. By momentum conservation, a $K \neq 0$ exciton cannot decay by emission of only a single photon. It can nevertheless decay by phonon-assisted transition. Phonon-assisted excitation of such an exciton followed by phonon-assisted recombination constitutes a two-phonon RRS process. Energy conservation requires

$$\omega_\ell - \omega_s = \omega(\Gamma(\kappa)) + \omega(\Gamma'(\kappa)) \tag{4}$$

From Eq. (1), the two-phonon RRS cross-section is given by

$$\sigma_R = [\alpha(\Gamma)\gamma_{rad}(\Gamma'(\kappa)) + \alpha(\Gamma')\gamma_{rad}(\Gamma(\kappa))]/\gamma_{tot}(\kappa) \tag{5}$$

where $\alpha(\Gamma)$ is the absorption cross-section for phonon-assisted excitation of the 1s exciton at $\omega_{1s}(\kappa)$ and $\gamma_{rad}(\Gamma(\kappa))$ is the phonon-assisted radiative decay rate of the exciton. It can easily be shown that for dispersionless phonon,

$$\alpha(\Gamma) \propto [\omega_\ell - \omega_{1s}(o) - \omega(\Gamma)]^{\frac{1}{2}} \text{ if } \omega_\ell > \omega_{1s}(o) + \omega(\Gamma)$$

$$= 0 \text{ otherwise.} \tag{6}$$

In Eq. (5), $\gamma_{rad}(\Gamma)$ is directly proportional to the square of the phonon-assisted transition matrix element and should be nearly independent of $\omega_{1s}(\kappa)$. However, since $\kappa \neq 0$, the total exciton decay rate γ_{tot} is no longer a constant but is a strong function of $\omega_{1s}(\kappa)$. Physically, excitons with higher energies have more lower energy states to decay into and hence the decay rate is higher. Taking into account only the relatively stronger exciton decay mechanisms, we can write[2]

$$\gamma_{tot}(\kappa) = A + B[\omega_\ell - \omega_{1s}(o) - \omega(\Gamma_{12}^-)] + C[\omega_\ell - \omega_{1s}(o) - 3\omega(\Gamma_{12}^-)]$$

$$+ \text{ ---} \tag{7}$$

where A, B, C are constants; the first term is due to scattering by defects and impurities, direct radiation, and optical-phonon-assisted radiation; the second term is due to emission of an LA phonon; the third term is due to simultaneous emission of two Γ_{12}^- phonons; etc. Equation (7) shows that γ_{tot} increases as ω_ℓ increases. Thus from Eqs. (5) - (7), we expect that σ_R for two-phonon RRS should first increase sharply at $\omega_\ell \sim \omega_{1s}(o) + \omega(\Gamma)$ and then drops off more gradually with increasing ω_ℓ. A number of sharp lines in Fig. 2 correspond to two-phonon Raman modes. Experimental results on RRS by two Γ_{12}^- phonons at 220 cm^{-1} near the absorption edge are presented in Fig. 6 as an example. The results on other two-phonon modes are similar. The dashed curve in Fig. 6 was obtained by calculating $\sigma_R(\omega_\ell)$ from Eqs. (5) - (7) using appropriate values for A, B, and C.[2] It is seen that the theoretical curve is in good agreement with the experimental results.

The agreement between theory and experiment gives the following affirmative information. 1) The observed optical absorption edge is indeed due to phonon-assisted excitation of 1s yellow excitons. 2) Various steps at the absorption edge are identified as due to phonon-assisted transitions involving various phonon modes. 3) The two-phonon modes in the Raman spectrum are unambiguously identified through their resonant behavior. 4) The relative coupling strengths between exciton and various phonons can be deduced from the intensity ratios of the different two-phonon modes. 5) The exciton lifetime (= $1/\gamma_{tot}$) is an energy-dependent quantity.

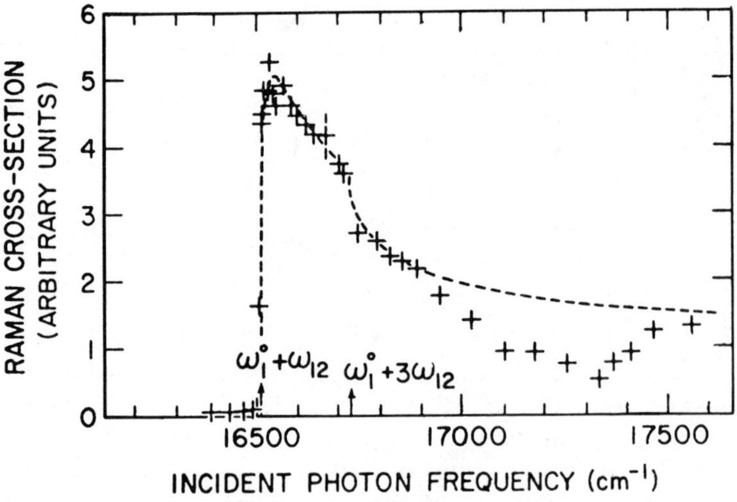

Fig. 6. The Raman cross-section of the $2\Gamma_{12}^-$ mode of Cu_2O plotted as a function of incident photon energies. The broken line is a plot of Eq. (5). (after Ref. 2).

It is also possible to deduce phonon dispersion from two-phonon RRS.[4] Equations (3) and (4) show that the Raman shift $(\omega_\ell - \omega_s)$ is a function of κ and hence ω_ℓ. If the phonons emitted are dispersionless, then $\omega_\ell - \omega_s$ is independent of ω_ℓ. This is true for most of the two-phonon modes in Cu_2O. However, the $\Gamma_{15}^{-(1)}$(LO) mode is fairly dispersive, and therefore the frequency of the $\Gamma_{12}^- + \Gamma_{15}^{-(1)}$(LO) mode has a clear dependence on ω_ℓ, as shown in Fig. 7. From Eqs. (3) and (4), if the exciton effective mass M is known, then since Γ_{12}^- is dispersionless, the dispersion of $\Gamma_{15}^{-(1)}$(LO) can be easily calculated. With $M = 3m_0$ where m_0 is the electron mass, we have found good agreement on the dispersion of $\Gamma_{15}^{-(1)}$(LO) between our measured value and the theoretical value of Carabatos and Prevot.[10]

In Fig. 7, there are another three strongly dispersive modes which split off from the $2\Gamma_{12}^-$ mode as $\omega_\ell \gtrsim \omega_{1s}(o) + \omega(\Gamma_{12}^-)$. These are modes connected with $2\Gamma_{12}^-$ + acoustic phonons.[4]

RRS OF $2\Gamma_{12}^-$ + ACOUSTIC PHONONS IN Cu_2O

The exciton at $\omega_{1s}(\kappa)$ excited by Γ_{12}^- - phonon-assisted absorption can first decay into a lower excitonic state at $\omega_{1s}(\kappa')$ by emitting one or several acoustic phonons and then radiatively recombine with emission of another Γ_{12}^- phonon. The RRS cross-section for $2\Gamma_{12}^- + TA(\omega_{TA})$ is

$$\sigma_R(\omega_\ell, 2\omega_{\Gamma_{12}^-} + \omega_{TA}) = \alpha(\Gamma_{12}^-)[\gamma_{TA}(\kappa,\omega_{TA})/\gamma_{tot}] [\gamma_{rad}(\Gamma_{12}^-)/\gamma_{tot}] \quad (8)$$

where $\gamma_{TA}(\kappa, \kappa')$ is the exciton decay rate from $\omega_{1s}(\kappa)$ to $\omega_{1s}(k')$ through emission of a transverse acoustic phonon and can be calculated.[4] With ω_ℓ fixed, Eq. (8) describes the Raman line-shape of the $2\Gamma_{12}^-$ + TA mode. The solid curve in Fig. 7 is a plot of $[\sigma(\omega_\ell, 2\omega_{\Gamma_{12}^-} + \omega_{TA})]_{max}$ versus ω_ℓ, obtained from Eq. (8) by assuming $M = 3m_0$. It agrees well with the experimental data.

Similarly, we can find $\sigma_R(\omega_\ell, 2\omega_{\Gamma_{12}^-} + \omega_{LA})$ for the $2\Gamma_{12}^-$ + LA(ω_{LA}) mode. The solid curve Y in Fig. 7 is a theoretical plot of $[\sigma_R(\omega_\ell, 2\omega_{\Gamma_{12}^-} + \omega_{LA})]_{max}$ versus ω_ℓ, assuming again $M = 3m_0$.

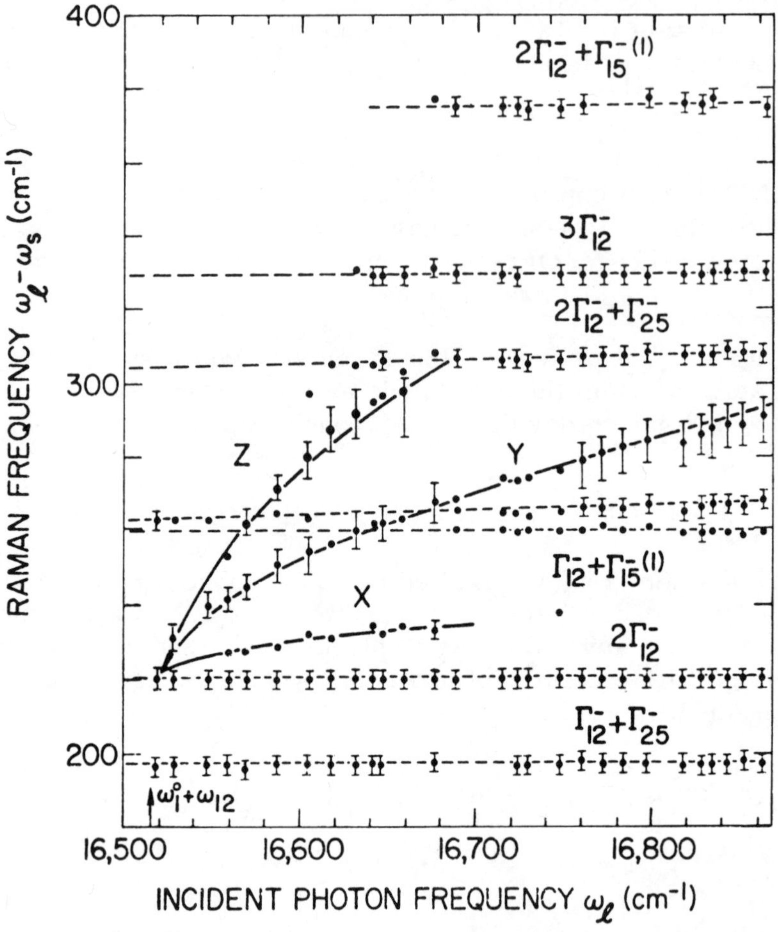

Fig. 7. Raman frequency shifts of all the observed Raman modes of Cu_2O between 190 and 400 cm^{-1} as a function of incident photon energy ω_ℓ. The broken curves are drawn for clarity. The solid curves are theoretical curves discussed in the text. The vertical bars over the experimental points indicate the half-widths of the corresponding Raman peaks. The position of the absorption edge at $\omega_{1s}(o) + \omega(\Gamma_{12}^-)$ is indicated by an arrow. (after Ref. 4).

It has a stronger dependence on ω_ℓ since the LA phonon has a larger dispersion than TA. We also found experimentally

$$\frac{\sigma_R(2\omega_{\Gamma_{12}^-} + \omega_{LA})}{\sigma_R(2\omega_{\Gamma_{12}^-} + \omega_{TA})} = \frac{\gamma_{LA}(\omega_{LA})}{\gamma_{TA}(\omega_{TA})} \simeq 45.$$

This actually corresponds to the ratio of the square of the matrix elements for exciton -LA and exciton -TA phonon coupling.

Because of the stronger exciton -LA -phonon coupling, even the $2\Gamma_{12}^- + 2LA$ mode was detectable. The calculated $[\sigma_R(\omega_\ell, 2\omega_{\Gamma_{12}^-} + 2\omega_{LA})]_{max}$ versus ω_ℓ is shown as the solid Z curve in Fig. 7. The quantitative agreement between theory and experiment in all these cases suggests that our choice of the exciton effective mass $M = 3m_0$ is indeed correct.

RRS OF THREE AND FOUR PHONONS IN Cu_2O.[4]

The exciton at $\omega_{1s}(K)$ can also first decay into a lower excitonic state by emitting one or several optical phonons and then radiatively recombine with emission of another optical phonon. For example, the RRS cross-section for the $3\Gamma_{12}^-$ mode is

$$\sigma_R(3\Gamma_{12}^-) = \alpha(\Gamma_{12}^-)[\gamma_p(\Gamma_{12}^-)/\gamma_{tot}][\gamma_{rad}(\Gamma_{12}^-)/\gamma_{tot}] \tag{9}$$

where $\gamma_p(\Gamma_{12}^-)$ is the exciton decay rate from $\omega_{1s}(K)$ to $\omega_{1s}(K')$ through emission of a Γ_{12}^- phonon. Since the lowest exciton state is at $\omega_{1s}(o)$, we must have $\omega_{1s}(K) \geq \omega_{1s}(o) + \omega(\Gamma_{12}^-)$. Then from Eq. (3), we find $\omega_\ell \geq \omega_{1s}(o) + 2\omega(\Gamma_{12}^-)$ as the threshold frequency for observing RRS of the $3\Gamma_{12}^-$ mode. Figure 8 shows the RRS enhancement of a number of 3-phonon modes. In each case, the Raman cross-section increases sharply at a predicted threshold frequency.

RRS by four Γ_{12}^- phonons has also been observed. The RRS cross-section is

$$\sigma_R(4\Gamma_{12}^-) = \alpha(\Gamma_{12}^-)[\gamma_p(2\Gamma_{12}^-)/\gamma_{tot}][\gamma(\Gamma_{12}^-)/\gamma_{tot}]. \tag{10}$$

Again, we expect $\gamma_p(2\Gamma_{12}^-) \neq 0$ only if $\omega_\ell \geq \omega_{1s}(o) + 3\omega(\Gamma_{12}^-)$. This threshold condition was indeed observed as shown in Fig. 8.

TWO-PHONON RRS AROUND THE EXCITED YELLOW EXCITON STATES.[4]

We can also have RRS with ω_ℓ or ω_s near resonance with the np excited states of the yellow exciton. Electric-dipole excitation of the np exciton states is allowed. RRS can occur, for example, with the excitation of an exciton in the np states followed by emission of two odd-parity phonons and a Stokes photon. Since the excitonic states are sharp, the RRS enhancement is also expected to be sharp. Figure 9 shows an example of two-phonon RRS with ω_ℓ near resonance with the np yellow exciton states. The experimental results agree well with the solid curve in the figure obtained theoretically from the prescribed scattering mechanism given in the inset.[4]

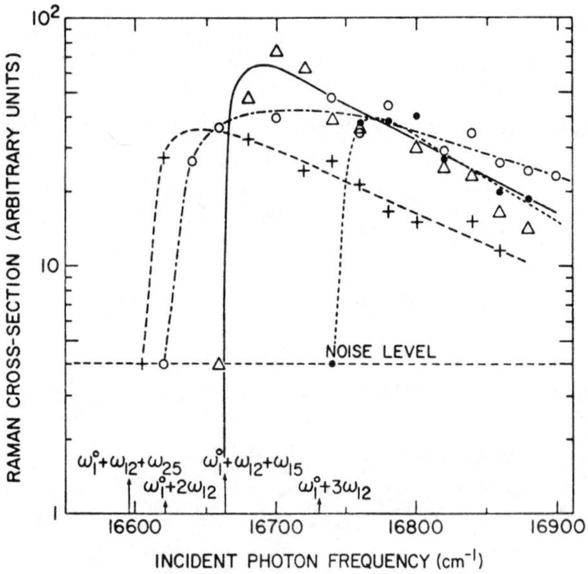

Fig. 8. Raman cross-sections of three- and four-phonon modes of Cu_2O plotted as a function of incident photon frequencies: ...+....+....$2\Gamma_{12}^- + \Gamma_{25}^-$ mode, -•-●-●-$3\Gamma_{12}^-$ mode Δ Δ Δ $2\Gamma_{12}^- + \Gamma_{15}^{-(1)}$ mode and ...●...●....$4\Gamma_{12}^-$ mode. The solid line is a theoretical curve.

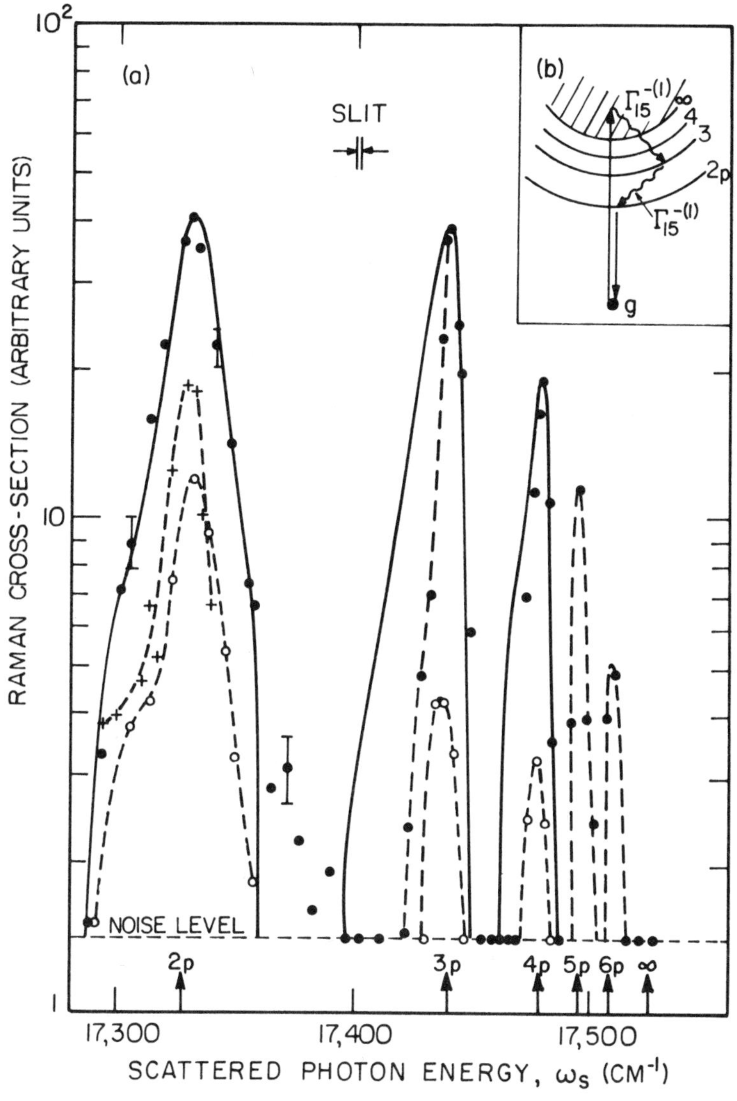

Fig. 9. (a) Absorption spectrum of Cu_2O measured at ~ 5^oK. The dashed curve represents the background absorption due to phonon-assisted transitions. (b) Raman cross-section of the $\Gamma_{15}^{-(2)}(L) + \Gamma_{12}^{-}$ ($770\ cm^{-1}$) modes of Cu_2O obtained at ~ 10^oK as a function of incident photon energies. The solid curve is the theoretical curve [Eq. (6)]. (c) Schematic diagram of the dominant resonant Raman process at $\omega_\ell \sim \omega_{3p}$. g stands for the ground state; β, for an allowed exciton. (after Ref. 4)

TWO-MAGNON RRS IN MnF_2.[5]

The principles of two-phonon RRS near the phonon-assisted absorption edge in Cu_2O can be applied to two-magnon RRS in MnF_2. Figure 10 shows the absorption spectrum of MnF_2 between 18400 and 18500 cm^{-1}. We shall discuss here only the case where the polarizations are perpendicular to the $\hat{c}$-axis. In the figure, E_1 and E_2 have been identified as the exciton lines and σ_1 and σ_2 as the corresponding magnon sidebands. The exciton and magnon dispersion curves are sketched in the inset of Figure 10.

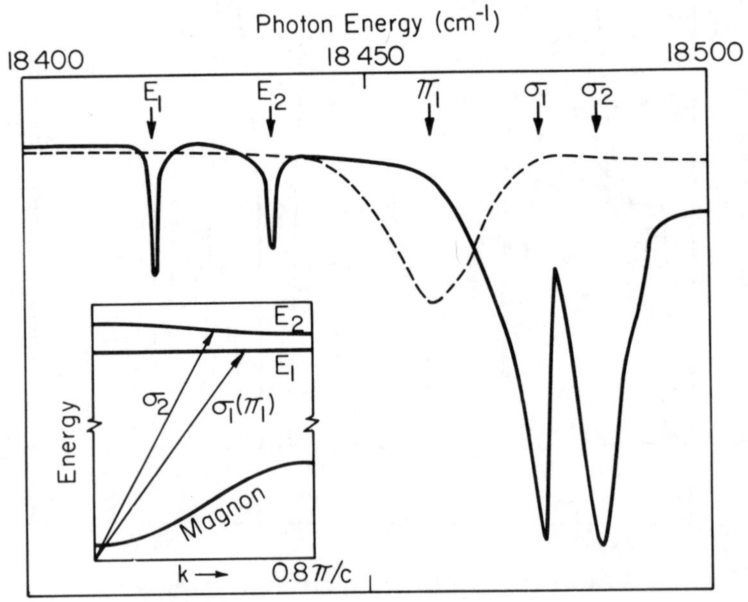

Fig. 10. Absorption spectrum of MnF_2 at 1.6 K between 18400 and 18500 cm^{-1}. The solid and the dashed curves are for polarizations perpendicular and parallel to the $\hat{c}$ axis, respectively. The inset is a sketch of the relevant energy levels. (after Ref. 5)

In the two-magnon RRS process, an exciton at $\omega_{1s}(k)$ is excited by magnon-assisted optical absorption followed by emission of a Stokes photon and another magnon. We therefore expect to observe resonant enhancement when ω_ℓ falls within the magnon sidebands σ_1 and σ_2. From energy conservation, we have

$$\begin{aligned}\omega_\ell &= \omega_{ex}(k) + \omega_{mag}(k) \\ &= \omega_s + 2\omega_{mag}(k) \qquad (11)\end{aligned}$$

This indicates that the Raman shift $\omega_\ell - \omega_s$ now depends on ω_ℓ and is a linear function of ω_ℓ if the excitonic dispersion can be neglected. This linear relation is represented in Fig. 11a by the dashed lines for the two excitons E_1 and E_2. They describe a portion of the two-magnon Raman shift data very well. The disagreement between theory and experiment away from the magnon sidebands is due to contribution from the non-resonant two-magnon Raman scattering.[11] As expected, the Raman shift determined from Eq. (11) is only valid when the resonant contribution dominates. Deviation on the high energy side close to a peak of a magnon sideband is presumably due to the strong dominance of the resonance enhancement near the zone edge.

It is easy to show that the two-magnon Raman cross-section is given by[5]

$$\sigma = \sigma_R + \sigma_{NR} \qquad (12)$$

where the resonant part is related to the nonresonant part by

$$\sigma_R \propto d\sigma_{NR}(\omega_\ell - \omega_s = 2\omega_\ell - 2\omega_{1s})/d\omega_s. \qquad (13)$$

The solid curves in Fig. 10b are obtained from Eqs. (12) and (13) using the experimental lineshape of $d\sigma_{NR}/d\omega_s$ and with σ_R normalized to its peak value. They describe the experimental data very well.

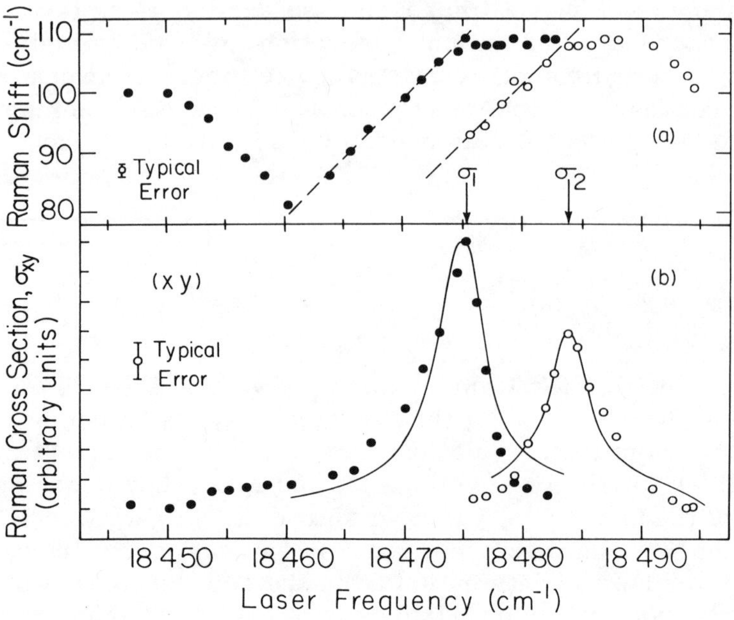

Fig. 11. (a) Two-magnon Raman shift and (b) two-magnon Raman cross-section as functions of the excitation frequency ω_ℓ. The exciting and the scattering radiations are polarized along $\hat{y}$ and $\hat{x}$, respectively ($\hat{x}$, $\hat{y} \perp \hat{c}$). (after Ref. 5)

CONCLUSION

We have used RRS in Cu_2O and in MnF_2 as examples to illustrate how powerful the resonant Raman technique is. It is shown that from RRS measurements, one can have a better understanding of the phonon or magnon structure of the crystal, and in addition, obtain information about the excitonic properties and exciton-phonon (magnon) coupling. The ideas and basic principles discussed here can be extended to electron-hole pairs and to other elementary excitations. For example, one-phonon and two-phonon RRS near the absorption edge of a crystal can presumably help in determining whether the energy gap is direct or indirect; and in some magnetic crystals, one may expect to observe RRS by a magnon-phonon near a phonon or magnon sideband.

This work was sponsored by the U.S. Energy Researcn and Development Administration.

REFERENCES

1. See, for example, E. F. Gross, J. Phys. Chem. Solids 8, 172 (1959); Adv. Phys. Sci. (Moscow) 76, 432 (1962).
2. P. Y. Yu, Y. R. Shen, Y. Petroff, and L. Falicov, Phys. Rev. Lett. 30, 283 (1973).
3. A. Compaan and H. Z. Cummins, Phys. Rev. Lett. 31, 41 (1973).
4. P. Y. Yu and Y. R. Shen, Phys. Rev. Lett. 32, 373 (1974); Phys. Rev. Lett. 32, 939 (1974); Phys. Rev. (1975, to be published).
5. N. M. Amer, T. C. Chiang, and Y. R. Shen, Phys. Rev. Lett. 34, 1454 (1975).
6. J. Reydellet, M. Balkanski, and D. Trivich, Phys. Stat. Solidi 52b, 175 (1972).
7. R. J. Elliot, Phys. Rev. 124, 340 (1961).
8. Y. R. Shen, Phys. Rev. B9, 622 (1974).
9. M. V. Klein, Phys. Rev. B8, 919 (1973).
10. C. Carabatos and B. Prevot, Phys. Stat. Solidi 44, 70 (1971).
11. P. A. Fleury, S. P. S. Porto, and R. Loudon, Phys. Rev. Lett. 18, 658 (1967).

SYMMETRY EFFECTS IN RESONANCE SCATTERING*

Joseph L. Birman

Physics Department, City College of the CUNY
New York, N.Y. 10031 USA

A brief review is given of a recently obtained result that the elements of the phenomenological scattering tensor for inelastic scattering of light by elementary excitations in a crystal can be expressed as a known product form in: Clebsch-Gordan Coefficients, times reduced matrix elements, thereby making maximal use of the symmetry of the system. Recent experimental work on Quadrupole-Dipole Resonance Raman Scattering by phonons in Cu_2O due to Professor H. Z. Cummins' group at CCNY illustrates and confirms the extraction of symmetry and dynamical information in a definitive and beautiful fashion.

THE SCATTERING TENSOR AND CLEBSCH-GORDAN COEFFICIENTS

In analyzing the scattering of radiation from condensed matter it is generally most useful to begin by constructing a phenomenological scattering tensor. At a later stage a microscopic theory can be developed. This procedure has been followed

*Supported in part by NSF grant # DMR74-21991 AO1, AROD grant #DAHCO4-75-G-0052 and FRAP-CCNY grant #10573N.

in recent work on solids[1] and fluids.[2]

The calculation of non-zero components of the scattering tensor for a crystal of given symmetry can be greatly simplified by utilizing a result recently given[1b] relating elements of the scattering tensor to the relevant Clebsch-Gordan Coefficients: this permits maximum factorization of the scattering tensor into a symmetry (Clebsch-Gordan) part times a reduced matrix element.

In this paper we shall briefly review that work and give some recent experimental results[3] obtained at City College on Cu_2O which confirm the predictions of the theory. Analogous work aimed at rationalizing calculation of the tensor in fluids is in progress.

Let the unit polarization vector of the incident radiation be $\hat{\epsilon}_1$ (cartesian components $\epsilon_{1\beta}$, $\beta = 1, 2, 3$) and that of the scattered radiation be $\hat{\epsilon}_2$ (components $\epsilon_{2\alpha}$, $\alpha = 1, 2, 3$). In an inelastic first-order scattering event (Raman or Brillouin) in which a single excitation of symmetry $(j\sigma)$ is produced, the amplitude of scattering is proportional to

$$\epsilon_{2\alpha} P^{(1)}_{\alpha\beta} (j\sigma) \epsilon_{1\beta} \tag{1}$$

When the frequency of incident, or scattered radiation is close to an absorption, resonance scattering occurs. Under this condition, symmetry breaking mechanisms can play a significant role in the scattering. For example, let the incident frequency be close to a quadrupole transition, and let $\vec{k}_1$ be the propagation vector of the incident radiation. Then "forbidden" wave vector dependent scattering can occur, producing an excitation of particular symmetry $(j\sigma)$, with amplitude proportional to

$$\epsilon_{2\alpha} P^{(1)}_{\alpha\beta\gamma} (j\sigma) \epsilon_{1\beta} k_{1\gamma} \tag{2}$$

in the case of first-order scattering. In Eqs. (1) and (2) the objects $P^{(1)}_{\alpha\beta}$, $P^{(1)}_{\alpha\beta\gamma}$ are the scattering tensors.

The scattering tensor $P^{(1)}_{\alpha\beta} (j\sigma)$ carries with it two essentially different pieces of information: cartesian indices $(\alpha\beta)$ and the symmetry species $(j\sigma)$ of the excitation produced; the tensor must transform accordingly. Let S be the rotational part of a crystal symmetry element of the crystal symmetry group G. A polar vector $\vec{r}$ transforms under S as:

$$r'_\mu = \sum_\lambda S_{\mu\lambda} r_\lambda = \sum_\lambda D^{(v)}(S)_{\mu\lambda} r_\lambda \tag{3}$$

Suppose the excitation produced is a normal mode $Q(^j_\sigma)$ e.g., a phonon, and transforms as:

$$Q(^j_\sigma)' = \sum_\tau D^{(j)}(S)_{\tau\sigma} Q(^j_\tau) \tag{4}$$

where $D^{(j)}$ is the jth irreducible representation and σ the row. The tensor $P^{(1)}_{\alpha\beta}(j\sigma)$ transforms as:

$$P^{(1)}_{\alpha\beta}(j\sigma) = \sum_{\lambda\mu} \sum_\tau S_{\alpha\lambda} S_{\beta\mu} D^{(j)}(S)^*_{\sigma\tau} P^{(1)}_{\lambda\mu}(j\tau) \tag{5}$$

The tensor given in Eq. (2) transforms as:

$$P^{(1)}_{\alpha\beta\gamma}(j\sigma) = \sum_{\lambda\mu\nu} S_{\alpha\lambda} S_{\beta\mu} S_{\gamma\nu} D^{(j)}(S^{-1})_{\tau\sigma} P^{(1)}_{\lambda\mu\nu}(j\tau) \tag{6}$$

Equations (5) and (6) can be formally solved once one recognizes how to utilize the Clebsch-Gordan Coefficients, i.e., how to recouple the reducible products of matrix elements into fully reduced - irreducible components. First consider the situation where $P^{(1)}_{\alpha\beta}$ is symmetric in (α, β). This is usually the case for inelastic (Raman) scattering by phonons far from resonance. We should then symmetrize in Eq. (5) by substituting

$$S_{\alpha\lambda} S_{\beta\mu} \rightarrow \tfrac{1}{2}(S_{\alpha\lambda} S_{\beta\mu} + S_{\beta\lambda} S_{\alpha\mu}) \equiv [D^{(v)}(S)_{(2)}]_{\alpha\beta\lambda\mu} \tag{7}$$

where $D^{(v)}_{(2)}$ is the symmetrized Kronecker square of the representation $D^{(v)}$ by which a polar vector transforms. Let U be the unitary matrix which fully reduces $D^{(v)}_{(2)}$, and assume each irreducible component $D^{(\ell)}$ in the reduction appears only once. Then

$$[D^{(v)}(S)_{(2)}]_{\alpha\beta\lambda\mu} = \sum_{\ell nn'} U_{\alpha\beta\ell n} D^{(\ell)}(S)_{nn'} U^{-1}_{\ell n'\lambda\mu} \tag{8}$$

Here ℓ refers to the irreducible representation, and n is the row. Substitute Eqs. (7) and (8) into Eq. (5), sum over all group elements in G, and use the orthonormality relation for irreducible representations to obtain

$$\ell_j P^{(1)}_{\alpha\beta}(j\sigma) = \sum_{\lambda\mu\tau} U_{\alpha\beta j\sigma} U^{-1}_{j\tau\lambda\mu} P^{(1)}_{\lambda\mu}(j\tau) \tag{9}$$

This is easily solved by choosing

$$P^{(1)}_{\lambda\mu}(j\tau) = c(j) U_{\lambda\mu j\tau} \tag{10}$$

where c(j) is the "reduced matrix element": it is characteristic of the irreducible representation j only. Equation (10) demonstrates that under the cited condition (no multiplicity) elements of the scattering tensor are the product of a Clebsch-Gordan Coefficient times a reduced matrix element. The second order tensor and also tensors for "morphic" effects transform as known sums of products of the Clebsch-Gordan Coefficients.[4]

The first order tensor for symmetry-broken resonance scattering, e.g., Eq. (2) obeys quite different symmetry. For example, in general different modes $(j\sigma)$ will participate in forbidden scattering than in allowed scattering. The strategy of solving Eq. (6) is the same as earlier: an additional Clebsch-Gordan matrix will arise here, however; because of the extra factor of $S_{\gamma\nu}$ arising from the transformation of components $k_{1\gamma}$. In dealing with forbidden scattering it proves useful to begin by symmetrizing on the index pair $(\beta\gamma)$, i.e., using "correct linear combinations" of $\epsilon_{1\beta} k_{1\gamma}$ --this also permits a priori incorporation of the transversality condition. Thus if we establish the correspondence $(\beta\gamma) \rightarrow A$ where A is the row index of irreducible representation $D^{(a)}$ then the "contracted" form of the scattering tensor for forbidden scattering (photon wave vector dependent) to produce an excitation $(j\sigma)$ is given by the correspondence

$$P^{(1)}_{\alpha A}(j\sigma) \rightarrow P^{(1)}_{\alpha\beta\gamma}(j\sigma) \tag{11}$$

It easily follows that $P^{(1)}_{\alpha A}(j\sigma)$ obeys an equation like Eq. (5), not symmetrized on αA so that

$$P^{(1)}_{\alpha A}(j\sigma) = c(j)\,V_{\alpha A j\sigma} \tag{12}$$

where V is the unitary Clebsch-Gordan matrix which reduces the direct product $D^{(v)} \otimes D^{(a)}$, analogously to Eq. (8). The uncontracted scattering tensor then follows from Eq. (11).

The computation of the elements of the scattering terms is now straightforward. For first order forbidden scattering the necessary input is simply the table of Clebsch-Gordan Coefficients for the crystal point groups.[5] For higher order scattering or morphic effects space group Clebsch-Gordan Coefficients may be needed.[4, 6]

QUADRUPOLE-DIPOLE RESONANCE SCATTERING IN Cu_2O

Recently Professor Cummins and his associates at The City College of New York investigated resonance Raman Scattering by single phonons in Cu_2O for incident photon frequency close to the 1S yellow exciton line. As is well known owing to the work of Gross and collaborators in Leningrad, Nikitine and collaborators in Strasbourg, and Elliott in Oxford, this optical 1S yellow transition is electric quadrupole allowed and corresponds to the representation Γ25+ of the cubic group O_h. For Raman scattering one can then use the phenomenological expression Eq. (2) for the amplitude of EQ-ED Raman scattering where $\vec{k}_1$ is the incident photon at ω_1 = 1s Y. The allowed phonon symmetry species $(j\sigma)$ are easily obtained from Eq. (3) by forming the direct product of the vector representation Γ15- (from $\hat{\epsilon}_2$) with Γ25+ (from the symmetrized combination of $\epsilon_{1\beta}k_{1\gamma}$ to obtain

$$\Gamma 2- \oplus \Gamma 12- \oplus \Gamma 15- \oplus \Gamma 25- = (j\sigma)\ . \tag{13}$$

All of these normally forbidden odd parity phonons are permitted in the EQ-ED Raman scattering. Using the result Eq. (11) and Eq. (12) we can in a straightforward fashion obtain the elements of the scattering matrix. The case $(j\sigma)$ = 12- is important: it is the most intense of the normally forbidden phonons to be active in the EQ-ED scattering for incident frequency near the 1s Y exciton. Using tables of the Clebsch-Gordan Coefficients for point group O_h we can determine the non-zero coupling coefficients for coupling each row of Γ15- with each row of Γ25+ to obtain one of the rows of Γ12-. In the notation of Eq. (12)

$$P^{(1)}_{xyz}(12-;1) = P^{(1)}_{xzy}(12-;1) = -P^{(1)}_{yxz}(12-;1)$$

$$= -P_{yzx}(12-;1) = \sqrt{3}\,b$$

and for the other partner

$$P_{xyz}(12-;2) = P^{(1)}_{xzy}(12-;2) = P^{(1)}_{yxz}(12-;2)$$

$$= P^{(1)}_{yzx}(12-;2) = b$$

$$P^{(1)}_{zxy}(12-;2) = P^{(1)}_{zyx}(12-;2) = -2b.$$

Here b is a reduced matrix element, i.e., a constant. All scattering tensors are equally straightforwardly found for other $(j\sigma)$.

Among the aspects of the theory which can be tested is the predicted dependence on angle of rotation of the forbidden scattering by each phonon. Professor Cummins and his coworkers have dramatically verified the predictions of the theory by measuring the angular dependence of the normalized intensity of Raman scattering from Γ12-, Γ25-, and Γ15-(LO), Γ15-(TO) phonons and comparing with experiment. The result is shown in Fig. 1: dots are experiment, and solid curve the predictions of the theory based on calculating non-vanishing matrix elements as described above. Further details are contained in a recently published paper by the group of Professor Cummins.[3]

Another dramatic verification of the symmetry predictions relates to a series of time-reversed Stokes-anti-Stokes experiments. The gist of these experiments is conveyed in Fig. 2, also from the work of Professor Cummins' group.[3] It gives the relation between time-reverse and/or in/out reversal times rotation by π about a principle direction (x+y) in the crystal. The predicted ratio of scattering for a time reverse pair is 7.1 ± 0.6. That measured is 6.6 ± 0.4 and 7.2 ± 0.4 in different configurations. Equally good agreement was found for theory (0.91) and

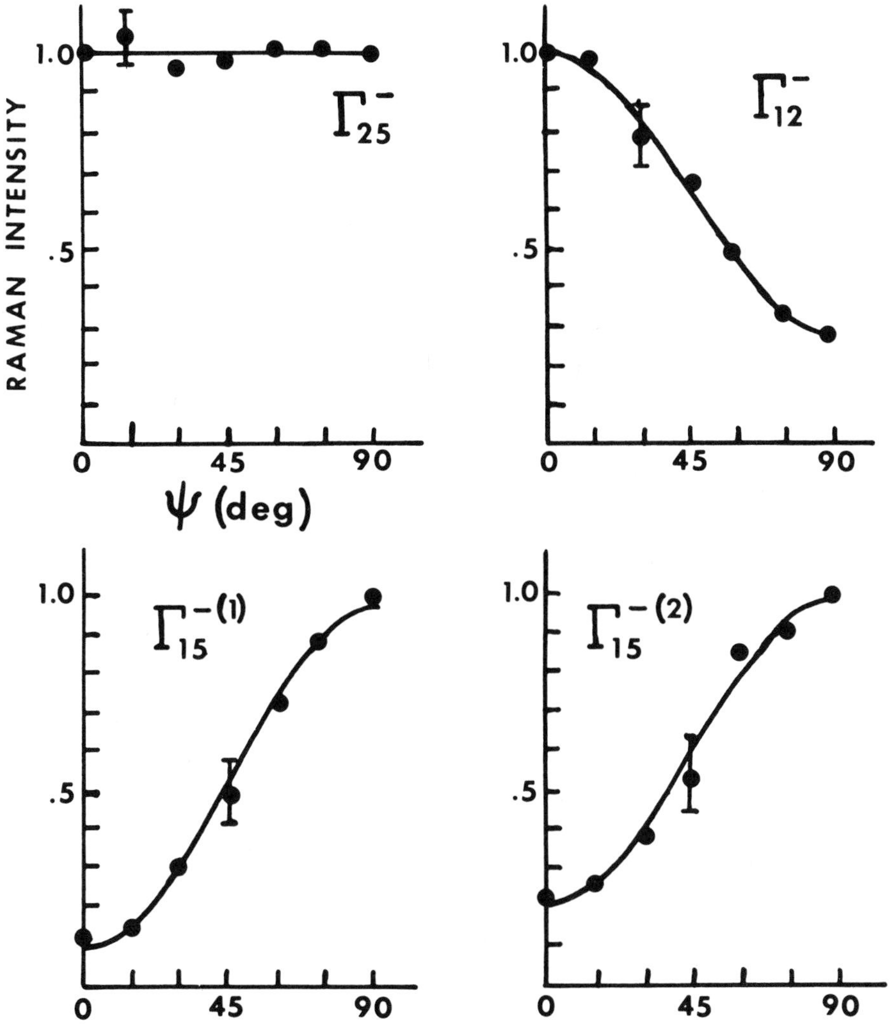

Fig. 1. Angular dependence of normalized intensity of Raman scattering from phonons of indicated symmetry--the laser was at resonance with the 1s Yellow exciton frequency. Dots are experimental points; solid curve is the theory. (See ref. 3).

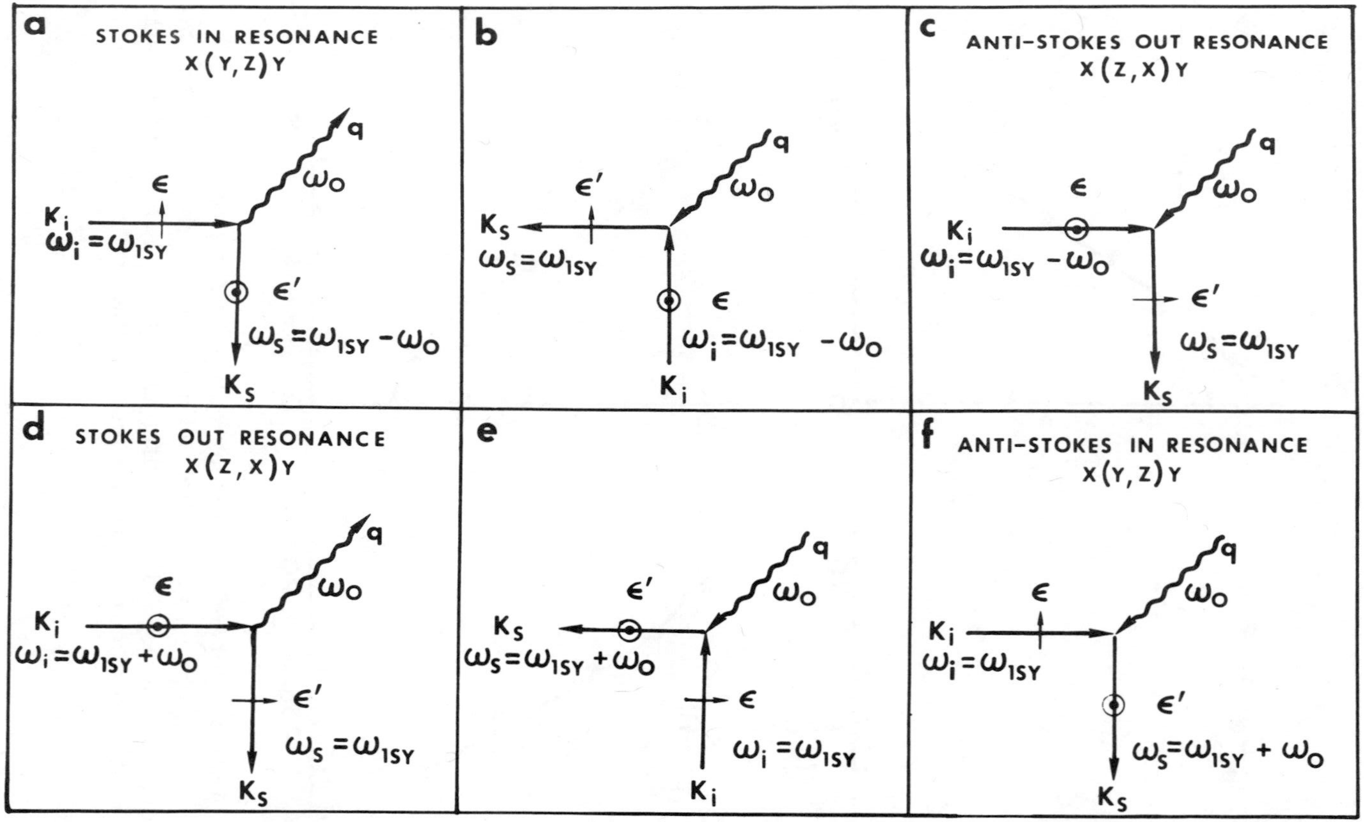

Fig. 2. Schematic of various time-reverse, in-out, and rotational transform processes. Several of these have been tested, and theoretical and experimental ratios are in excellent agreement. (See ref. 3).

experiment (0.96 + 0.05) ratios of in/out resonances.[3]

The work of Cummins et al[3] has not only verified the symmetry predictions of the theory, and thereby confirmed the electric quadrupole-electric dipole (EQ-ED) model of the scattering. In addition they determined the incident frequency dependence of relative intensity of scattering by phonons of each species. This will permit determination of such microscopic parameters as exciton-phonon and exciton-phonon coupling strengths for excitons in the different states of the Yellow series. This work is in progress.

In summary, the very careful study of the detailed properties of one-phonon Raman scattering at a forbidden resonance gives excellent hope of enabling us to unravel all aspects of resonance Raman scattering including separation of symmetry effects and dynamical effects.

REFERENCES

1. a) J. L. Birman, "Theory of Crystal Space Groups and Infra-red and Raman Lattice Processes of Insulating Crystals" Handbuch der Physik 25/2b (1974) Springer-Verlag; b) J. L. Birman and R. Berenson, Phys. Rev. B9, 4512, 4518 (1974).
2. a) V. S. Vikhrenko, Sov. Phys. Usp. 17, 558 (1975); b) B.Ya Zel'dovich, Sov. Phys. JETP 36, 39 (1973).
3. A. Z. Genack, H. Z. Cummins, M. A. Washington, A. Compaan, Phys. Rev. 12, 2478 (1975).
4. R. Berenson (to be published).
5. G. F. Koster, J. O. Dimmock, R. G. Wheeler, H. Statz, "Properties of the Thirty-Two Point Groups" (MIT Press 1963).
6. R. Berenson and J. L. Birman, J. Math. Phys. 16, 227 (1975); R. Berenson, I. Itzkan, J. L. Birman ibid, 16, 236 (1975).

LIGHT ABSORPTION AND RAMAN SCATTERING BY PHONONS BOUND TO IMPURITY CENTERS

E. I. Rashba and A. B. Zimin

L. D. Landau Institute for Theoretical Physics
Academy of Sciences of the USSR
Moscow, USSR

In this paper we discuss the optical properties of one specific type of local vibration modes. A distinctive feature of these modes is the following. They are caused by the electron-phonon interaction in the impurity centre rather than by the difference in masses, or in force constants, as usual. These modes are named dielectric modes.

The energy spectrum is schematically shown in Fig. 1. In both parts of this figure on the left the electronic spectrum is shown, and on the right the vibration spectrum. Here ω_o is the frequency of lattice phonons in the perfect crystal, and ω_i the frequencies of the local modes.

When the electron-phonon coupling is weak in some sense, the frequencies ω_i are close to ω_o. In this case it is convenient to consider dielectric modes as quantum states of a phonon bound to the electronic impurity centre. We shall use this terminology below. Then $\lambda_i = \omega_o - \omega_i$ is the binding energy of the phonon with the impurity centre.

These bound states were originally considered for the resonance situation, when some electronic excitation energy of the centre nearly coincides with ω_o:

$$\epsilon_i - \epsilon_o \approx \omega_o \tag{1}$$

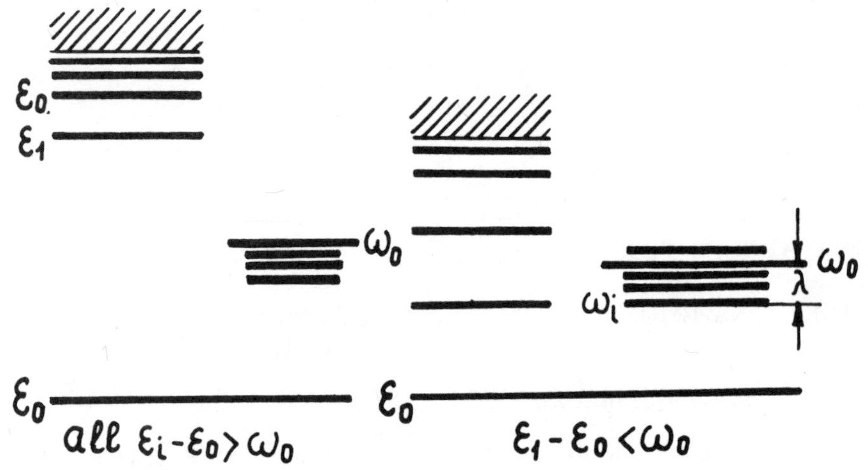

Fig. 1

This special case was theoretically analyzed in Ref. 1-3. Experimentally bound states were observed from some impurities in Si,[4] for donors in GaP,[3] and for acceptors in CdS.[5] It is interesting to note that a rather complicated spectrum including several bound states was found. In GaP one bound state of s-type and one bound state of p-type were observed; the assignment of four phonon bands in CdS is tentative yet. It is very significant to stress that the resonance condition, Eq. (1), is strongly violated both in GaP and CdS: the impurity centres in these crystals are far from resonance. For instance, the ionization energy R of acceptors in CdS is $R \approx 4\omega_0$.

We describe here the principal results of the general theory which is applicable to the centres with arbitrary electronic spectrum and arbitrary ratio of ionization energy R to ω_0. The central problem is to overcome the diffuclties which arise in perturbation theory due to summation over the infinite number of electronically excited states of the centre. In the resonance situation it is possible to retain only the finite number of resonance states, but in the general case the whole electronic spectrum should be included. Our theory describes the energy spectrum of bound states, as well as the spectra of optical absorption and Raman scattering.

The energy spectrum of the coupled electron-phonon system may be found either from the electron Green function, or from the phonon one. Let us first follow the first way. We have to consider the one-particle electron Green function near the threshold of emission of the ω_0 phonon, that is, for the frequencies $\omega \approx \epsilon_0 + \omega_0$, ϵ_0 being the ground level of the electronic subsystem.

In Fig. 2 a succession of dangerous diagrams is shown, which must be summed up for the weak electron-phonon coupling case. These diagrams include the maximum number of dangerous sections at the fixed number of vertices, every one-phonon section being a dangerous one. This succession was discussed as applied to various aspects of the phonon bound state problem in Ref. 6.

For an impurity centre the summation of this succession is equivalent to the solution of the eigenvalue problem.[7]

$$\lambda\{(\lambda + H - \epsilon_0)^2 - \omega_0^2\}\varphi = 2(\lambda + H - \epsilon_0)\bar{A}\varphi \tag{2}$$

for functions φ_i. This equation will be of principal significance for us below. The functions φ_i are closely connected to the wave functions of the phonon bound to the impurity centre. Here H is the electronic Hamiltonian of the impurity centre when the electron-phonon interaction is neglected. λ is the binding energy of the phonon with the impurity centre. The operator A represents the electron-phonon interaction and equals

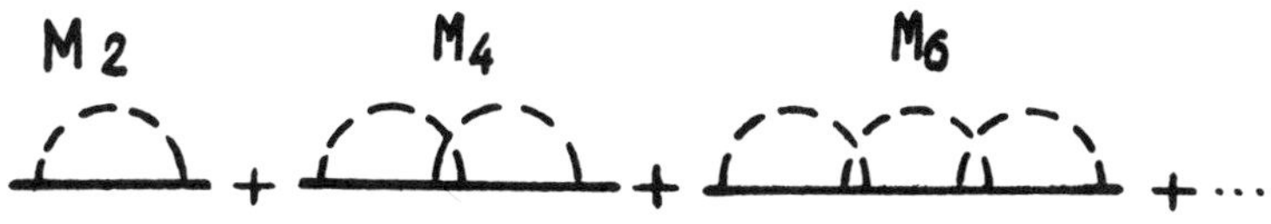

Fig. 2

$$A(\vec{r}, \vec{r}') = \psi_o(\vec{r}) V(\vec{r} - \vec{r}') \psi_o(\vec{r}'), \tag{3}$$

where $\psi(\vec{r})$ is the eigenfunction of H corresponding to the eigenvalue ϵ_o,

$$V(\vec{r} - \vec{r}') = \sum_{\vec{q}} |c_{\vec{q}}|^2 e^{i\vec{q}\vec{r}} \tag{4}$$

and $c_{\vec{q}}$ are the coefficients in the electron-phonon interaction Hamiltonian. For polarization interaction

$$c_{\vec{q}} \sim \frac{1}{q}, \quad V(\vec{r}) = \frac{e^2 \omega_o}{2r} \left(\frac{1}{\kappa_o} - \frac{1}{\kappa_\infty}\right); \tag{5a}$$

$V(\vec{r})$ shows the Coulomb behaviour. For the non-polarization interaction

$$c_{\vec{q}} = \text{const}, \quad V(\vec{r}) \sim \delta(\vec{r}). \tag{5b}$$

$\bar{A}$ is the part of operator A acting in the subspace $(\varphi, \psi_o) = 0$.

The investigation of the equation for φ allows one to find the general structure of the energy spectrum.[7] Here modes arise corresponding to all the values of the angular momentum. We emphasize that for each value of the angular momentum an infinite number of modes arise. When all $\epsilon_i - \epsilon_o > 0$, then all the eigenvalues λ_i are positive. This means that all the frequencies ω_i are lower than ω_o. When some $\epsilon_i - \epsilon_o < \omega_o$, the equal number of $\lambda_i < 0$, that is, the equal number of modes with $\omega_i > \omega_o$, arise.

The binding energies for some first states may be estimated as

$$|\lambda_i| \sim 0.1\,\alpha\omega_o(\omega_o/R)^{\frac{1}{2}}, \tag{6}$$

where α is the Fröhlich coupling constant.

The very significant point is that the functions φ_i allow one to calculate the parameters of the optical spectra connected with the impurity-phonon complexes: the absorption spectra, as well as the Raman-scattering spectra.

For example, the one-particle Green function, represented by the series shown in Fig. 2, allows one to find the intensities of local phonon satellites in the absorption spectrum of the impurity excitons.[8] The intensities of these satellites are determined by

$$I(\omega) \sim \Sigma \left| \left((\omega - H)^{-1} \bar{A}_{\varphi_i} \right)_{\vec{k}=0} \right|^2 \quad \delta(\omega - \omega_i). \tag{7}$$

The index $\vec{k} = 0$ shows that the zero-momentum Fourier component of the corresponding function must be taken. For example, for the isotopic impurities in molecular crystals the Green function of the impurity Frenkel exciton $(\omega - H)^{-1}$ is well known; thus, the intensities in the absorption spectrum are easily determined by functions φ_i.

As applied to semiconductors the absorption and Raman spectra of shallow neutral impurity centres are the most interesting. The lattice absorption at frequencies ω_i corresponds to creation of the bound phonons and may be described by the contribution to the polarization operator from the diagrams shown in Fig. 3a. Here D is the full phonon Green function, and the

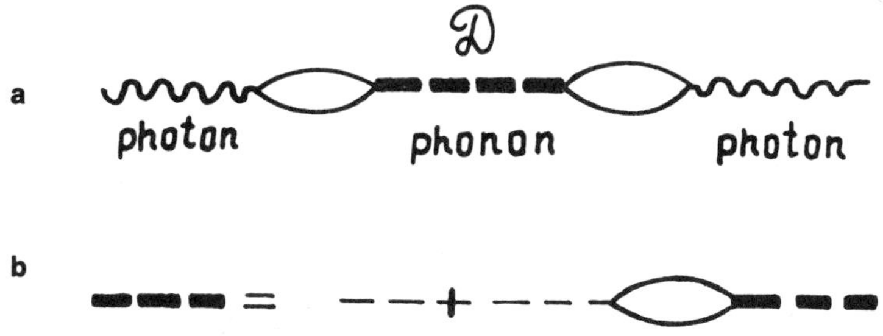

Fig. 3

electronic bubbles standing at both sides of it involve the summation over the whole electronic spectrum of the impurity centre. The equation for D is shown in Fig. 3b. It is important that by calculating the electronic bubbles in the diagrams of Fig. 3a and 3b the nonrenormalized electronic Green functions may be used. This is due to the fact that the quantum state in which the ground level of all impurity centres is occupied must be taken as the vacuum state here, and for such vacuum state the electron Green function has no singularities near $\epsilon_o + \omega_o$. Thus, the central problem here is the summation of electronically excited states over the infinite succession.

It can be shown that the equation for D may be reduced to the equation for φ discussed above. As a result, we can find the phonon wave function Φ_i, corresponding to the i-th bound state:

$$\Phi_i(\vec{q}) \sim c_{\vec{q}} (\psi_0 | e^{-i\vec{q}\vec{r}} | \varphi_i); \tag{8}$$

it is given here in the momentum representation. For the non-polarization interaction $c_{\vec{q}}$ = const and

$$\Phi_i(\vec{r}) \sim \psi_0(\vec{r}) \varphi_i(\vec{r}) . \tag{9}$$

Thus, φ_i is connected with the phonon wave function Φ_i by a very simple relation. For polarization phonons this relation is more complicated.

The intensity of light absorption found from the polarization operator of Fig. 3a is

$$I(\omega) \sim \sum_i |\langle \psi_0 | \vec{r} | \varphi_i \rangle|^2 \delta(\omega - \omega_i) \tag{10}$$

So, the intensity of the i-th band is completely determined by the matrix element $\langle \psi_0 | \vec{r} | \varphi_i \rangle$ and, thus, by the φ_i function. It is very interesting to note that the functions ψ_0 and φ_i entering into the matrix elements are the solutions of two utterly different equations: $H\psi_0 = \epsilon_0 \psi_0$ and Eq. (2). The oscillator strength may be estimated as

$$f_i \sim \alpha(\omega_o/R)^{5/2} \quad . \tag{11}$$

The Raman scattering is described by this diagram (Fig. 4), where the vertices correspond to the Raman process in a perfect lattice. The intensity of scattering corresponding to creation of bound phonon states is determined by the same matrix elements $\langle\psi_o|\vec{r}|\varphi_i\rangle$. The ratio of the intensity of this scattering to that produced by the lattice phonons may be estimated as $n_i a_b^3$; here n_i is the impurity concentration, and a_b is the radius of the impurity centre. Thus, the impurity scattering per one centre when compared with the perfect lattice scattering per one cell is stronger by a factor $\sim(a_b/a)^3$, where a is the lattice constant. This gigantic scattering intensity results from coherent lattice vibrations in the volume $\sim a_b^3$ around the impurity centre, and thus has the same origin as the gigantic oscillator strengths for impurity excitons.[9]

So, we have a quite general and effective theory of the phonon-impurity complexes, which allows to express the energy levels and intensities of optical transitions in terms of functions φ_i and of the basic equation (2) for them.

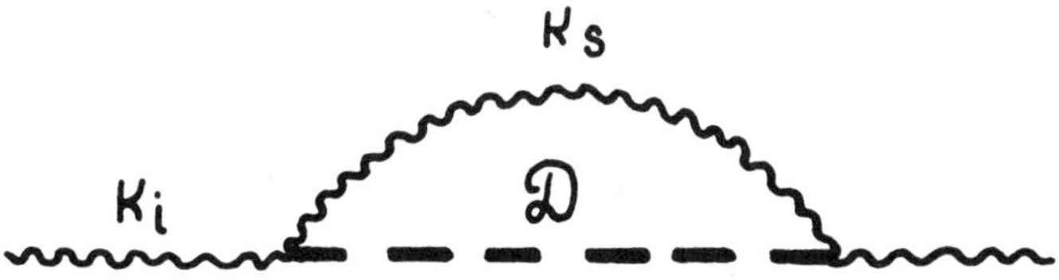

Fig. 4

REFERENCES

1. Sh. M. Kogan, R. A. Suris, Zh. Eksp. Teor. Fiz. 50, 1279 (1966) Sov. Phys. - JETP, 23, 850 (1966).
2. S. Rodriguez, T. D. Schultz, Phys. Rev. 178, 1252 (1969).
3. P. J. Dean, D. D. Manchon Jr., J. J. Hopfield, Phys. Rev. Lett., 25, 1027 (1970).
4. A. Onton, P. Fisher, A. K. Ramdas, Phys. Rev. Lett. 19, 781 (1967). H. J. Hrostowski and R. H. Keiser, J. Phys. Chem. Sol. 4, 148 (1958).
5. D. C. Reynolds, C. W. Litton, T. C. Collins, Phys. Rev. B4, 1868 (1971).
6. I. B. Levinson, E. I. Rashba, Usp. Fiz. Nauk 111, 683 (1973) Sov. Phys.-Uspekhi 16, 892 (1974); Rep. Progr. Phys. 36, 1499 (1973).
7. E. I. Rashba, ZhETF Pis. Red. 15, 577 (1972) Sov. Phys.-JETP Letters 15, 411 (1972); Izv. AN SSSR, ser. fiz. 37, 619 (1973).
8. E. I. Rashba, A. B. Zimin, ZhETF 66, 1479 (1974).
9. E. I. Rashba, Opt.i Spektr, 2, 568 (1957); E. I. Rashba, G. E. Gurgenishvili, Fiz. Tverd. Tela, 4, 1029 (1962) Sov. Phys.-Solid State 4, 759 (1962).

Section VII

Scattering from Electronic Excitations and Spin-Flip Processes

SOME FEATURES OF LIGHT RAMAN SCATTERING

BY BULK POLARITONS

V. L. Strizhevskii, V. I. Kislenko,
F. N. Marchevskii, Yu. N. Yashkir

T. G. Shevchenko Kiev State University
Kiev, USSR

In this work some essential general features of light Raman Scattering (RS) by bulk polaritons under stationary and nonstationary conditions are discussed. In the first Section the influence of the scattering polariton motion on the RS intensity is considered. Section 2 is devoted to the analysis of some aspects of the influence of the noise nature of light polariton RS on the time dependence of scattering intensity under the essentially nonstationary conditions of excitations. Finally, in the third Section the results of the investigation of nonstationary active spectroscopy of spontaneous RS by polaritons are presented.

THE INFLUENCE OF SCATTERING POLARITON MOTION

The specificity of the polariton character of scattering excitations lies, in particular, in their motion: the polariton group velocity $v_p \neq 0$. As a consequence, the finite-state density of the system "crystal+field" (appearing in the calculation of the scattering integral intensity) is determined both by the Stokes wave group velocity v_s and by v_p. To analyze the influence of v_p-finiteness it is sufficient to take into account that the scattering probability W in a transparent noncentrosymmetrical crystal per unit time into the unit solid angle determining the integral (with respect to the frequency) scattering intensity in the fixed direction (the scattering being excited by a plane monochromatic pump wave), is proportional to $\delta[\omega_s(\vec{k}_s) + \omega_p(\vec{w}) - \omega_\ell]$ (ω is the frequency,

$\vec{w} = \vec{k}_\ell - \vec{k}_s$, $\vec{k}_{\ell,s}$ are wave vectors, the indices ℓ, s and p here and below correspond to the pump, Stokes and polariton waves respectively). On the other hand, fixed-angle scattering intensity is obviously proportional to

$$\int_0^\infty W dk_s$$

(it should be summarized by the variable, which determines the finite state of the system and it is not fixed by measurement conditions). For the calculation of the latter integral it is expedient to use the well-known correlation [1].

$$\delta[\omega_s(\vec{k}_s)+\omega_p(\vec{w})-\omega_\ell] = \sum_i \left| \frac{\partial}{\partial k_s}[\omega_s(\vec{k}_s)+\omega_p(\vec{k}_\ell-\vec{k}_s)]_{k_s=k_s^i} \right|^{-1} \delta(k_s-k_s^i)$$

where the index i enumerates the roots k_s of the equation $\omega_s(\vec{k}_s) + \omega_p(\vec{w}) - \omega_\ell = 0$, and the derivative by k_s is calculated at the fixed direction of the vector $\vec{k}_s$. It is not difficult to show that

$$\frac{\partial}{\partial k_s}[\omega_s(\vec{k}_s) + \omega_p(\vec{w})] = \gamma_s v_s - v_p \cos\psi$$

where $\gamma_s \approx 1$ is the cosine of the angle between group and phase velocities of the Stokes wave, ψ is the angle between $\vec{k}_s$ and $\vec{w}$. Accordingly the scattering efficiency σ contains the factor $M = |1 - \xi|^{-1}$ where

$$\xi = v_p \cos\psi / v_s \ ; \ \sigma = \sigma_o M \ ,$$

$$\sigma_o = \frac{\kappa_p \sigma_f S_f [1 + A(1-\kappa_p^2]^2}{\epsilon_\infty (1-\kappa_p^2)^2 + S_f} \ , \quad \kappa_p = \frac{\omega_p}{\omega_f}$$

The expression for σ_o is presented for the case, when ω_p is in the vicinity of the frequency ω_f of a polar mechanical phonon; the dielectric permittivity in this region is approximated by the expression $\epsilon_p = \epsilon_\infty + S_f/(1-\kappa_p^2)$, S_f is the oscillator strength, ϵ_∞ is the high-frequency limit of ϵ_p (with respect to ω_f); σ_f is the phonon scattering efficiency; A is the electron-deformation parameter determined in such a way that the value $A(1 - \kappa_p^2)$ is the ratio of the nonresonant contribution from the corresponding nonlinear polarizability to the resonant one.

Further details may be found in [2, 3]. The factor M was not taken into account in a number of works (see [4] and others). This is correct when ω_p is near to ω_f and, correspondingly, $v_p \approx 0$, $\xi \simeq 1$, $M \simeq 1$. But usually (in the region of small scattering angles) ω_p is essentially remote from ω_f, and M differs greatly from 1. And moreover, cases of the group synchronism are possible (near "critical" scattering angles [2, 3], when $\xi = 1$ and $M \to \infty$). A correct expression for σ may be found in this case after accurate integration of the scattering spectrum [2].

The influence of the factor M on scattering intensity for a number of real crystals is studied (theoretically and experimentally), in particular, in the work [2]. Therefore, we do not here further discuss this question. It must be noted, that this influence appears, in general, to be very essential, and it determines both quantitative and qualitative peculiarities of the scattering intensity.

Scattering polariton motion influences also the width of scattering polariton lines. Indeed, the scattering spectrum width is determined by the rate of decrease of the gain factor g (which describes the interaction of waves) with the increase of wave mismatch $\Delta\vec{k} = \vec{k}_\ell - \vec{k}_s - \vec{k}_p$ (here $\vec{k}_p$ is a polariton wave vector). In the nonresonant region of polariton frequencies we have $g = a/[(\Delta\vec{k})^2 + (\alpha_p/2)^2]$ [5, 6], where α_p is the energetic absorption coefficient at frequency ω_p, and the value a does not depend on $\Delta\vec{k}$; it is supposed that $g \ll \alpha_p$ (usually this condition is well satisfied). The spontaneous RS intensity is proportional to g. It is obvious that g is essentially different from zero, if $|\Delta k| \leq \alpha_p/2$. Wave mismatch $\Delta\vec{k}$ may be represented approximately as proportional to frequency mismatch $\Delta\omega$. As a result, a scattering line halfwidth Δ is expressed as a product $\Delta = \mu\alpha_p$, where $\mu = |\partial(\Delta k)/\partial\omega|^{-1}$. The factor α_p (in the vicinity of an isolated phonon vibration with the frequency $\omega_f - i\gamma/2$) is proportional to the product of phonon damping γ and phonon strength ν (determined according to [4]). The value μ^{-1} is the rate of wave mismatch growth by the growth of frequency mismatch; it depends on the group mismatch of Stokes and polariton waves (the same may be said about the factor M introduced above). As a result it appears that $\Delta = \gamma\nu M$. Indeed, this formula is correct far away from "critical" scattering angles, because in their vicinity in the linear (with respect to $\Delta\omega$)

approximation we have $M \to \infty$.

Thus, a scattering line halfwidth may be represented as a product of three factors, each factor having a clear physical sense. In particular, one may say that the factor $M = \mu / v_p \cos \theta_p$ (θ_p is the angle between $\vec{k}_p$ and the normal to the nonlinear slab surface) is determined by the rate of wave mismatch growth with frequency mismatch $\Delta\omega$.

RS BY POLARITONS UNDER NONSTATIONARY CONDITIONS

Suppose now that polariton RS is excited by a plane pump wave with a time-dependent amplitude (for example, by some short pulse, the duration τ_ℓ of which is comparable or essentially less than the life-time τ_0 of scattering polaritons). The value τ_0 may be determined as the inverse halfwidth of the stationary polariton RS line and it depends on the scattering angle. The calculation of the time dependence of the scattering intensity B for this case is carried out in works [7, 8] (the most interesting and important "local" nonstationarity caused by the finiteness of τ_0 is taken into account, while "wave" nonstationarity caused by the difference of group velocities v_ℓ and v_s was neglected, the difference between v_s and v_p being naturally taken into account). Here we dwell briefly on only one consequence from [7, 8] connected with the behavior of the scattering intensity at $\tau_\ell \ll \tau_0$, because the results published by different authors do not correlate well, the question being in differences based on physical reasons. This question was already discussed partially in [8], and therefore we shall touch upon only some aspects.

In the work [9] on the example of phonon RS it is asserted in particular, that in the region $\tau_\ell < \tau_0$ the value B decreases $\tilde{\tau}_\ell = \tau_\ell / \tau_0$ times due to the weakening of the effective noise sources which are responsible for RS. We present, however, qualitative arguments which testify that the above-mentioned decrease is absent and that the value B follows the change of the pump intensity $I_\ell(t)$, having the time to reach the stationary state at every momentary value of $I_\ell(t)$. In fact, the condition of the proportionality between B(t) and $I_\ell(t)$ amounts to the assumption that the relaxation time τ' in the subsystem (with which the scattering subsystem is in contact and which is the reason for energy dissipation of the latter* is small compared to τ_ℓ (and, obviously, $\tau' \ll \tau_0$), so at every momentary value of the scattering excitation coordinates in the system the "partial" (in the sense of [10]) equilibrium state had the time to set in (due to quick relaxation of the dissipating subsystem), the energy passing from the pump to the dissipative subsystem.** Further, accidental (Langevin) forces, which may formally be regarded as the reason for scattering fluctuations [11], behave as white noise, i.e., the corresponding spectral power density $k(\omega)$ is constant throughout the spectrum. From the formal point of view the above-mentioned condition is sufficient for $B(t) \sim I_\ell(t)$ (compare with [7, 8]).

In the problem of RS by polaritons the role of initial noise sources is played by fluctuations of the medium electric polarization (the medium is supposed to be nonmagnetic), and here $k(\omega) \sim \epsilon''_p(\omega)$ [12]. In the nonresonant polariton region the value $\epsilon''_p(\omega)$ changes essentially in an interval of the order of the phonon frequency ω_f. At the same time the scattering line halfwidth is obviously of the order of τ_ℓ^{-1}. Hence, if $\tau_\ell^{-1} \ll \omega_f$, then the change of $\epsilon''_p(\omega)$ through the scattering spectrum may be neglected. From this it is already clear that $B(t)$ follows $I_\ell(t)$ without any lag (as a matter of fact, the situation does not differ in principle from the stationary case), so any reason for effective sources weakening by a factor of τ_ℓ are absent. Naturally, this refers to the stimulated RS too, though in this case the direct proportionality between $B(t)$ and $I_\ell(t)$ is absent due to the influence of the polariton coherent component.

All these considerations are represented in [7, 8] in quantitative form. Phenomenological expressions for the intensity of scattering by polaritons and optical phonons (or by isolated molecule vibrations) appear to be identical and therefore the corresponding features for these cases are the same.

NONSTATIONARY ACTIVE SPECTROSCOPY OF SPONTANEOUS RS BY POLARITONS AND PHONONS

The method of observing spontaneous RS of some auxiliary (probe) wave by coherent polaritons prepared in an independent

---

*And, hence, it is the reason of scattering excitation damping.
**A corresponding time parameter of this passing is τ'.

process (active spectroscopy of the spontaneous RS by coherent polaritons [13, 14]) may be used under nonstationary conditions [15] too. This puts on the agenda the theoretical description of this process. It may be developed by the synthesis of the calculation method of the active spectroscopy of spontaneous RS under stationary conditions developed in [13] and the calculation method of the nonstationary coherent polariton field (excited in stimulated RS) presented in [7, 8].

The statement of the problem is as follows: In a plane-parallel slab of the thickness L a plane pump wave excites (due to stimulated RS) coherent Stokes and (nonresonant) polariton waves with due regard to the possible external signal at the Stokes frequency $\omega_s = \omega_\ell - \omega_p$. A probe wave (all values related to it are marked with the index π) is spontaneously scattered by generated coherent polaritons, with waves $\omega_{\sigma,a} = \omega_\pi \mp \omega_p$ arising. Let us represent fields at frequencies $\omega_{\ell,\pi}$ in the form

$$\vec{E}_{\ell,\pi}(\vec{r},t) = \vec{e}_{\ell,\pi} A_{\ell,\pi}(t - \frac{z}{V}) \exp [i(\vec{k}_{\ell,\pi}\vec{r} - \omega_{\ell,\pi} t)] + c.c.$$

The axis z is directed along the normal to the slab surfaces. To avoid unessential complications let us neglect the difference between the z-components of group velocities of waves $\omega_{\pi,\ell,s,\sigma,a}$ (v is the common value of these components, $\vec{e}_{\ell,\pi}$ are unit polarization vectors).*

For short we omit the intermediate calculations and present at once the final result for surface brightness of the output slab surface at frequency ω_σ:

$$B_\sigma(L,\xi) = B_\sigma^0(1+\bar{n}_p)G_\sigma(\xi)L\Big[1 + \frac{2L}{\pi}\int_{\xi_0}^{\xi}\Big(1 + \frac{N_s(\xi')}{1+\bar{n}_p}\Big)G_s(\xi')I_1^2(\zeta)\zeta^{-2}e^{(\xi'-\xi)\gamma}d\xi'\Big], \tag{1}$$

*The corresponding criterion is $|v_i^{-1} - v_j^{-1}|L \ll \tau_k$, where τ_k is the time duration of the ω_k - wave impulse, the indices i, j, k take on values π, ℓ, s, σ, a.

where

$$N_s(\xi) = B_s^e(\xi)/B_s^0, \; B_{\sigma,s}^0 = \frac{\hbar\omega_{\sigma,s}^3 n_{\sigma,s}^2}{8\pi^3 c^2}, \; \bar{n}_p = \left(e^{\frac{\hbar\omega_p}{k_0 T}} - 1\right)^{-1},$$

$$\xi = t - \frac{L}{V}, \; \gamma = \frac{1}{\tau_0}, \; \zeta = \left[\frac{2L}{\pi}\int_{\xi'}^{\xi} G_s(\xi_1)d\xi_1\right]^{\frac{1}{2}}, \; G_{s,\sigma}(\xi) = \frac{\pi}{2}\gamma g_{s,\sigma}(\xi).$$

Here I_1 is the modified Bessel function of the first order, $g_{\sigma,s}(\xi)$ are the gain factors in the line centre of stimulated RS of wave ω_π (with the generation of radiation ω_σ) and of the wave ω_ℓ (with the generation of radiation ω_s) respectively. These values are determined by well-known expressions [6] and they are proportional to the momentary values of intensities $I_{\pi,\ell}(\xi)$; B_s^e is the surface brightness of the external Stokes signal on the input slab surface; $\xi_0 = t_0$ is the initial time moment taken arbitrarily before the arrival of impulses to the input surface. The formula (1) describes both polariton and purely vibrational cases, if we use appropriate expressions for g.

The formula (1) is sufficiently general because it takes into account both coherent and noise components of the wave ω_p, the coherent component being generated with regard to a possible signal $N_s(\xi)$ at the frequency ω_s. Let us first analyze the case of the weak pump determined by the condition $\zeta \ll 1$. Here

$$B_\sigma(L\xi) = B_\sigma^0(1+\bar{n}_p)G_\sigma(\xi)L + B_\sigma^0 G_\sigma(\xi)\frac{L^2}{2\pi}\int_{\xi_0}^{\xi} N_s(\xi')G_s(\xi')e^{(\xi'-\xi)\gamma}d\xi'$$

The first term corresponds to the usual spontaneous RS by noise polaritons (it is omitted below) and the second item corresponds to spontaneous RS by coherent polaritons due to the signal N_s. Let τ_0 be small compared to the overlap time τ_1 of the signal and pump pulses. Then the quasistationary regime is realized and

$$B_\sigma = B_\sigma^0 G_\sigma(\xi)G_s(\xi)\cdot N_s(\xi)\frac{L^2}{2\pi\gamma}$$

If one knows the shape of two (of the three) impulses (G_σ, G_s, and N_s) and the dependence $B_\sigma(\xi)$ is measured, then the time shape of the third impulse may be studied.

Inversely, if $\tau_0 >> \tau_1$, then at an extremely short pump pulse, when one can put $G_s(\xi) = G_s^0\delta(\xi - \xi_\ell)$ (G_s^0 does not depend on ξ) we get (at $\xi \geq \xi_\ell$)

$$B_\sigma(L, \xi) = B_\sigma^0 N_s(\xi_\ell) G_s^0 C_\sigma(\xi) \exp [(\xi_\ell - \xi)\gamma] \frac{L^2}{2\pi}$$

Here at $\tau_\pi \geq \tau_0$ the dependence $B_\sigma(\xi)$ is determined mainly by the factor $\exp [(\xi_\ell - \xi)\gamma]$, so that a first-hand scanning of the phonon damping is realized (see Fig. 1a*). At $\tau_\pi << \tau_0$ the dependence $B_\sigma(\xi)$ is determined by the factor $G_0(\xi)$ with the peak value proportional to $f = \exp [(\xi_\ell - \xi_\pi)\gamma]$, so that one can determine τ_0 by varying the moment of the probe pulse entrance (see Fib. 1b, compare with [15]).

Finally, if values $\tau_{\ell,\pi,0}$ are of the same order, then the change of the polariton coherent component (in the process of impulse interaction) plays an essential role. In this case one does not succeed in obtaining any simple analytical expressions, and we present a numerical illustration (Fig. 2). It corresponds to the shape of impulses I_π, I_ℓ and N_s of the form

$$\exp [-4(\xi - \xi_{\pi,\ell,s})^2 / \tau^2_{\pi,\ell,s}]$$

The case 2a corresponds to $\tilde{\tau}_\ell = \tilde{\tau}_s = 0.25$, $\tilde{\tau}_\pi = 1$, $\xi_\pi = \xi_\ell = \xi_s$ ($\tilde{\tau}_{\pi,\ell} = \tau_{\pi,\ell} / \tau_0$, $\tilde{\tau}_s = \tau_s^e / \tau_0$, τ_s^e is the duration of $N_s(\xi)$). It is seen, in particular, that there is a delay of the impulse $B_\sigma(\xi)$ relative to the impulses $I_{\ell,\pi}(\xi)$, $N_s(\xi)$. The case 2B corresponds to $\tilde{\tau}_\ell = \tilde{\tau}_s = \tilde{\tau}_\pi = 0.25$, $\xi_\ell = \xi_s$ and to different ξ_π (curves 1, 2, 3 correspond to $\tilde{\xi}_\pi^{(1)} = \tilde{\xi}_\ell - 0.6$, $\tilde{\xi}_\pi^{(2)} = \tilde{\xi}_\ell$, $\tilde{\xi}_\pi^{(3)} = \tilde{\xi}_\ell + 0.6$, $\tilde{\xi}_{\ell,\pi} = \xi_{\ell,\pi} / \tau_0$).

We shall consider now the case of strong pumping ($\zeta >> 1$).

---

*In Figs. 1, 2 for the intensities relative units are used, these units being unconnected to each other for different waves. At the same time, impulse shapes are presented without distortions.

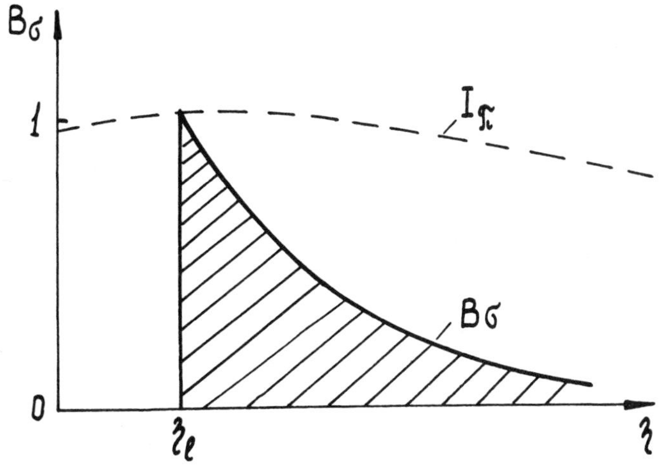

Fig. 1a

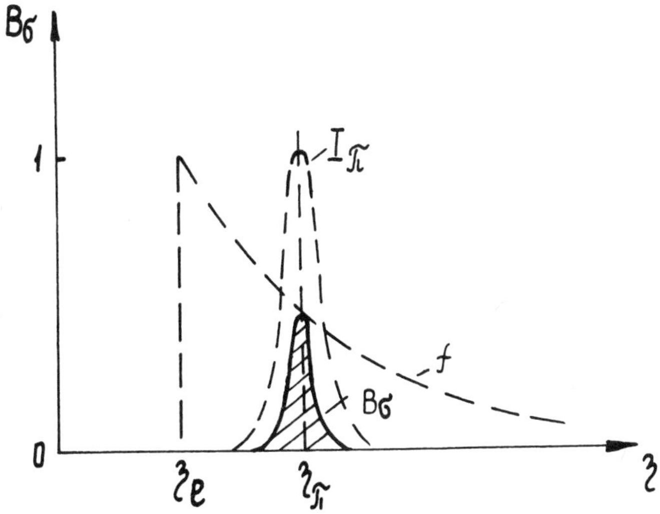

Fig. 1b

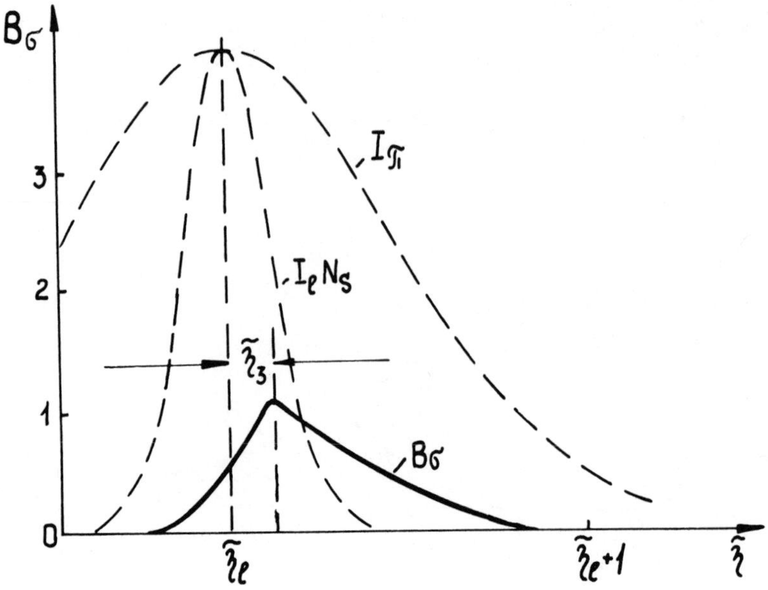
B_σ
3
2
1
0
I_π
$I_\ell N_s$
$\tilde{\eta}_3$
B_σ
$\tilde{\eta}_\ell$
$\tilde{\eta}_\ell+1$
$\tilde{\eta}$

Fig. 2a

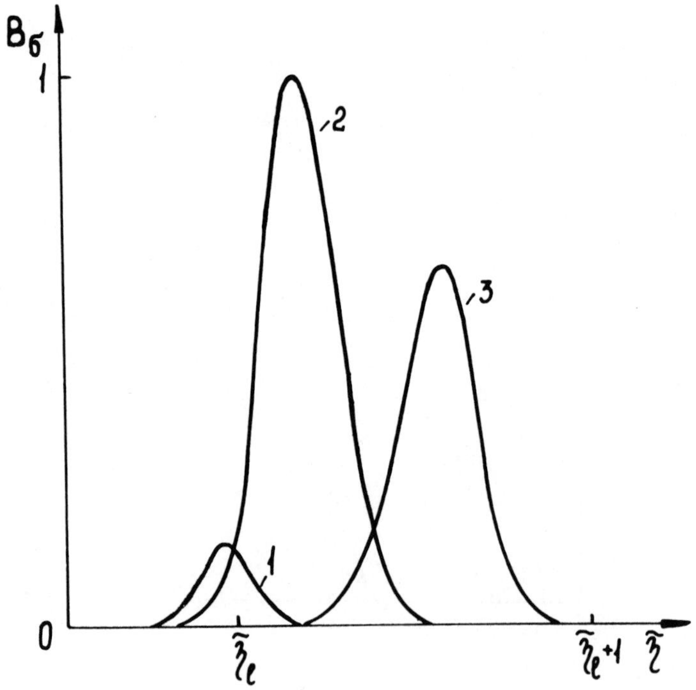
B_σ
1
0
1
2
3
$\tilde{\eta}_\ell$
$\tilde{\eta}_\ell+1$
$\tilde{\eta}$

Fig. 2b

Here

$$B_\sigma(L\xi) = B_\sigma^0 G_\sigma(\xi)\frac{L^2}{\pi^2}\int_{\xi_0}^{\xi}[1+\bar{n}_p+N_s(\xi')]G_s(\xi')\zeta^{-3}e^{2\zeta+(\xi'-\xi)\gamma}d\xi'$$

Though coherent polariton excitation is more effective in this case, the time dependence of the intensity B_σ (after the biharmonic impulse* cessation) appears to be analogous to that discussed above. The analysis of the dependence $B_\sigma(\xi)$ may be carried out similarly to the previous case.

REFERENCES

1. A. S. Davydov, Kvantovaja mechanika, M., Fizmatgiz, 1963.
2. V. L. Strizhevskii, Yu. N. Yashkir, Phys. Stat. Sol. (b), 69, N6 (1975).
3. V. V. Obukhovskii, H. Ponath, V. L. Strizhevskii, Phys. Stat. Sol., 41, 837, 847 (1970).
4. R. Loudon, LSSS (Springer Verlag, N.Y.) p.25, (1969); E. Burstein, S. Ushioda, A. Pinczuk, J. F. Scott, ibid., p.43; J. F. Scott, S. Ushioda, ibid., p.57; S. Ushioda, A. Pinczuk, E. Burstein, D. L. Mills, ibid., p.347.
5. V. L. Strizhevskii, V. V. Obukhovskii, H. Ponath, Zh. Exper. Teor. Fiz., 61, 537 (1971).
6. V. L. Strizhevskii, Zh. Exper. Teor. Fiz. 62, 1446 (1972).
7. F. N. Marchevskii, V. L. Strizhevskii, J. Raman Spectr. 3, 7, 15 (1975).
8. F. N. Marchevskii, V. L. Strizhevskii, Preprint ITR-75-9P, AN USSR, 1975.
9. N. Bloembergen, M. J. Colles, J. Reintjes, C. S. Wang, Ind. J. Pure Appl. Phys. 9, 874 (1971).
10. L. D. Landau, E. M. Lifshitz, Statistical Physics M, Fizmatgiz, 1959.
11. M. Lax, Phys. Rev. 145, 110 (1966); H. Haken, W. Weidlich, Z. Phys. 89, 1 (1966); H. Sannerman, Z. Phys. 188, 480 (1965).
12. L. D. Landau, E. M. Lifshitz, Electrodynamics of the continuous media, M., Fizmatgiz, 1959.
13. V. L. Strizhevskii, Yu. N. Yashkir, Collection "Physical

*pump + input Stokes signal.

foundations of the registration and treatment of the information by the laser radiation" (Theses of the Third All-Union Conference "Physical Foundations of the information transmission by the laser radiation") Kiev, 1973, p.21; V. L. Strizhevskii, Yu. N. Yashkir, Kvantovaja elektronika, M., 2, N5 (1975).
14. D. N. Klyshko, Kvantovaja electronika, M., 2, 265 (1965).
15. L. Laubreau, D. Von der Linde, W. Keiser, Opt. Communs. 7, 193 (1973).

ELECTRONIC RAMAN SCATTERING

Miles V. Klein

Dept. of Physics and Materials Research Laboratory
University of Illinois at Urbana-Champaign
Urbana, Illinois 61801 USA

This paper briefly reviews some aspects of electronic Raman scattering in simple semiconductors. Emphasis is placed on the strength of the scattering as predicted by the f-sum rule. The mechanism for light scattering by free and bound electrons in a many-valley semiconductor is outlined. Data are presented from two systematic studies of the dependence of the spectrum on donor concentration, from Doehler et al on Ge(As) and from Jain and Klein on Si(P). The gross features of the spectra may be understood in terms of a new sum rule, valid for all concentrations. It shows that the strength of the light scattering is a direct manifestation of the short-range, central-cell, donor potential.

This paper will be concerned solely with electronic Raman scattering in semiconductors. I shall first review some aspects of electronic Raman scattering in crystals with simple, non-degenerate band structures, for example GaAs.[1] Then I shall present some new results on free and bound electrons in many-valley semiconductors.

A scattering diagram for a Stokes scattering of a photon by a free electron is shown in Fig. 1. The relevant wave-vectors, angular frequencies, and polarization unit vectors are labelled by $\vec{k}$, ω, and $\vec{e}$, respectively. The cross-section for this process is

$$\frac{d\sigma}{d\Omega} = \frac{\omega_s}{\omega_L} r_0^2 (\vec{e}_L \cdot \vec{e}_s)^2 \sim 10^{-26} \text{ cm}^2 \tag{1}$$

where $r_0 = e^2/mc^2$ is the Thomson radius. The scattering is from density fluctuations, which couple to the square of the vector potential:

$$H' \propto r_0 \rho_q^\dagger \vec{A} \cdot \vec{A} \; . \tag{2}$$

The cross-section for scattering from a collection of many electrons becomes

$$\frac{d^2\sigma}{d\omega\, d\Omega} = r_0^2 (\vec{e}_L \cdot \vec{e}_S)^2 S(q, \omega)\, \omega_S / \omega_L \tag{3}$$

where the structure factor is given by

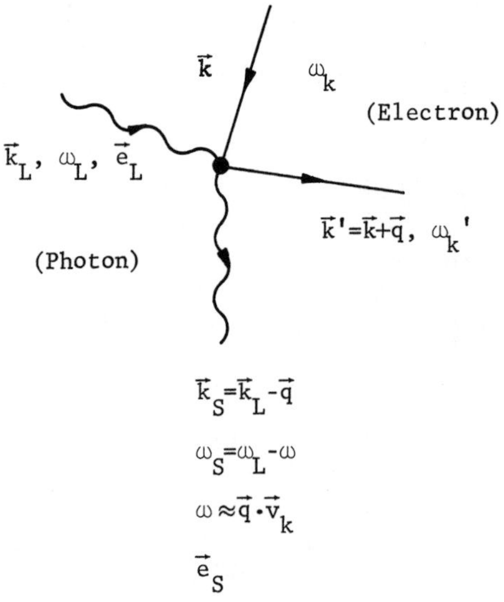

Fig. 1. Diagram illustrating the kinematics of scattering of light by a free electron and introducing the notation used in this paper.

$$S(q,\omega) = \sum_f \mathrm{Av}_i \, |(f\,|\,\rho_q\,|\,i)|^2 \, \delta(\omega_{fi} - \omega) \tag{4}$$

where ρ_q is the density-fluctuation operator

$$\rho_q = \Sigma_j \, e^{i\vec{q}\cdot\vec{r}_j} \tag{5}$$

The parallel polarization selection rule results from the $(\vec{e}_S \cdot \vec{e}_L)$ factor. The overall strength of the scattering from N electrons must obey the f-sum rule, which leads to

$$\frac{1}{N}\int_0^\infty \frac{d^2\sigma}{d\omega\, d\Omega}\, \frac{\omega\, d\omega}{(1+n_\omega)} = r_o^{\,2} (\vec{e}_L \cdot \vec{e}_S)^2 \, \frac{\omega_S}{\omega_L}\, \frac{\hbar q^2}{2m} \tag{6}$$

Here $n_\omega = [\exp(\hbar\omega/k_B T) - 1]^{-1}$. The sum rule and Eq. (6) are valid when there are no velocity-dependent forces, but there may be static potentials. The effects of electron-electron Coulomb interactions are included.

In a simple semiconductor with a single type of carrier in a single band the above results still hold with m in r_o and m in Eq. (6) replaced by a scalar effective mass, provided that the appropriate resonance enhancement factor is inserted if $\hbar\omega_L$ is close to the band gap E_G.

By comparing Eqs. (1) and (6) we see that the sum rule (6) can be interpreted as saying that the average of the cross-section per electron times the excitation frequency is $r_o^{\,2}\hbar q^2/2m$. For scattering of visible laser light $\hbar^2 q^2/2m$ is about 10^{-2} meV, well below the energy of plasmons or of Doppler shifted electrons at the Fermi surface. Thus the average cross-section per electron is of order

$$r_o^{\,2}\hbar q^2/(2m\omega_p) << r_o^{\,2}$$

or (7)

$$r_o^{\,2}\hbar q^2/(2mqv_f) << r_o^{\,2}$$

In polar semiconductors the longitudinal optic phonon will couple to the plasmon. The plasmon part of the resulting coupled modes will still scatter light via the ρA^2 mechanism. There is an additional non-interfering scattering mechanism in which the optical susceptibility is modulated by the macroscopic electric field of the longitudinal coupled modes and by the atomic displacements of the phonon part of the modes. The paper by Mirlin et al at this symposium is a good example of recent work on these coupled modes.

For a simple semiconductor plasma the plasmon contribution to the cross section essentially exhausts the sum-rule, Eq. (6). In this case the contribution to light scattering via the ρA^2 mechanism from electrons being singly excited out of the equilibrium distribution is strongly suppressed by screening. Such excitations have a characteristic excitation frequency given by the Doppler shift qv_f or qv_{rms}. Solid state effects provide additional scattering mechanisms for such excitations, and these do not become screened. The most important of these solid state mechanisms is scattering via spin density fluctuations. This mechanism depends on the spin-orbit splitting of the top of the valence bands. If Δ, the amount of the splitting, is small compared with E_G, then the perturbation in Eq. (2) is replaced by

$$H' \propto r_o |\vec{e}_L \times \vec{e}_S| \left[\frac{\hbar\omega_L}{E_G}\,\frac{\Delta}{E_G}\right] (\rho^{\dagger}_{q\uparrow} - \rho^{\dagger}_{q\downarrow}) \tag{8}$$

where $\rho_{q\uparrow}$ and $\rho_{q\downarrow}$ are density fluctuation operators for spin up and spin down electrons. The spin quantization axis is taken along $\vec{e}_L \times \vec{e}_S$. The perturbation (8) yields a sum rule in which the right hand sides of (6) and (7) are multiplied by

$$\left[\frac{\hbar\omega_L \Delta}{E_G^2}\right]^2 < 1 \tag{9}$$

and $\vec{e}_L \cdot \vec{e}_S$ is replaced by $\vec{e}_L \times \vec{e}_S$. The mean cross-section per electron is thus still very much less than r_o^2.

Consider now a bound electron on a donor in a simple semiconductor. Within the effective mass approximation, the interdonor Raman transitions will still obey Eq. (6) (with m* substituted for m). The situation is quite different for electrons on donors in a many-valley semiconductor such as Si or Ge. The valley degeneracy is partly split by the short range, "central cell" potential, and the 1s orbitals of the hydrogenic ground state combine to form a "valley-orbit split" multiplet. The fully symmetric ground state has the only orbital that is nonzero at the central cell, and its energy is lowered below the hydrogenic effective mass ground state energy by the attractive central cell potential that essentially only it experiences. To a first approximation the other orbitals retain the effective mass energy. Raman scattering from the ground state to one or more excited states of the multiplet is strongly Raman-allowed.

The mechanism for scattering of light by electrons in a many-valley semiconductor is provided by a generalization of the ρA^2 term for a single carrier.[2-6] If $\vec{\mu}_\ell$ is the (m/m*) tensor for the ℓth valley, then the kinetic energy for an electron in that valley is $\frac{1}{2}m\vec{p}\cdot\vec{\mu}_\ell\cdot\vec{p}$. When we make the replacement $\vec{p}\rightarrow\vec{p}+e\vec{A}/c$, and collect the bilinear terms in A we find for a collection of electrons that we must make the replacement

$$(\vec{e}_L\cdot\vec{e}_S)r_o\rho_q^{\dagger}\rightarrow r_o\hat{\rho}_q^{\dagger} \tag{10a}$$

where

$$\hat{\rho}_q=\Sigma_\ell\,\alpha^\ell\rho_\ell(q) \tag{10b}$$

and

$$\alpha^\ell=\vec{e}_L\cdot\overleftrightarrow{\mu}_\ell\cdot\vec{e}_S\,. \tag{10c}$$

$\rho_\ell(q)$ denotes the density fluctuation operator for electrons in valley ℓ.

The quantity $\hat{\rho}(q)$ is an intervalley density fluctuation operator. When it is inserted into Eq. (4) a form factor $\hat{S}(q,\omega)$ is obtained:

$$\hat{S}(q,\omega)=\sum_f Av_i\,|(f|\hat{\rho}(q)|i)|^2\,\delta(\omega_{fi}-\omega) \tag{11}$$

This gives the cross-section for light scattering:

$$\frac{d^2\sigma}{d\Omega d\omega} = r_o^2 \frac{\omega_S}{\omega_L} \hat{S}(q,\omega) \tag{12}$$

For a dilute system of neutral donors the spectrum of intervalley fluctuations is just that provided by the valley-orbit split ground state multiplet. This was first observed for Si(P) by Wright and Mooradian, who observed a strong line of E symmetry at 105 cm^{-1}.[7] We are interested here in how this line, or rather how the spectrum of intervalley fluctuations, evolves as the doping is increased and the semiconductor passes into the metallic impurity band regime. That strong valley-orbit light scattering would persist into the metallic regime was first shown by Colwell and Klein in SiC(N).[8] They found that above the transition the spectrum was a continuum that started at zero energy.

A systematic study of this effect as a function of doping was first performed by Doehler et al on Ge(As).[9] Figure 2 was taken from their paper, and it shows how the valley-orbit line at 36 cm^{-1} for isolated As donors broadens asymmetrically. For donor concentrations n_d of 1.7×10^{17} cm^{-3} and higher this donor line seems to appear superimposed on a continuum that begins at zero energy shift and is monotonically decreasing in intensity. The critical concentration for the metal-semiconductor transition in Ge(As) is about 3×10^{17} cm^{-3}. The broadened valley-orbit line appears to persist well into the metallic regime.

K. Jain and the author have made similar measurements on Si(P) having donor concentrations n_d in the range 7×10^{16} to 5×10^{18} cm^{-3}. The spectra were excited with 5 W of unpolarized 1.064μ Nd:YAG laser light. Figure 3(a) for $n_d = 7 \times 10^{16}$ cm^{-3} shows a single sharp line at 105 cm^{-1} caused by the 1s(A_1) to 1s(E) valley-orbit transition of the donor electron. Our results are summarized in Fig. 3. For this system the metal-semiconductor transition occurs for $n_d \sim 3 \times 10^{18}$ cm^{-3}. It appears that the residual effect of the broadened valley-orbit line has vanished at the transition and above. (Figs. 3(i) and (j).) This behavior is not consistent with that for Ge(As) shown in Fig. 2.[10]

For both Ge(As) and Si(P) the polarization properties of the electronic continuum that is observed at the highest concentrations studied is the same as that of the valley-orbit line and is the same as that predicted for the intervalley fluctuation mechanism discussed above. We leave detailed discussions of

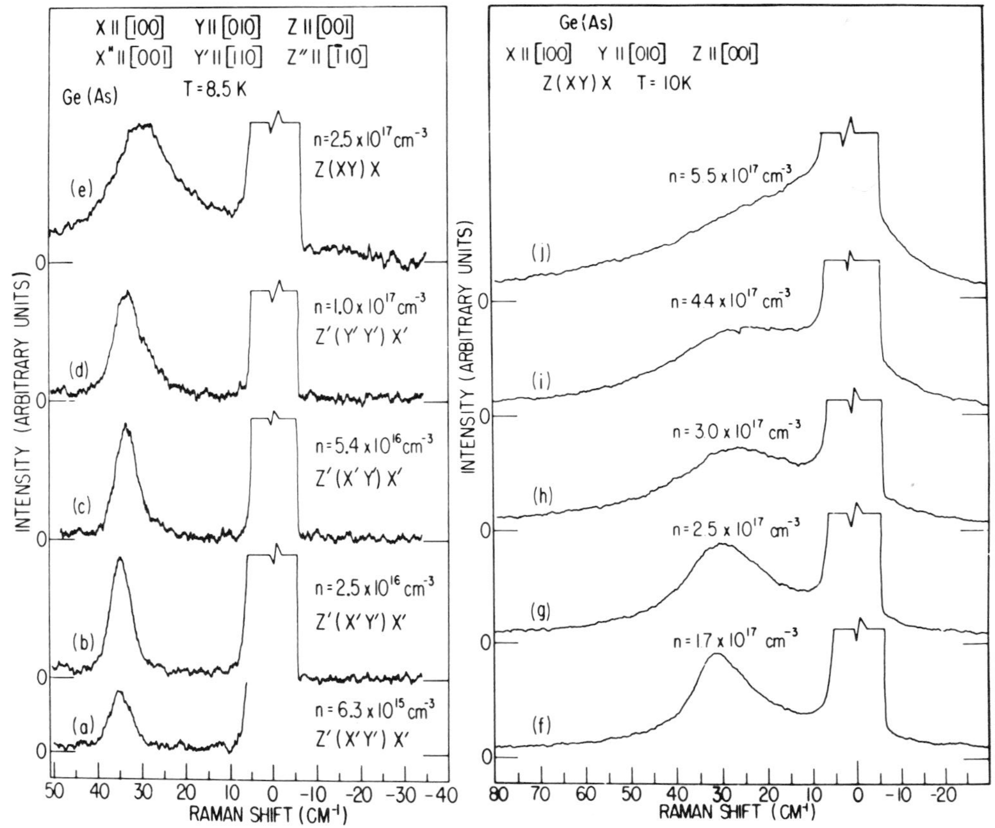

Fig. 2. Temperature-dependence of the Raman spectra of Ge(As). Excitation was provided by a 2.1 μ ABC:YAG laser. From Doehler et al, Ref. 9.

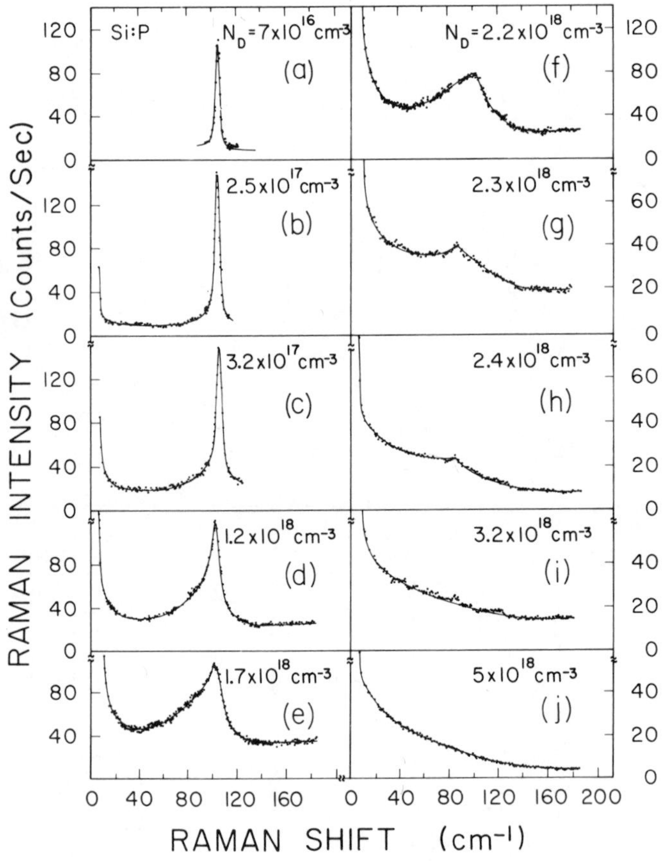

Fig. 3. Stokes Raman spectra of Si(P) for various donor concentrations. Due to laser beam heating the actual sample temperature, as determined from the ratio of Stokes and anti-Stokes strengths, increased from 20 K for sample (a) to about 50 K for samples (h-j). The small peak at 84 cm^{-1} in spectra (g) and (h) is a non-lasing fluorescence line from the Nd:YAG laser.

the spectra and their significance for the various theories of the metal-insulator transition for future work. Here we wish only to establish a general property of the spectra that is connected with a new sum rule.[11]

The new sum-rule may be derived in a manner similar to the "f-sum rule" obeyed by $S(q, \omega)$ for a single-component plasma. By arguments similar to those of Nozières and Pines[12] extended to finite temperatures, we find

$$\int_0^\infty \frac{\hat{S}(q,\omega)\omega\, d\omega}{1+n_\omega} = \Sigma_{\ell\ell'} \alpha^{\ell}\alpha^{\ell'} \langle [[\rho_\ell(q)^\dagger, H], \rho_\ell{}'(q)] \rangle_T , \tag{13}$$

where

$$H = V_c + V_{ee} + T + V_{sr} \tag{14}$$

is the total Hamiltonian, and $\langle \cdots \rangle_T$ denotes a thermal expectation value in the equilibrium ensemble. V_c, the Coulomb potential due to the donors, commutes with $\rho_\ell(q)^\dagger$, as does V_{ee}, the electron-electron Coulomb potential. The kinetic energy is

$$T = \sum_{j\ell} \frac{1}{2m} \vec{p}_j \cdot \overleftrightarrow{\mu}_\ell \cdot \vec{p}_j . \tag{15}$$

The short range potential energy is given by a sum over donor sites R of the form:

$$V_{sr} = - V_0 \sum_R F(R)^\dagger F(R) \tag{16}$$

where V_0 is the strength of the central cell potential, and $F(r)$ is the field operator associated with the effective mass envelope function in the A_1, ground, multiplet. The contributions of T and V_{sr} to the right hand side of Eq. (13) may be computed directly. The respective contributions give two independent terms:

$$\int_0^\infty \frac{\hat{S}(q,\omega)\omega d\omega}{1+n_\omega} = N \sum_{\ell=1}^{v} \frac{(\alpha^\ell)^2}{v} \frac{\hbar}{2m} (\vec{q} \cdot \overleftrightarrow{\mu}_\ell \cdot \vec{q})$$

$$+ \frac{N}{2v^2} \Sigma_{\ell\ell'} (\alpha^\ell - \alpha^{\ell'})^2 V_0 p_{cc} . \tag{17}$$

Here N is the number of electrons, v is the number of valleys, and

$$p_{cc} = N^{-1} \sum_R \langle F(R)^\dagger F(R) \rangle_T \tag{18}$$

is the average number of electrons in the central cell of a donor. The valley-orbit splitting for an isolated donor is given by

$$\hbar\omega_{VO} = V_0 |F(R)|^2 \tag{19}$$

Thus we may write

$$V_0 = \hbar\omega_{VO}/p_{cc}^{is} , \tag{20}$$

where p_{cc}^{is} is the value of p_{cc} for an isolated neutral donor in its $1s(A_1)$ ground state. Equations (17) and (18) state the sum rule for VO Raman scattering.

Our experiment on Si(P) was conducted with $\vec{e}_1 \| (110)$, $\vec{e}_2 \| (1\bar{1}0)$, $\vec{k}_1 \| (001)$, and $\vec{k}_2 \| (110)$. Equations (17) and (18) then give

$$N^{-1} \int_0^\infty \frac{\hat{S}(q,\omega)\omega d\omega}{1+n_\omega} = \frac{(\mu_\parallel - \mu_\perp)^2}{6} \left[\frac{\hbar q^{*2}}{2m} + \frac{\omega_{VO} p_{cc}}{p_{cc}^{is}} \right] . \tag{21}$$

The parameter q^* is given by

$$q^{*2} = (\vec{q} \cdot \overleftrightarrow{\mu}_\ell \cdot \vec{q}) \tag{22}$$

for valleys ℓ along $\pm x$ or $\pm y$ cube axes; $\mu_\perp$ and $\mu_\parallel$ are principal values of μ. For the conditions of our experiment, we find

$$\frac{\hbar q^{*2}}{2m} = 0.12 \text{ cm}^{-1} \tag{23}$$

Since $\omega_{VO} = 105 \text{ cm}^{-1}$, the second, or intervalley scattering, term in (21) will dominate the sum-rule as long as p_{cc}/p_{cc}^{is} is greater than about 10^{-2}.

We have computed values for the left-hand side of Eq. (21) for our Raman data of Fig. 3. The strength of the 520 cm^{-1} Raman-active k = 0 phonon mode was used as an internal calibration.

The results are shown in Fig. 4. The scatter is probably due to variations in relative response of the photomultiplier tube and due to difficulties in drawing a consistent base line. There seems to be a decrease of about a factor of four in the integrated spectrum, in the sense of (21), over the concentration range of our measurements. According to Eq. (21), this implies that p_{cc} has dropped to $\frac{1}{4}$ of its value for isolated donors.

There are two explanations for the decrease of p_{cc}. Laser beam heating of the samples during the measurements for Fig. 3 and for Fig. 4 caused the temperature of the crystals to rise from about 20 K for $n_d = 0.6 - 3.2\text{x}10^{17}$ cm^{-3} to 50 K for $n_d = 3.2 - 5.0\text{x}10^{18}$ cm^{-3}. Thermal activation could account for a decrease in p_{cc}, since the ground-state $1s(A_1)$ orbital is the only one with appreciable amplitude at the donor sites. For instance, if one assumes that there are five other orbitals 80 cm^{-1} above the $1s(A_1)$ orbitals, at T = 50 K one estimates a value of p_{cc}/p_{cc}^{is} of 0.6 from this effect. The other major cause of the decrease in central-cell occupation is probably due to delocalization of the $1s(A_1)$ orbitals as they interact with one another at high donor concentrations. As bonding-type molecular orbitals are formed in

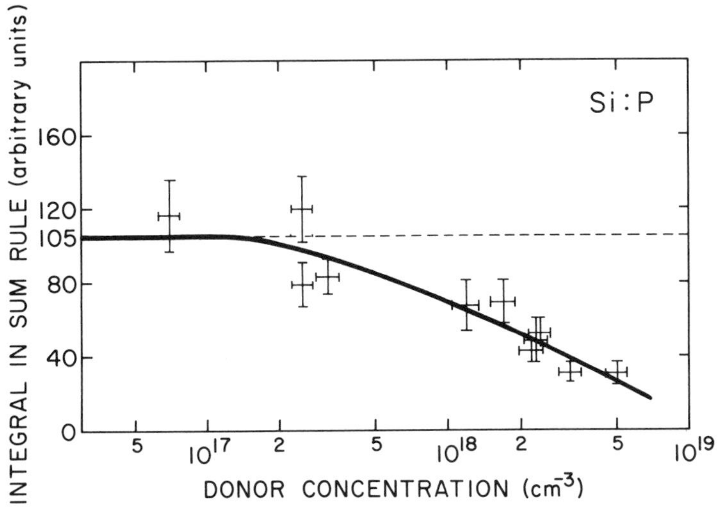

Fig. 4. Dependence of the left-hand side of Eq. (21) on donor concentration. The strength of the 520 cm^{-1} zone-center optic phonon was used for calibration.

the region between donors, the occupation probability in the central cells will decrease. Careful, temperature-dependent studies will be necessary to separate these two causes of the decrease in p_{cc}.

We conclude that the strength, in the senses of the sum rule, of the valley-orbit Raman line is due to the short-range, central cell potential. The average cross-section per electron is of order r_0^2. This is true for both localized orbitals in the dilute concentration regime and for delocalized orbitals in the impurity band regions. If the short-range potential could be "turned-off", the strength of the scattering would decrease by several orders of magnitude and would obey the usual q^2-dependent sum-rule.

REFERENCES

*This work was supported in part by the National Science Foundation under Grant GH-37757 and in part by the Advanced Research Projects Agency under contract DAHC 15-73-G-10.

1. The subject of electronic Raman scattering in semiconductors will be covered in a review article by the author to appear in Raman Scattering, M. Cardona, ed. (Springer Verlag, to be published).
2. P. M. Platzman, Phys. Rev. 139, A379 (1965).
3. S. V. Gantsevich, V. L. Gurevich, V. D. Kagan, and R. Katilius in Light Scattering in Solids, edited by M. Balkanski (Flammarion, Paris, 1971) p. 94.
4. S. V. Gantsevich, V. L. Gurevich, and R. Katilius, Zh. Eksp. Teor. Fiz. 57, 503 (1969); Soviet Phys. JETP 30, 276 (1970).
5. N. Tzoar and E. N. Foo in Light Scattering in Solids, edited by M. Balkanski (Flammarion, Paris 1971) p. 119.
6. E. N. Foo and H. Tzoar, Phys. Rev. B6, 4553 (1972).
7. G. B. Wright and A. Mooradian, Phys. Rev. Lett. 18, 608 (1967).
8. P. J. Colwell and M. V. Klein, Phys. Rev. B6, 498 (1972).
9. J. Doehler, P. J. Colwell, and S. A. Solin, Phys. Rev. Lett. 34, 584 (1975); J. Doehler, Phys. Rev. B (to be published).
10. Note added at the time of writing: Acting upon a suggestion of Prof. S. A. Solin, we have examined more closely the temperature-dependence of the more heavily doped samples

used in parts (h-j) of Fig. 3. When the temperature is lowered from 50 to 30K, a knee is found at 85 cm^{-1} for $n_d = 3.2x10^{18}$. It moves to 55 cm^{-1} for $n_d = 5x10^{18}$. Thus Si(P) is not as different from Ge(As) as a comparison of Figs. 2 and 3 would suggest. These results are discussed in more detail in Ref. 11.

11. K. Jain and M. V. Klein, in Proc. Third International Conference on Light Scattering in Solids, edited by M. Balkanski (Flammarian, Paris, to be published); K. Jain, S. Lai, and M. V. Klein, to be published.
12. P. Nozières and D. Pines, Phys. Rev. 109, 741 (1958).

THEORY OF SPIN-FLIP LINE SHAPE IN CdS

P. A. Wolff,* J. G. Ramos‡ and S. Yuen

Department of Physics
Massachusetts Institute of Technology
Cambridge MA USA

INTRODUCTION

Spin-flip light scattering is a process in which an electron in a crystal (generally a semiconductor) scatters light with change of its spin direction. Usually such experiments are performed in a static magnetic field. The scattered frequency is then shifted from the incident frequency by the electron spin resonance frequency. The latter varies linearly with applied field--hence, spin-flip scattering is a magnetically tunable Raman process. This feature of spin-flip scattering was paramount in the minds of early workers in the field, and provided much of the impetus for subsequent research. An elegant device--the spin-flip laser--resulted from this work.

To date, it has been less well appreciated that the spin-flip process also provides fundamental information concerning electron dynamics in semiconductors--dynamics both of the electron spin motion, and of its translational motion through the crystal lattice. To exploit this fact, one must know how to relate the

---

*Work sponsored by the Air Force Office of Scientific Research under AFOSR Contract/Grant No. AFOSR-71-2010.

‡Permanent address: Instituto de Fisica "Gleb Wataghin" Universidade Estadual de Campinas, Campinas, SP, Brasil.
Fellowship from FAPEST (Brasil).

properties of the spin-flip line (such as position, intensity, line width, line shape, etc.) to parameters which characterize electron motion. The purpose of this paper is to discuss these relationships and, in particular, to express the spin-flip line shape in terms of microscopic parameters. The most important of these is the spin diffusion constant which cannot easily be measured in other ways. In particular, practically nothing is known about spin diffusion in the vicinity of the metal-insulator transition in semiconductors.

We will test our ideas concerning the spin-flip line shape by comparing the theory to experiments on n-CdS. CdS has a large spin-flip cross-section, and the line shape has been carefully measured. For this case the theory works well. It accounts in detail, and without adjustable parameters, for most of the data concerning angular and temperature variation of the spin-flip linewidth. In the low temperature range, where discrepancies are found, they are understandable in terms of the effects of the electron-electron interaction. Neither the data nor the theory are yet good enough to characterize this region--but, it is clearly an area for further study.

In spin-flip measurements, the magnetic tunability of the scattering aids the experimentalist by giving a unique signature to the spin-flip signals--a resonance whose position in frequency is known, and varies linearly with magnetic field. In this, and other respects, spin-flip scattering is akin to EPR. Spin-flip scattering and EPR measure the same basic property of the conduction electron system--the transverse spin susceptibility, $\chi^{+}(k, \omega)$. EPR experiments determine essentially the $k = 0$ value of $\chi^{+}(k, \omega)$, whereas spin-flip scattering can be used to determine $\chi^{+}(k, \omega)$ over a range of k. In this sense, spin-flip scattering is a more powerful technique than EPR.

FORMULATION OF THEORY

For n-type III-V semiconductors, the spin-flip matrix element has the form[1]

$$M_{\text{spin-flip}} = D\left(\frac{e^2}{mc^2}\right) \langle f \,|\, (\vec{\sigma} \cdot \vec{\epsilon}_o \times \vec{\epsilon}_1)\, e^{i\vec{k}\cdot\vec{r}} \,|\, i \rangle \,, \tag{1}$$

where $\vec{k} = (\vec{k}_0 - \vec{k}_1)$; $\vec{k}_0$ and $\vec{k}_1$ are wave vectors of the incident and scattered light waves; $\vec{\epsilon}_0$ and ϵ_1 are the corresponding polarization vectors; and D is a dimensionless constant whose value is determined by a sum of interband matrix elements.[1] The states "i" and "f" are both in the conduction band. We regard D as a known parameter--evaluated either from theory or experiment. D has a resonance,[1] as the photon energy approaches the direct band gap, which can be used to enhance spin-flip cross-sections. In CdS there is a particularly fortunate coincidence of the 4880 Å line of the Ar^+ laser with a strong, interband transition. Huge spin-flip cross-sections--of order 10^{-18} cm^2/ster.--have been reported[2] with such excitation.

The theory of spin-flip scattering has not yet been worked out in detail for II-VI semiconductors. These crystals are more complicated--for the present purpose--than III-V semiconductors because they are hexagonal. The uniaxial crystal field lifts the degeneracy (required by symmetry in cubic crystals) at the $k = 0$ point of the valence band. In CdS the splitting is small--about .05 eV. Hence, if the photon frequencies are well away from resonance, the spin-flip matrix element has the same form as that for a III-V semiconductor. Spin-flip polarization selection rules consistent with this interpretation are observed with green (5145 Å) Ar^+ laser light scattering in CdS.

On the other hand, if the incident light is resonant with a particular interband transition, this channel dominates the spin-flip matrix element and gives it a different form from that indicated by Eq. (1). Such a situation occurs with blue light (4880 Å line of the Ar^+ laser) illumination of CdS. As was first pointed out by Thomas and Hopfield,[2] this frequency nearly coincides with that of a bound exciton transition, of the exciton associated with the uppermost ($J = 3/2$, $m_J = \pm 3/2$) valence band. The spin-flip process is dominated by the virtual transition passing through this level, and the spin-flip matrix element takes the form:

$$M_{\text{spin-flip}} = D\left(\frac{e^2}{mc^2}\right) \langle f \,|\, (\vec{\sigma} \cdot \hat{c})(\hat{c} \cdot \vec{\epsilon}_0 \times \vec{\epsilon}_1)\, e^{i\vec{k}\cdot\vec{r}} \,|\, i \rangle , \qquad (2)$$

where $\hat{c}$ is a unit vector in the c-axis direction. Spin-flip selection rules consistent with Eq. (2) were observed by Thomas and Hopfield with 4880 Å radiation.

Thus, depending upon the laser frequency, the spin-flip cross-section in CdS can take a variety of forms. For our purpose, it is sufficient to note that the matrix element has the general form

$$M_{\text{spin-flip}} = D\left(\frac{e^2}{mc^2}\right) \langle f \,|\, (\vec{\sigma} \cdot \vec{\alpha})\, e^{i\vec{k}\cdot\vec{r}} \,|\, i \rangle \ , \tag{3}$$

where $\vec{\alpha}$ is a vector that does not involve the electronic variables. We anticipate that spin-flip matrix elements will generally have this form. The essential point is that, in spin-flip scattering with momentum transfer $\vec{k} = (\vec{k}_0 - \vec{k}_1)$, light couples to the electronic variable $\vec{\sigma} e^{i\vec{k}\cdot\vec{r}}$ which is the kth Fourier component of the spin density. Our subsequent considerations will be based on this fact and will not require a knowledge of the vector $\vec{\alpha}$.

So far, the discussion has been concerned with single particle spin-flip scattering. In deriving Eq. (1), for instance, one imagines a semiconductor crystal containing a single conduction electron in otherwise intrinsic material. In practice, a semiconductor always contains many carriers--often sufficiently many that carrier-carrier interactions play a central role in determining its properties. These interactions are particularly important at low temperatures, and can lead to a Mott (metal-insulator) transition in appropriately doped crystals. We believe that spin-flip scattering will prove to be an especially powerful tool for studying such transitions.

Many body effects in light scattering have been discussed by Hamilton and McWhorter.[3] Their prescription for treating the problem is a straightforward extension of Eq. (3). In an interacting electron system they assume that light couples to the conduction electron spin density via a spin-flip interaction of the form:

$$H_{sf} = D\left(\frac{e^2}{mc^2}\right) \sum_j \left[(\vec{\sigma}_j \cdot \vec{\alpha}) e^{i\vec{k}\cdot\vec{r}_j}\right] \ , \tag{4}$$

where the j-sum runs over all conduction band electrons. The matrix elements of H_{sf}, taken between many-body states for the conduction electrons, then determine the scattering rate. This prescription includes the effects of many-body interactions in the initial and final states, but ignores their influence on the virtual, intermediate states which appear in the sum which determines D. Generally speaking, it is a good approximation to ignore many-body effects in virtual states since electrons spend only a short time in them.

Henceforth, we use Eq. (4) to describe spin-flip scattering. This interaction involves the operator

$$\vec{\sigma}_{-\vec{k}} = \sum_j (\sigma_j e^{i\vec{k}\cdot\vec{r}_j}) \quad , \tag{5}$$

the Fourier transform of the electron spin density. A straightforward calculation, paralleling van Hove's[4] well-known analysis, yields the following expression for the Stokes portion of the spin-flip spectrum:

$$\left.\frac{d^2\sigma}{d\omega\, d\Omega}\right|_{\text{Stokes}} = 2\pi \,|D|^2 \left(\frac{e^2}{mc^2}\right)^2 \left(\frac{\omega_1}{\omega_0}\right) S^+(\vec{k},\omega) \quad , \tag{6}$$

where

$$S^+(\vec{k},\omega) \equiv \int_{\infty}^{\infty} e^{i\omega t} \langle \sigma^-_{\vec{k}}(t)\sigma^+_{-\vec{k}}(0) \rangle \frac{dt}{2\pi} \tag{7}$$

is the Fourier transform of the spin-spin correlation function. The angular brackets in this expression indicate a thermal average over the exact, many-body states of the interacting, conduction electron system. At equilibrium, the spin-spin correlation function is related to the transverse spin susceptibility by the fluctuation-dissipation theorem:[5]

$$\left. S^+(\vec{k},\omega) \right|_{\text{Equil.}} = \frac{-4\,\text{Im}\,\chi^+[(\vec{k},\omega)]}{[1-e^{-\beta\vec{\omega}}]} \quad , \tag{8}$$

where

$$\chi^+(\vec{k},\omega) = -\frac{i}{4}\int_0^{\infty} e^{i\omega t} \langle [\sigma^-_{\vec{k}}(t), \sigma^+_{-\vec{k}}(0)] \rangle dt \tag{9}$$

is the susceptibility. The manipulations leading from Eq. (3) to Eqs. (6), (7), (8) and (9) are quite standard, so we have not discussed them in detail. The final result is important, because it demonstrates that spin-flip light scattering experiments measure the k-dependent spin susceptibility. This conclusion is implicit in several analyses[6] of the spin-flip problem, but has not yet been fully exploited to study electron dynamics in semiconductors such as CdS.

THE SPIN SUSCEPTIBILITY

The central problem in calculating the spin-flip spectrum is that of determining the transverse spin susceptibility, $\chi^+(\vec{k},\omega)$. We are interested in this function for k-values typical of light scattering experiments. In the CdS work mentioned above, the wave vector transfer (k) was of order $5\text{x}10^5\ \text{cm}^{-1}$. The corresponding length, $\ell = (2\pi/k) \simeq 10^{-5}$ cm, is large compared to the average interelectron spacing, the effective Bohr radius of electrons in CdS, and the mean free path. Under such conditions, the spin magnetization (even in the presence of strong many-body effects) can be described by a Bloch equation of the form:

$$\frac{\partial M^+}{\partial t} + i\omega_s M^+ + \frac{(M^+ - \chi_o H^+)}{T_2} - D_s \nabla^2 M^+ = i\mu g M_o H^+ . \tag{10}$$

Here T_2 is the transverse spin-relaxation time, D_s the spin diffusion constant, and H^+ the perturbing field that excites the spin system. If H^+ is a plane wave, Eq. (10) has the solution:

$$M^+ = \frac{(-\omega_s + \frac{i}{T_2})\chi_o H^+}{[(\omega-\omega_s) + i(\frac{1}{T_2} + D_s k^2)]} , \tag{11}$$

implying that

$$\chi^+(\vec{k},\omega) \equiv \frac{M^+(\vec{k},\omega)}{H^+(\vec{k},\omega)} = \frac{(-\omega_s + \frac{i}{T_2})\chi_o}{[(\omega - \omega_s + i(\frac{1}{T_2} + D_s k^2)]} . \tag{12}$$

The structure factor becomes

$$S^{+}(\vec{k}, \omega) = \frac{4}{\pi(1 - e^{-\beta\omega})} \left[\frac{\gamma \chi_o}{(\omega - \omega_s)^2 + \gamma^2} \right] \tag{13}$$

where

$$\gamma \equiv [\frac{1}{T_2} + D_s k^2] . \tag{14}$$

The main features of this spectrum have been confirmed by experiments of Scott, Damen and Fleury[7] (SDF). Their line shapes are Lorentzian, centered at the spin resonance frequency of electrons in CdS. The k^2 variation of the line width was tested by measuring the dependence of linewidth on scattering angle (θ). Equation (14) implies that the linewidth (γ) has angular variation:

$$\gamma = [\frac{1}{T_2} + 4 D_s k_o^2 \sin^2(\theta/2)] . \tag{15}$$

SDF's data fit Eq. (15) to within experimental error. The agreement suggests that our general picture of the spin-flip line shape is correct, and that spin diffusion is indeed responsible for the k^2 variation of the line width. This feature of spin-flip scattering enables one, even in an interacting electron system, to make a direct measurement of the spin diffusion constant.

SPIN DIFFUSION--COMPARISON WITH EXPERIMENT

Electron motion in semiconductors is most often described as a single particle process, in which electron-electron interactions are ignored. This picture is valid at elevated temperatures, or in heavily doped (degenerate) samples even near T = 0. When the single particle picture is applicable, the spin diffusion (D_s) and particle diffusion (D) constants are equal. Moreover, the particle diffusion constant is related to the mobility (μ) via the Einstein relation. Thus, whenever many-body effects are unimportant, there is a direct relationship between the spin-flip line width and a standard transport coefficient.

The preceding argument implies that mobility data can be used to estimate high temperature spin-flip linewidths in CdS. At low temperatures, however, the single particle picture fails in n-CdS samples of the doping level ($n \simeq 5 \times 10^{17}$/cc) studied by

SDF. This sample has carrier concentration only slightly higher than that at which a transition from metallic to insulating behavior takes place at low temperatures in CdS (Mott transition).[8] Nevertheless, in the absence of a theory for this transition, we have attempted to analyze the SDF data by assuming $D = D_S$, and relating D to the measured mobility (μ) via the Einstein relation. This procedure is suspect in two ways--D_S need not equal D in an interacting system; there may also be many-body corrections to the Einstein relation.[9]

We now compare the measured spin-flip linewidths, with those calculated on the assumption $D_S = D$. To estimate D, we use the Einstein relation and mobility data[8] for n-CdS. The classical Einstein relation has the form $D = (kT/e)\mu$. This result has been generalized to degenerate and interacting systems by Kubo.[9] He finds

$$D = \left(\frac{n\mu}{e}\right)\left(\frac{\partial n}{\partial \zeta}\right)^{-1}, \tag{16}$$

where ζ is the chemical potential. This result can also be rewritten in terms of a compressibility (κ) via the relation

$$n^2\kappa = \left(\frac{\partial n}{\partial \zeta}\right). \tag{17}$$

To calculate κ we assume the electrons are a non-interacting gas. κ is then evaluated, as a function of temperature, from tabulated integrals of Fermi functions. The mobility data of Toyotomi and Morigaki enable us to determine D via Eq. (16) and the linewidth via Eq. (15) and the assumption $D = D_S$. Figure 1 shows a comparison of measured and calculated linewidths determined in this way for an $n = 5\text{x}10^{17}$/cc CdS crystal. The agreement is good. Both μ and $\partial n/\partial \zeta$ for this sample are temperature dependent. At higher temperatures, μ is constant and the nearly linear temperature variation of the linewidth follows from the classical Einstein relation $D = (kT\mu/e)$. On the other hand, at low temperatures the Einstein coefficient saturates and the linewidth variation is caused by the temperature variation of μ, which is quite strong. This simple theory also accounts for spin-flip linewidths measured in more heavily doped n-CdS samples.[10]

At the lowest temperatures ($T < 20^{\circ}$K) the theory outlined above predicts values of D_S <u>larger</u> than those observed in the $n = 5\text{x}10^{17}$/cc sample. The discrepancy at 4°K is about a factor 7 -- well outside experimental error. There are two possible

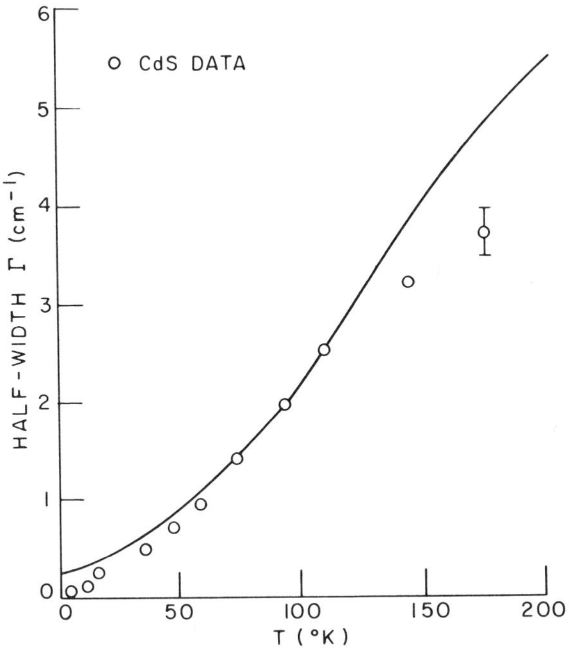

Fig. 1. Spin-flip linewidth vs. temperature for n-CdS.

sources of this discrepancy: (i) Impurity banding, which modifies $\partial n/\partial \zeta$ [see Eq. (16)]. This effect amounts to a modification of the single electron dynamics by the impurity potential.
(ii) Many-body effects. Electron-electron interactions can modify the Einstein relation [Eq. (16)], or cause the assumption $D = D_S$ to fail. In either case, our analysis would be incorrect.

Preliminary calculations, based on theories of Matsubara and Toyozawa, Kane, Cyrot-Lackmann and Lukes,[11] lead us to the conclusion that impurity banding effects cannot account for the seven-fold discrepancy between theory and experiment. Thus, we tentatively ascribe this discrepancy to electron-electron interaction. Further studies will be required to resolve this question. In any case, it seems clear that spin-flip scattering can give important information concerning this region of semiconductor behavior.

One of the authors (P.A.W.) has had extensive discussions with S. Geschwind concerning spin-flip scattering. We are grateful for his advice and encouragement concerning this problem.

REFERENCES

1. Y. Yafet, Phys. Rev. 152, 858 (1966); P. L. Kelley and G. B. Wright, D. L. Kelley and S. H. Groves, Procs. of Intl. Conf. on Light Scattering Spectra of Solids (Springer-Verlag, New York, 1969).
2. D. G. Thomas and J. J. Hopfield, Phys. Rev. 175, 1021 (1968).
3. D. C. Hamilton and A. L. McWhorter, Procs. Intl. Conf. on Light Scattering Spectra of Solids (Springer-Verlag, New York, 1968).
4. L. van Hove, Phys. Rev. 95, 249 (1954).
5. D. N. Zubarev, Soviet Physics--Uspekhi 3, 320 (1960).
6. R. W. Davies and F. A. Blum, Phys. Rev. B3, 3321 (1971); R. W. Davies, Phys. Rev. B7, 3731 (1972); S. Yuen, P. A. Wolff and B. Lax, Phys. Rev. B9, 3394 (1974).
7. P. A. Fleury and J. F. Scott, Phys. Rev. B3, 1979 (1971); J. F. Scott, T. C. Damen and P. A. Fleury, Phys. Rev. B6, 3856 (1972).
8. S. Toyotomi and K. Morigaki, Jour. Phys. Soc. Japan 25, 807 (1968). See also N. F. Mott and E. A. Davis, Electronic Processes in Non-Crystalline Materials (Clarendon Press, Oxford, 1971).
9. R. Kubo, Reports on Progress in Physics, Vol. XXIX (1966), Page 255.
10. S. Geschwind, R. Romestain, G. E. Devlin and P. A. Wolff, Proc. XII Intl. Conf. on Phys. of Semiconductors, (B.G. Teubner, Stuttgart, 1974).
11. T. Matsubara and Y. Toyozawa, Prog. Theor, Phys. 26, 739 (1961); E. O. Kane, Phys. Rev. 131, 79 (1963); T. Lukes, K.T.S. Somaratma and K. Tharmalingam, J. Phys. C; Solid State Phys. 3, 1631 (1970); F. Cyrot-Lackmann and J. P. Gaspard, J. Phys. C; Solid State Phys. 7, 1829 (1974).

INTERACTION BETWEEN THE LOW FREQUENCY BRANCHES OF A CRYSTAL'S ENERGY SPECTRUM NEAR PHASE TRANSITIONS

Yu. A. Popkov, V. V. Eremenko, V. I. Fomin, and A. P. Mokhir

Physico-Technical Institute of Low Temperatures
Academy of Sciences of the Ukrainian SSR
Kharkov, USSR

INTRODUCTION

Interactions between various excitations in solids have long been a problem attracting physicists' interest from both theoretical and experimental viewpoints. Recent studies of phase transitions permitted a number of new phenomena to be predicted and found, and the known ones to be understood better. Here, not of the least importance, has been Raman scattering spectroscopy which was revived by the development of laser techniques. It suffices to mention studies of polaritons, Fermi-resonance, Jahn-Teller effect, magnon interaction under two-magnon scattering, interactions between soft optic modes and acoustic vibrations. These and other phenomena have been discussed at great length in the literature [1-3] and at international conferences on light scattering in solids [4, 5].

During most phase transitions the frequency of some excitation (soft optic phonon modes under structural transitions, magnons under magnetic ordering, rotons in superfluid helium, etc.) turns to zero. A strong temperature dependence of such frequencies near a phase transition causes various branches of the energy spectrum to cross and provides a good means for studying interactions between the appropriate quasi-particles.

Note that a soft mode always has full symmetry in the ordered phase and is therefore Raman-active.

This paper discusses the behavior of soft optic modes, their interaction with acoustic vibrations, and the temperature dependence of two-magnon light scattering in $KMnF_3$ studies by methods of inelastic light scattering (Raman and Mandelstam-Brillouin). Besides, an anomalous behavior of the optic phonon in $CoCO_3$ under antiferromagnetic ordering was considered, which is related to the specific characteristics of the exciton and magnon spectra for $CoCO_3$.

SOFT MODES IN $KMnF_3$

At room temperature $KMnF_3$ has the perovskite structure O_h^1, where the first order Raman spectrum is forbidden. As the temperature falls, two structural phase transitions lowering the lattice symmetry occur. The first transition ($T_{c1} = 187.6^oK$) is attributed [6] to the soft mode Γ_{25} at the [111] boundary of the Brillouin zone (BZ), and the second one ($T_{c2} = 102^oK$) to the soft mode M_3 at the [110] boundary. Critical temperatures vary greatly in the literature and we here present sufficiently accurate results from Mandelstam-Brillouin scattering experiments on our samples. The displacement of fluorine ions in the neighboring pseudocubic cells under these transitions is shown in Fig. 1. The angles φ_1 and φ_2 are the order parameters.

Below 187.6^oK the crystal $KMnF_3$ has the tetragonal structure D_{4h}^{18} with twice the formula (two molecules) in the unit cell [7]. Since the transition reduces the BZ, the point R (the [111] boundary of the cubic phase) falls into the centre of a new zone, and additional phonon modes appear at $k = 0$, including those allowed in the first-order Raman spectrum. In particular, the triply-degenerate soft mode Γ_{25} splits into $A_{1g} + E_g$. This was studied in detail for $SrTiO_3$ which has a similar phase transition at 105^oK [8]. The $KMnF_3$ structure below T_{c2} is not understood clearly yet, although it is most probably the rhombic one D_{2h}^{17} with four times the formula in the cell, which is supported by a comparison of Raman spectra with group-theoretical analysis [9, 10]. Under this transition the point M (the [110] boundary of the cubic zone) falls into the BZ centre, new phonon modes appear at $k = 0$, and the soft mode M acquires full symmetry (corresponds to the A_{1g} representation). The lower-

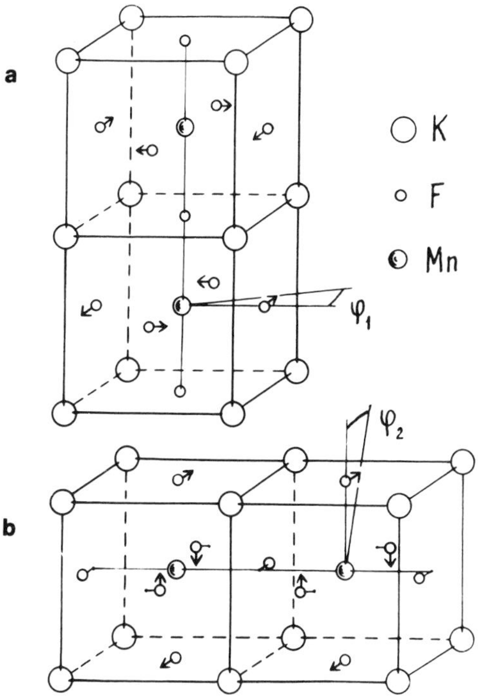

Fig. 1. Perovskite structure and fluorine ion shift in $KMnF_3$ under phase transitions.

(a) O_h^1 D_{4h}^{18} transition due to soft mode Γ_{25} condensation.

(b) D_{4h}^{18} D_{2h}^{17} with M_3 mode condensation.

ing of the lattice symmetry splits the E_g vibration into $B_{2g} + B_{3g}$. Thus, at low temperatures in the low frequency region the $KMnF_3$ Raman spectra must display four lines originating from Γ_{25} and M_3.

The behavior of the soft modes is shown in Fig. 2. Below T_{c1} one observes the E_g mode (according to the polarization measurement), its splitting at 102°K and, finally, all the four lines at low temperatures. A jump-like change in the soft mode

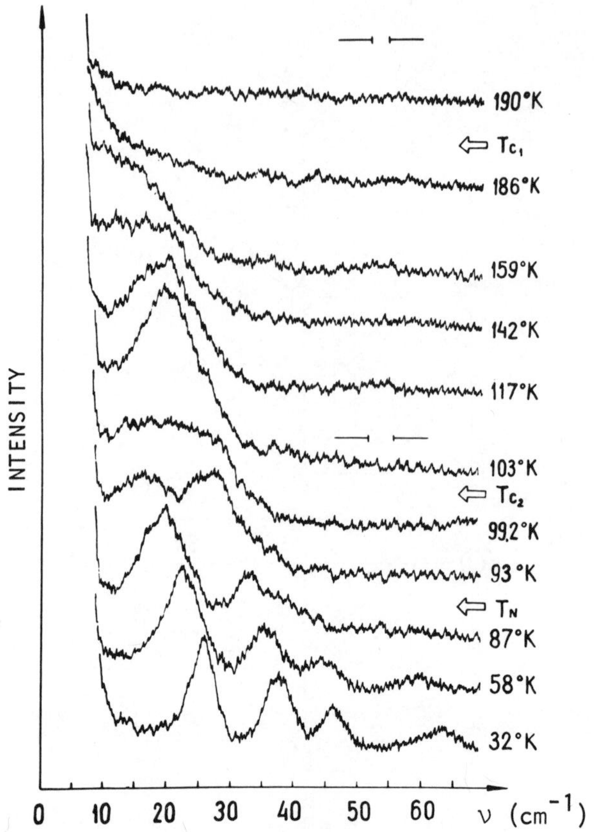

Fig. 2. Low frequency region of $KMnF_3$ Raman spectrum at various temperatures. Shown are (xx+xz) spectra excited by a 6328 Å line He-Ne laser.

frequencies is observed at $T_N = 87^oK$ appropriate to the $KMnF_3$ transition to an antiferromagnetic state [11]. This behavior is the evidence of an essential change in the lattice parameter under magnetic ordering which may be accompanied by symmetry lowering. For Raman spectra this is not so important, however, since even in the rhombic phase degeneracy of all vibrations is lifted.

The presence of degrees of freedom in magnetically ordered crystals due to magnetic moment vibrations gives rise to a new mechanism of light scattering involving spin waves (magnons) [3,12]. Both one-magnon scattering with spin wave excitation at q = 0 and two-magnon scattering involving mainly spin waves appropriate to the BZ boundary, where their density of states is maximum, were observed experimentally in a number of compounds. Because of the peculiar magnetic structure and the absence of spin-orbit interaction in the Mn^{2+} ground state, the $KMnF_3$ spectrum shows only two-magnon light scattering, which was studied earlier at low temperature [5, p. 371; 13]. Since this process is determined by the short-range magnetic order, which is also preserved at $T > T_N$, the two-magnon scattering line can be observed in the paramagnetic region with a non-zero frequency of the maximum.

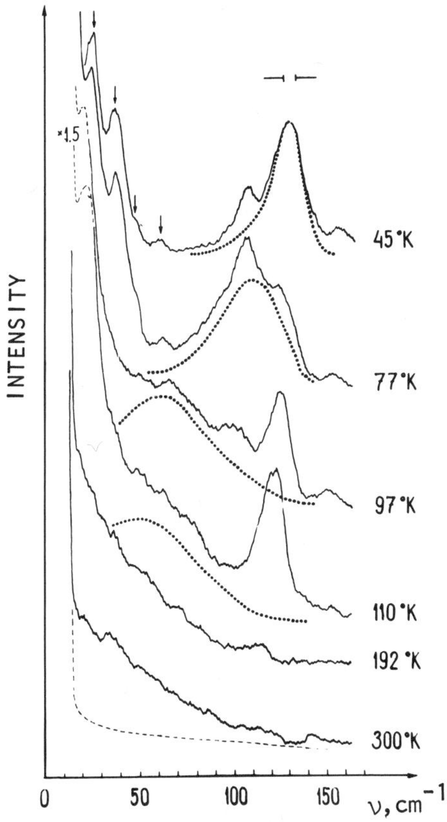

Fig. 3. Raman spectra of $KMnF_3$ at various temperatures excited by a 4880 Å line Ar laser. Arrows point the position of soft phonon modes. Dotted line indicates the shape of a two-magnon scattering line. The dashed curve represents the instrument response function.

This paper reports on the spectrum measurement over a wide temperature range. The results obtained are shown in Fig. 3. The two-magnon scattering line is observed down to room temperatures, though it becomes smeared with increasing temperature. The temperature dependence of the frequencies of the maxima for all of the above lines are illustrated in Fig. 4. The phonon origin at 100 cm^{-1} and higher was considered earlier [9,15]. The E_g frequency at 105-170°K is well described by the dependence $\nu = A(T_c - T)^{\frac{1}{2}}$, where A = 2.3 cm^{-1} $deg^{-\frac{1}{2}}$ and T_c = 187.6°K.

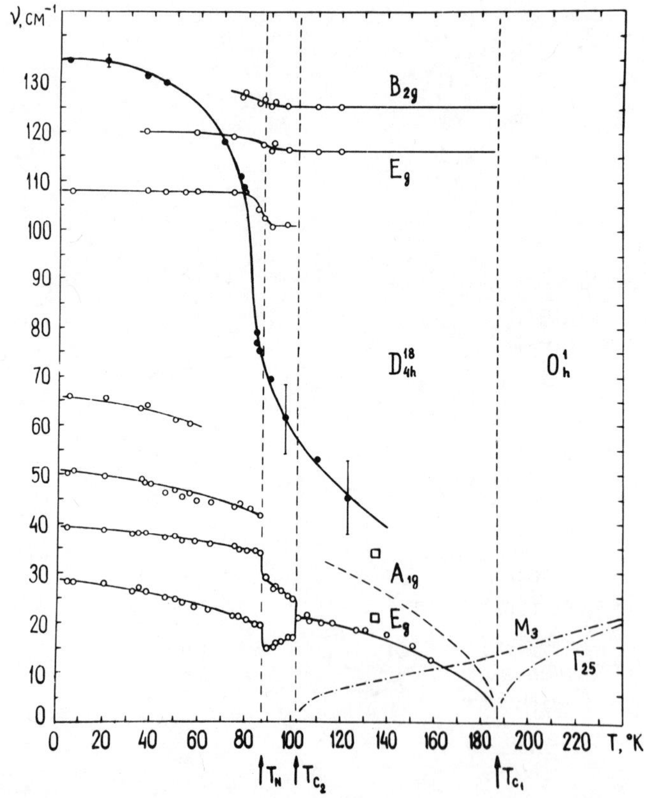

Fig. 4. Frequency behavior of phonon (open circles) and two-magnon (solid circles) light scattering in $KMnF_3$. Dash-dot lines and squares stand for inelastic neutron scattering data [14]. The dashed curve shows the calculated frequencies of soft mode A_{1g}.

As seen in Fig. 3 and 4, at above 90°K the two-magnon scattering line is within the frequency region of the expected A_{1g} position. The observed 62 cm^{-1} line at 100°K was interpreted as an A_{1g} mode [9]. The symmetry considerations say that these excitations may interact. The A_{1g} mode decays into two boundary magnons, which likely accounts for its low intensity. As far as we know, this process has not been studied yet. Note that in the spectrum of isomorphous $SrTiO_3$ the A_{1g} mode intensity is much higher than the E_g one [8].

INTERACTION OF OPTICAL AND ACOUSTIC MODES IN A $KMnF_3$ CRYSTAL

Figure 5 shows the temperature behavior of frequencies of longitudinal acoustic (LA) phonons. The Mandelstam-Brillouin light scattering was measured on samples of various orientations using a He-Ne laser [16, 17]. All the three phase transitions at 187.6°K, 102°K (structural) and 87°K (magnetic) are seen clearly. Under both structural transitions the LA phonon behavior is similar: a sharp change when approaching the critical temperature and then a frequency jump, which proves these are first-order transitions. But in this region hysteresis was not observed (at least not wider than 0.2°K). Below T_N the crystal twins [15], so that the measurement was possible only in the phonon propagation along $[11\sqrt{2}]$ close to the body diagonal of a cubic cell.

The anomalous behavior of acoustic phonons during phase transitions was attributed to the interaction with soft optical phonons [1, 18-21]. The above results can be explained invoking both the resonance (linear) and nonlinear mechanisms of interaction. The first one is operative below T_N where the soft mode is Raman-active, and predicts a sharp change in the LA phonon frequencies at the critical point (Fig. 5, dash-dot lines). The agreement with the experiment was obtained for the following elastic constants in the tetragonal phase (in 10^{11} dyne/cm^2):

$C_{11} = 11.21$ $\quad C_{12} = 3.8$ $\quad C_{13} = 4.55$

$C_{33} = 10.46$ $\quad C_{44} = 2.34$ $\quad C_{66} = 2.56$

The calculation involved the elastic constants of the cubic phase [16] (the same units):

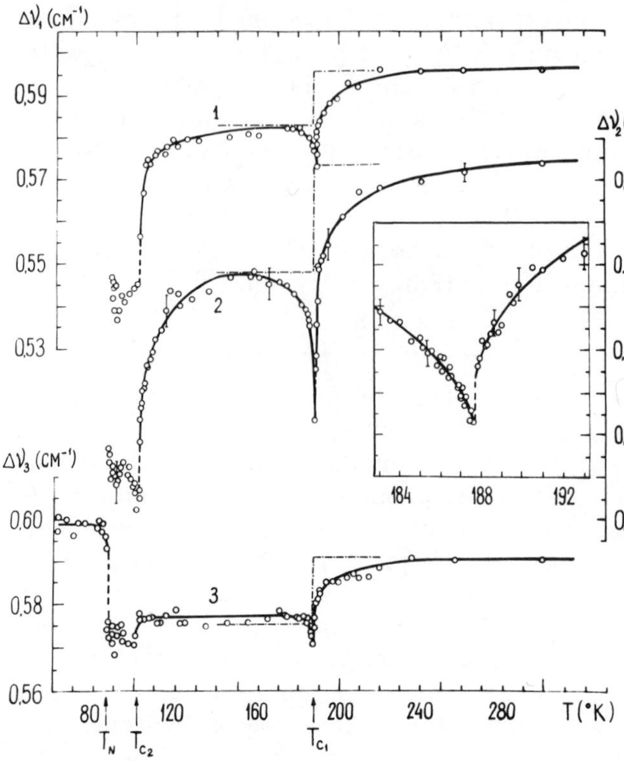

Fig. 5. Temperature dependence of LA phonon frequency at various propagations: 1 - q||[110] ; 2 - q||[100] ; 3 - q||[11√2]. Dash-dot curves represent the calculation involving only resonance interaction between soft optic modes and acoustic vibrations [20].

$C^c_{11} = 11.46 \qquad C^c_{12} = 4.05 \qquad C^c_{44} = 2.56$

The V-shaped dependence of the LA phonon frequencies in the vicinity of the transitions is accounted for by the nonlinear mechanism of interaction which describes the acoustic phonon decay into a pair of boundary phonons of the soft optical branch. The phonon frequency change in the cubic phase must be proportional to $(T - T_c)^{-\frac{1}{2}}$ [19]. This readily fits the experiment through an appropriate choice of critical temperature. A qualitative comparison is, however, meaningless because of the uncertainty of T_c for the first-order phase transition.

Unfortunately, the lack of agreement between literature data does not permit a correct comparison of our results with those of ultrasonic measurements [22-24]. Nevertheless, it may be asserted that in the phase transition region one observes a noticeable dispersion in the sound velocity in the interval from ultrasonic (10^7 c/s) to hypersonic (10^{10} c/s) frequencies studied in Mandelstam-Brillouin scattering experiments.

INTERACTION BETWEEN OPTIC PHONONS AND EXCITON-MAGNON PAIRS IN $CoCO_3$ CRYSTALS

The $CoCO_3$ crystal has the rhombohedral calcite type structure D^6_{3d} with two formulae per unit cell. T_N = 18°K it becomes antiferromagnetic. The calcite structure allows five Raman-active lattice vibrations $A_{1g} + 4E_g$. Three of them (of the highest frequencies) correspond to internal vibrations of the Co_3^{2-} group and the rest to external vibrations. We have found all five lines, and their frequencies at 4.2°K are 222, 312, 724, 1088 and 1437 cm^{-1}.

Besides, there are lines due to electronic transitions in the Co^{2+} ion. Their intensity is similar to that of phonons owing to the presence of an unquenched orbital moment in the Co^{2+} ground state $^4T_{1g}$ (^4F). With the trigonal component of the crystalline field and the spin-orbit interaction involved, the $^4T_{1g}$ term splits into six Kramers doublets, the energy spacing between the extreme components being about 1000 cm^{-1}. All the transitions are allowed in the Raman spectrum and found experimentally. Their frequencies at 50°K are 163, 600, 918, 961 and 1024 cm^{-1}. The Kramers degeneracy is lifted by the exchange field in the magnetically ordered state. This spectrum and its change at T_n will be the subject of more detailed studies. Here we consider the anomalous behavior of the lowest frequency E_g phonon.

Figure 6 shows the temperature dependence of frequencies for low energy excitations, and Fig. 7 illustrates the temperature change in the half-width of two external lattice vibrations. It is seen that the 310 cm^{-1} line suffers no change, and the temperature dependences of the frequency and half-width of the other E_g phonon are anomalous below T_N. A similar behavior was

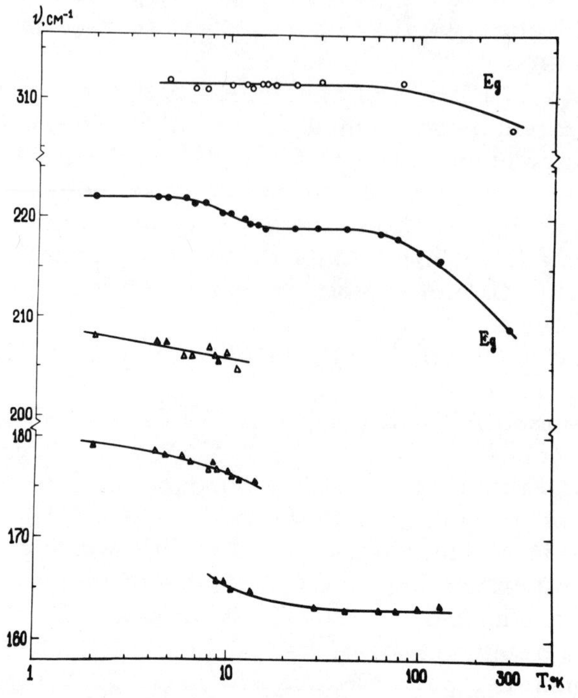

Fig. 6. Temperature dependence of frequencies of lowest electron transitions (triangles) and phonons (points) of Raman $CoCO_3$ spectrum.

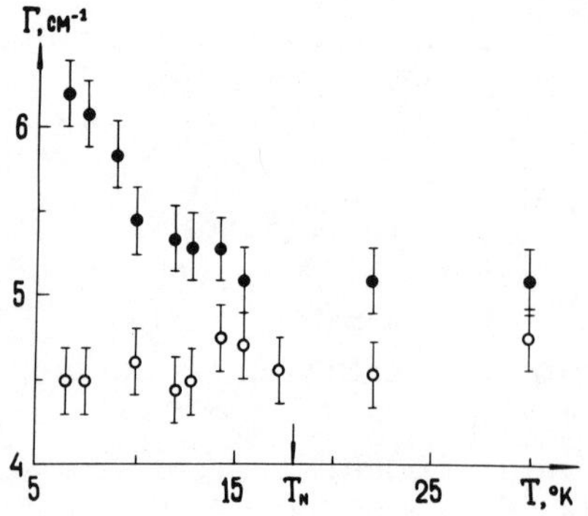

Fig. 7. Behavior of two E_g phonon half-widths with 220 cm^{-1} (solid circles) and 310 cm^{-1} (open circles) frequencies.

earlier found for CoF_2 from the Raman spectrum measurement [25], and in IR absorption [26]. To explain this effect, a number of mechanisms were proposed which involved phonon-exciton interactions [25-27]. Qualitative agreement between calculation and experiment was obtained [25] for the k = 0 phonon decay into two boundary excitations, exciton and magnon. The coupling constant was derived experimentally. The process is shown schematically in Fig. 8. For CoF_2 the numerical calculation was made, since the data on inelastic neutron and light scattering and IR absorption permit evaluation of dispersion and half-widths for all excitations involved.

$CoCO_3$ was not measured in this way, thus only an estimate is possible. The frequency of the phonon in question at 2^oK is 222 cm^{-1}. At this temperature a 207 cm^{-1} exciton transition is also observed. The exciton dispersion is assumed negligible. The estimate of magnon energy at the zone boundary may be derived from the ordering temperature and, independently, from the exciton line shift in the T_N region (Fig. 6). Both estimates are about 15 cm^{-1}. Thus, the total magnon + exciton energy at resonance is in agreement with the phonon frequency, which creates additional phonon relaxation below T_N. This results in the frequency change and the appropriate phonon line broadening in the light scattering spectrum. As temperature increases, the magnon frequency falls and the phonon is out of resonance. Note that the usual phonon line broadening due to anharmonic vibrations appears at about, and above, 50^oK.

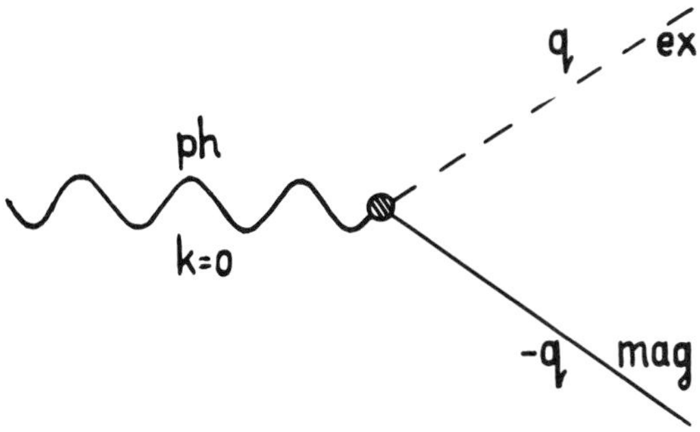

Fig. 8. Diagram of k = 0 phonon decay into exciton and magnon with wave vectors equal in magnitude and opposite in direction.

CONCLUSION

Several processes of interaction between various excitations in $KMnF_3$ and $CoCO_3$ have been described. They are due to a sharp change in the phonon or magnon energy near phase transitions. The interaction of soft optical and acoustic phonons are clearly understood, while the processes involving magnons in magnetically ordered crystals are relatively new. The anomalies in the phonon behavior in $CoCO_3$ due to its splitting into an exciton and a magnon below T_N may be calculated quantitatively using the dispersion data on low frequency excitations. The possibility of interaction between a soft optic phonon mode and two magnon excitation in $KMnF_3$ has been suggested. This process has not been treated theoretically yet, and is of undoubted interest.

The authors are indebted to N. Yu. Ikornikova and B. V. Beznosikov for single crystals of $CoCO_3$ and $KMnF_3$ kindly granted for our experiment.

REFERENCES

1. P. A. Fleury, J. Acoust. Soc. Am., 49, 1041 (1971); Comments Solid State Physics, IY, 167 (1972).
2. J. F. Scott, Rev. Mod. Phys., 46, 83 (1973).
3. Yu. A. Popkov, V. V. Eremenko, Fizika Kondensirovannogo sostoyania, FTINT AN UkSSR, Kharkov, vyp.27, p.2 (1973).
4. Proceedings of the International Conference on Light Scattering in Solids, edited by G. B. Wright, Springer-Verlag, New York, 1969.
5. Proceedings of the Second International Conference on Light Scattering in Solids, edited by M. Balkanski, Flammarion Sciences, Paris, 1971.
6. G. Shirane, V. J. Minkiewicz, A. Linz, Solid State Communs, 8, 1941 (1970).
7. V. J. Minkiewicz, Y. Fujii, Y. Jamada, J. Phys. Soc. Japan, 28, 443 (1970).
8. P. A. Fleury, J. F. Scott, J. M. Worlock, Phys. Rev. Letters, 21, 16 (1968).
9. D. J. Lockwood, B. N. Torrie, J. Phys. C., 7, 2729 (1974).
10. V. V. Eremenko, V. I. Fomin, Yu. A. Popkov, N. A. Sergienko, Fizika nizkikh temperatur, 1, No. 8, 1975.
11. A. J. Heeger, O. Beckman, A. M. Portis, Phys. Rev., 123, 1652 (1961).
12. P. A. Fleury, R. Loudon, Phys. Rev., 166, 514 (1968).

13. Yu. A. Popkov, V. I. Fomin, B. V. Beznosikov, Pisma v ZhETF, 11, 394 (1970).
14. V. J. Minkiewicz, G. Shirane, J. Phys. Soc. Japan, 26, 674 (1969).
15. Yu. A. Popkov, V. V. Eremenko, V. I. Fomin, Fiz. tverd. tela, 13, 2028 (1971).
16. Yu. A. Popkov, V. I. Fomin, L. T. Karchenko, Fiz. tverd. tela, 13, 1626 (1971).
17. V. I. Fomin, Yu. A. Popkov, ZhETF, to be published.
18. H. Thomas, K. A. Müller, Phys. Rev. Lett., 21, 1256 (1968).
19. E. Pytte, Phys. Rev., B1, 924 (1970).
20. J. C. Slonzewski, H. Thomas, Phys. Rev., B1, 3599 (1970).
21. F. Schwabl, Phys. Rev., B7, 2038 (1973).
22. K. S. Aleksandrov, L. M. Reshchikova, B. V. Beznosikov, Fiz. tverd. tela, 8, 3637 (1966); Phys. Status Solidi, 18, k17 (1966).
23. B. Okai, J. Yoshimoto, J. Phys. Soc. Japan, 34, 837 (1973).
24. K. Fossheim, D. Martinsen, A. Linz, Anharmonic Lattices, Structural transitions and melting, ed. by T. Riste, p. 141 (Noordhoff-Leiden, 1974).
25. R. M. Macfarlane, H. Morawitz, Phys. Rev. Lett., 27, 151 (1971).
26. S. J. Allen, H. J. Guggenheim, Phys. Rev., B4, 937 (1971).
27. D. L. Mills, S. Ushioda, Phys. Rev., B2, 3805 (1970).

Section VIII

Scattering by Superfluids

TRICRITICAL POINTS AND LIGHT SCATTERING IN He^3-He^4 MIXTURES*

Michael J. Stephen

Physics Department, Rutgers University

New Brunswick, New Jersey 08903 USA

The critical behavior of the thermodynamic quantities near tricritical points is determined in three dimensions. It is shown that the mean field theory is modified by logarithmic corrections and these logarithmic corrections are determined for the free energy, osmotic coefficient, coexistence line, order parameter, surface tension and coherence length. Recent light scattering data near the tricritical point in He^3-He^4 mixtures is discussed.

A tricritical point is a point in the space of the thermodynamic variables where three coexisting phases simultaneously become identical. In contrast, at an ordinary critical point two coexisting phases become identical. The term tricritical point was first coined by Griffiths.[1] Tricritical points have been observed in a variety of systems; He^3-He^4 mixtures, metamagnets, e.g., $FeCl_2$ where there are competing ferromagnetic and antiferromagnetic interactions and in mixtures of three or more liquids, e.g., ethanol, water and carbon dioxide.

---

*Supported in part by the National Science Foundation under Grant No. GH-38458.

The phase diagram of a He^3-He^4 mixture is shown in Fig. 1. As the He^3 concentration is increased, the λ temperature decreases and terminates at the tricritical point. At lower temperatures the normal-superfluid phase transition is of first order and is accompanied by phase separation.

The situation in the case of mixtures of three fluids is more complicated. The system requires four thermodynamic variables for its description, e.g., T and three chemical potentials μ, μ_2 and μ_3. The thermodynamic space is four-dimensional and at the tricritical point all four of these thermodynamic fields take on fixed values. The tricritical point is thus an isolated point in this space and can be hard to locate experimentally. Widom[2] has recently reviewed some of the experimental evidence in these systems. The He^3-He^4 mixtures are simpler because there is a symmetry between positive and negative values of the order parameter. This symmetry does not exist in ternary fluid mixtures.

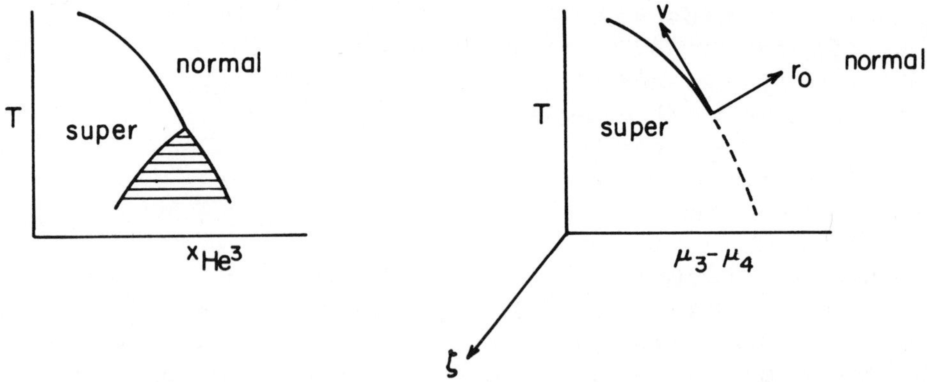

Fig. 1. Phase diagram of a He^3-He^4 mixture. In (a) the λ line decreases with increasing He^3 concentration and terminates at the tricritical point. It is more convenient to consider the phase diagram in the space of the thermodynamic fields (b). The first order line is shown dotted. ζ is a third thermodynamic field which couples to the order parameter. The scaling fields r_0, v are measured from the tricritical point and are normal and tangent to the critical line at the tricritical point respectively.

The Landau theory[3] of tricritical points is based on the free energy or mean field Hamiltonian expanded in powers of the order parameter s(x):

$$H/kT = \int d^3x \left[\tfrac{1}{2} r_o s^2(x) + \frac{v}{4!} s^4(x) + \frac{\omega}{6!} s^6(x) + (\vec{\nabla} s(x))^2 \right] \quad (1)$$

The parameters r_o and v vanish at the tricritical point (T_t, μ_{34t}) and are assumed to be linear functions of $T - T_t$ and $\mu_{34} - \mu_{34t}$ close to the tricritical point. The λ line is determined by $r_o = 0$, $v > 0$. So physically these variables are normal and tengential to the critical line at the tricritical point. The parameter $\omega > 0$ and is assumed constant. In the case of ternary fluid mixtures odd power terms also occur in the expansion (1) in powers of the order parameter s.

We want to determine the effects of fluctuations on the thermodynamic properties near a tricritical point. In the case of critical points it was pointed out by Larkin and Khmelnitski[4] that in more than four dimensions the mean field theory is not modified by the effects of fluctuations. In exactly four dimensions logarithmic corrections to mean field theory occur while in less than four dimensions the mean field exponents are changed. For the case of tricritical points it was first noted by Riedel and Wegner[5] that the anomalous dimension is three. This means that in more than three dimensions fluctuation effects can be neglected while in exactly three dimensions fluctuations give rise to logarithmic corrections to mean field theory. This is a very nice situation and enables us to solve for the tricritical behavior with logarithmic accuracy in three dimensions. The above results follow immediately from the Ginzburg[6] criterion in which we compare the mean square fluctuations in the order parameter with the square of the order parameter. Thus from Eq. (1) (with $v = 0$)

$$\langle s \rangle^2 \sim r_o^{\frac{1}{2}} \quad , \quad \langle \delta s^2 \rangle \sim \int \frac{d^d k}{r_o + k^2} \sim r_o^{\frac{d}{2} - 1}$$

and

$$\frac{\langle \delta s^2 \rangle}{\langle s \rangle^2} \sim r_o^{\frac{1}{2}(d-3)} \quad (2)$$

showing that the effects of fluctuations are small when $d > 3$.

The existence of a marginal dimension is very important in our present understanding of phase transitions. As mentioned above this idea was first developed by Larkin and Khmelnitski[4] who determined the logarithmic corrections to the mean field theory of uniaxial ferroelectrics (or ferromagnets) in three dimensions. (In leading order a uniaxial system for $d = 3$ is equivalent to an isotropic system with $d = 4$.) One important result of these authors was their prediction that the specific heat should diverge like $C_\pm = A_\pm (\ln|t|)^{1/3}$ above and below the transition with an amplitude ratio $A_-/A_+ = 4$. These predictions have recently been confirmed in some measurements of the specific heat of $LiTbF_4$ by Ahlers et al.[7] In my opinion these results are very important and give us confidence that the present field theoretic approach to the theory of phase transitions is valid.

We now confine the discussion to exactly three dimensions and determine the logarithmic corrections to the mean field theory of tricritical points. The free energy expansion (1) is regarded as an effective Hamiltonian and the order parameter s is generalized to be an n-component continuous spin variable $-\infty < s_i < \infty$. The small expansion parameter of the theory is the six point-vertex Γ_6. An examination of the perturbation series for this vertex shows that the leading logarithmic corrections arise from those graphs with three internal lines. These graphs may be summed by the usual parquet method[8, 9] which leads to the differential equation determining Γ_6:

$$\Gamma_6' = -\frac{3n+22}{15}\Gamma_6^2 F_3' \tag{3}$$

where a prime indicates differentiation with respect to the inverse susceptibility r and F_3 arises from the graphs with three internal lines

$$\begin{aligned} F_3(r) &= \frac{1}{(2\pi)^6}\int d^3k_1 d^3k_2 G(k_1)G(k_2)G(k_1+k_2) \\ &\simeq -\frac{1}{32\pi^2}\ln(r/k_c^2) \end{aligned} \tag{4}$$

where $G(k) = (k^2+r)^{-1}$ is the propagator and k_c is the momentum cut off (for convenience we set $k_c = 1$). Equation (3) is integrated with the boundary condition $\Gamma_0(r = k_c^2) = \omega$, the bare interaction, which gives

$$\Gamma_6 = \frac{\omega}{L} \tag{5}$$

where $L = 1 + (3n+22/480\pi^2)\omega \ln(1/r)$. Γ_6 is our small expansion parameter; for $r \to 0$, $\Gamma_6 \sim |\ln r|^{-1}$ and is small and independent of the bare interaction ω. We note that ω determines the size of the critical region which we define by $L > 1$.

The remaining vertices in the problem can be determined provided v is small. The exact condition is

$$\frac{v^2}{\omega} < r L^{2p-1} \tag{6}$$

where $p = 2(n+4)/3n+22$ (see below). From Fig. 1 we see that this condition requires that we approach the tricritical point along paths which do not lie close to the critical line. Using this condition the equations determining the four-point vertex Γ_4 and self-energy Σ are

$$\Gamma'_4 = -\frac{2(n+4)}{15} \Gamma_4 \Gamma_6 \Gamma'_3 \tag{7}$$

$$\Sigma' = \frac{n+2}{18} \Gamma_4^2 F'_3 . \tag{8}$$

Again using the boundary condition that these quantities reduce to their mean field values when $r = k_c^2$ we find

$$\Gamma_4 = \frac{v}{L^p} \tag{9}$$

and the inverse susceptibility

$$r = r_0 - \Sigma(r)$$

$$= r_0 - \frac{5}{6}\,\frac{n+2}{6-n}\,\frac{v^2}{\omega}\,(L^{1-2p} - 1) . \tag{10}$$

We now consider some thermodynamic quantities of interest. The free energy expanded in a Taylor series in powers of the order parameter $M = \langle s \rangle$ is given by

$$F(M) - F(0) = \tfrac{1}{2}\Gamma_2 M^2 + \frac{1}{4!}\Gamma_4 M^4 + \frac{1}{6!}\Gamma_6 M^6 \tag{11}$$

where $\Gamma_2 = r$ and is given in Eq. (10) and Γ_4 and Γ_6 are given by Eq.(9) and Eq. (5). This free energy difference, when $L > 1$, is of the form

$$F(M) - F(0) = r_o^{\frac{3}{2}} L^{\frac{1}{2}} f\left[\frac{v}{r_o^{\frac{1}{2}}} L^{\frac{1}{2}-p}, \frac{M}{r_o^{\frac{1}{4}}} L^{-\frac{1}{4}}\right]. \tag{12}$$

Other thermodynamic quantities of interest can be determined from the free energy (11) by differentiation (in differentiating (11) to logarithmic accuracy the logarithmic factors may be regarded as constant). In general it is found that the mean field expressions are modified by powers of logarithms. Some of the interesting results are:

(i) At the tricritical point $r_o = v = 0$, $r \sim \omega M^4$ and the order parameter varies with field as

$$\zeta \sim \omega M^5 L^{-1} \tag{13}$$

(ii) The first order coexistence line is determined by

$$r_o = \frac{5v^2}{2(6-n)\omega}\left[\frac{n+26}{12} L^{1-2p} - \frac{n+2}{3}\right] \tag{14}$$

(iii) The spontaneous order parameter on the coexistence line is

$$M_{oc}^2 = -\frac{15v}{\omega} L^{1-p} \tag{15}$$

(iv) The light scattering intensity in the ordered phase on the coexistence line is determined by the osmotic coefficient β which can be shown to be

$$\beta^{-1} \sim \frac{L^p}{|v|}. \tag{16}$$

Thus the light scattering intensity diverges with an exponent of 1 with a logarithmic correction. In the disordered phase the exponent is also unity but the logarithmic corrections are smaller.

Recently Watts and Webb[10] have measured the total intensity of light scattering approaching the tricritical point in the single phase region ($T > T_t$) and along each branch of the coexistence curve. The intensity of light scattering is proportional to the inverse osmotic coefficient β^{-1} (Eq. (16)). The exponent in each case was close to 1 although in the measurements along the coexistence curve in the ordered phase the critical region was found to be small and earlier measurements[11] gave a value of 1.6 for the exponent. The light scattering measurements of β^{-1} are in good agreement with those obtained from vapor measurements by Goellner and Meyer[12] and Alvesalo et al.[13] The experiments are not accurate enough to detect logarithmic corrections. The results (15) and (16) are also valid for ternary fluid mixtures for which $p = 2/5$ (i.e., $n = 1$).

(v) The surface tension σ may be shown to vary as

$$\sigma \sim v^2 L^{\frac{3}{2} - 2p} \tag{17}$$

and thus has exponent two with a logarithmic correction term.

(vi) The coherence length $1/\xi \sim vL^{\frac{1}{2}-p}$. An interesting experiment to measure the coherence length in ternary fluid mixtures has been suggested by Widom.[14] It involves measuring the thickness of the phase boundary between the two outer phases in the ternary mixture. The boundary diverges like ξ.

REFERENCES

1. R. B. Griffiths, Phys. Rev. Lett. 24, 715 (1970).
2. B. Widom, Fundamental Problems in Statistical Mechanics III, North Holland (to be published).
3. L. D. Landau and E. M. Lifshitz, Statistical Physics,

Addison-Wesley (1958).
4. A. I. Larkin and D. E. Khmelnitski, Zh. Eksp. Teor. Fiz. 56, 2087 (1969) (Sov. Phys. J.E.T.P. 29, 1123 (1969).
5. E. K. Riedel and F. J. Wegner, Phys. Rev. Lett. 29, 349 (1972).
6. V. L. Ginzburg, J. Exp. Theor. Phys. (USSR) 13, 243 (1943).
7. G. Ahlers, A. Kornblit and H. J. Guddenheim, Phys. Rev. Lett. 34, 1227 (1975).
8. M. J. Stephen, E. Abrahams and J. P. Straley, Phys. Rev. B12, 256 (1975).
9. M. J. Stephen, Phys. Rev. B12, 1015 (1975).
10. D. R. Watts and W. W. Webb, Proc. of the 13th International Conference on Low Temp. Phys. (Boulder, Colo., 1972) to be published.
11. D. R. Watts, W. I. Goldburg, L. D. Jackel and W. W. Webb, J. Phys. 33, Cl-155 (1972).
12. G. Goellner and H. Meyer, Phys. Rev. Lett. 26, 1534 (1971).
13. T. Alvesalo, P. Berglund, S. Islander, G. R. Pickett, and W. Zimmermann, Proc. of the 12th International Conference on Low Tem. Phys. (Kyoto, Japan, 1970).
14. B. Widom, Phys. Rev. Lett. 34, 999 (1975).

CRITICAL SCATTERING OF LIGHT BY A TWO-COMPONENT FLUID*†

Richard A. Ferrell

Department of Physics and Astronomy
University of Maryland
College Park, Maryland USA

A phenomenological approach to the problem of predicting the critical correlation function of a binary liquid is discussed. The main physical ideas are emphasized, especially the analogy with relativistic field theory. The principal mathematical tool is Cauchy's theorem and analytic continuation of the correlation function into the complex momentum plane. In this way the correlation function can be expressed by a dispersion relation involving a positive definite spectral function. The practical advantages of the method for fitting data are illustrated.

In this talk I wish to describe a phenomenological theory for the critical scattering of light by a two-component fluid near its critical point. My task here will be to discuss the general ideas on which the approach is based. I will not enter into the mathematical details, as these can be found in the published presenta-

---

*Work supported in part by the National Science Foundation and in part carried out under the auspices of the Center for Theoretical Physics and the Center of Materials Research.

†Talk given at the first US-USSR Symposium on the Scattering of Light by Condensed Matter, Moscow May 26-30, 1975.

tion of the method.[1] The approach exploits in an essential way an idea expressed by Lieb,[2] by Polyakov,[3] and by Migdal[4] some years ago.

Figure 1 shows how a two-component liquid separates into two phases when the temperature T is lowered and comes close to the critical temperature T_c. Although our phenomenological theory can be generalized to an arbitrary point in this phase diagram of temperature vs. concentration, in the form that it has so far been given it is limited to a line[5] just above the critical point, as shown in Fig. 1. This line terminates at the coexistence curve at the critical point characterized by the critical temperature T_c and the critical concentration c_c. Figure 2 shows the usual light scattering experiment. Light with incident momentum p_i is scattered by angle θ so that its final momentum is p_f. p_f is in a different direction but has essentially the same magnitude as p_i. The momentum difference is consequently given in terms of p_i and θ by

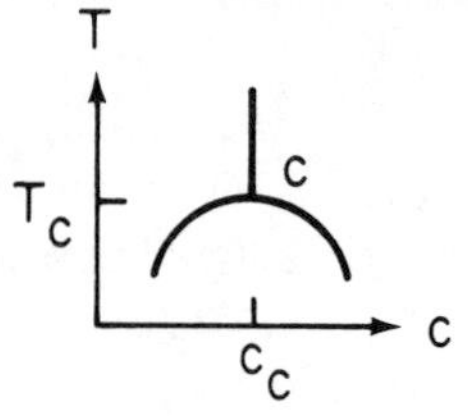

Fig. 1. Temperature-concentration phase diagram for a binary liquid. T_c and c_c are the critical temperature and concentration, respectively. The convex curve indicates the coexisting states of broken symmetry, while the vertical line shows the approach to the critical point (labeled "c"), along symmetrical states. It is only along this line that the phenomenological theory is valid.

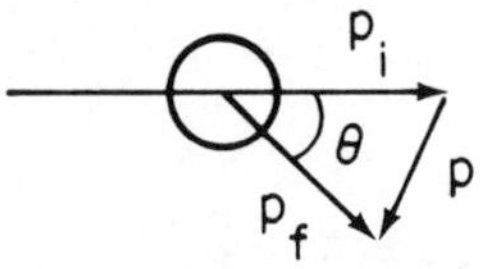

Fig. 2. Quasi-elastic scattering of light of initial and final momentum p_i and p_f, respectively. The scattering angle θ and momentum transfer p are related by Eq. (1).

$$p = 2\,p_i \sin \theta/2 \qquad (1)$$

What we want to characterize by our phenomenological theory is the dependence of the scattering either on the angle θ, or more conveniently, on the momentum transfer p. Figure 3 shows the answer to this question which is given by the theory of Ornstein and Zernike.[6] According to this theory the intensity is proportional to the Fourier transform of the correlation function

$$g_{O.Z.}(p^2) \propto \frac{1}{p^2 + \kappa^2} \qquad (2)$$

It is convenient to deal with the reciprocal of the correlation function

$$g^{-1}_{O.Z.}(p^2) \propto \kappa^2 + p^2 \qquad (3)$$

As shown in Fig. 3 this expression is characterized by a straight line which intersects the momentum-squared axis at the negative value $-\kappa^2$; κ is the reciprocal of the correlation length. κ decreases and finally goes to zero as $T \to T_c$. Such a sequence of values of κ is illustrated by the parallel lines of Fig. 3. In the limiting case $\kappa = 0$ the straight line passes through the origin.

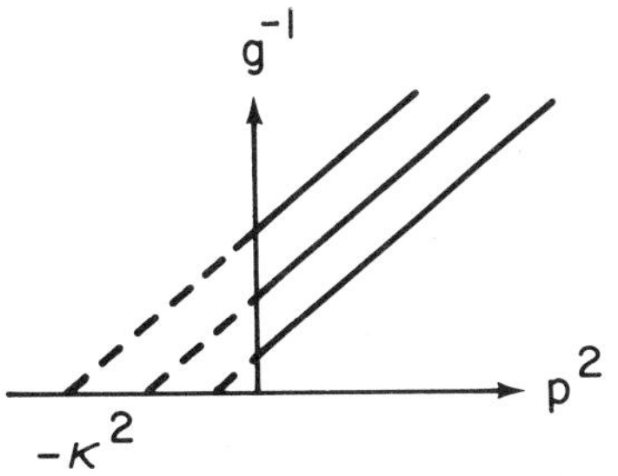

Fig. 3. Ornstein-Zernike approximation for the correlation function g (p^2) for three different values of κ^{-1}, the correlation length. The fact that the plot of $g^{-1}(p^2)$ vs. p^2 is a straight line is a result of the approximation. The correct plot is curved, as shown in Fig. 6.

This, however, is known to be an inaccurate representation of the critical correlation function, resulting from the simplifying approximations made in the Ornstein-Zernike theory. For this limiting case we know that the correct dependence of the correlation function is given by

$$g^{-1}(p^2) \propto p^{2-\eta} \tag{4}$$

The value of η has been determined experimentally both for single-component and two-component fluids. It is generally assumed that both types of fluids are characterized by the same universal value of η. The most recent determination is $\eta = 0.11 \pm 0.03$. This indicates that the neglect of η in Eq. (4) would give an error in the correlation function of the order of 11%. This is inacceptable at the present time for a theory which aspires to describe experiments which are now achieving an accuracy of the order of 1%.

It is convenient to consider p^2 as a complex variable and to study the function $g(p^2)$ at an arbitrary point in the complex p^2-plane. In other words, we generalize Fig. 3 where $g(p^2)$ is considered to be a function of p^2 both for physically significant positive values of p^2 as well as for unphysical negative values of p^2. It is natural to take the additional step of letting p^2 be a complex number. This generalization is represented in Fig. 4, where the pole in $g(p^2)$ is shown at $p^2 = -\kappa^2$ on the negative real axis. $g(p^2)$ is expected to be an analytic function of p^2 except at this pole and along the cut which begins at $-9\kappa^2$ and extends indefinitely to the left along the negative p^2 axis. According to a theorem of function theory, $g(p^2)$ is determined by its variation

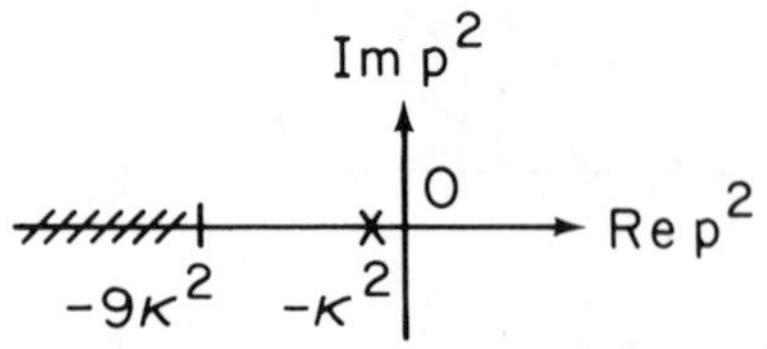

Fig. 4. Complex p^2-plane. Although the Ornstein-Zernike approximation for $g(p^2)$ gives the pole at $p^2 = -\kappa^2$, it omits the cut at $p^2 \leq -9\kappa^2$. The latter is responsible for producing the curvature in the Ornstein-Zernike plot shown in Fig. 6.

along any line in the complex p^2-plane. In our phenomenological theory we prefer to work with the reciprocal (so as to avoid the pole) and to characterize $g^{-1}(p^2)$ by its behavior along the cut. In particular, we determine it by means of a kind of dispersion relation entirely from $\mathrm{Im}\, g^{-1}(p^2)$, the so-called "spectral function", along the cut. This application of Cauchy's theorem is not only a mathematically advantageous procedure but also has an important physical significance. The work of Polyakov[3] and Migdal[4] has made it especially clear that the statistical mechanics problem of a phase transition in a three-dimensional Euclidean space is mathematically equivalent to a relativistic field theory in two space dimensions and one time dimension. The cut along the unphysical negative p^2 axis in the statistical mechanics problem corresponds to a quite ordinary time-like vector in the relativistic field theory problem. In fact, $p^2 < -9\kappa^2$ corresponds to the production of three particles, each of mass κ. The threshold for the production of five particles occurs at $-25\kappa^2$, and so forth. One of the main advantages to this reformulation of the problem of determining the critical correlation function is that $\mathrm{Im}\, g^{-1}(p^2)$ along the cut, by the field theory analogy, corresponds to a production rate and consequently is positive definite. This restricts the search of an appropriate approximate form for the correlation function to a positive definite spectral function which has, in fact, further general restricting features.

In addition to the threshold at $p^2 = -9\kappa^2$ the most important of these conditions is the asymptotic behavior of $\mathrm{Im}\, g^{-1}(p^2)$ for $p^2 \to -\infty$. $\mathrm{Im}\, g^{-1}(p^2)$ is also constrained in the way it approaches the asymptotic behavior. This is associated with the specific heat singularity. Further details on this can be found in the original publication.[1] Here I will limit myself to a discussion of the asymptotic condition, which is most readily obtained by considering the limiting case $\kappa = 0$.

The analytic continuation of p^2 to the negative axis gives from Eq. (4) an $\mathrm{Im}\, g^{-1}(p^2)$ proportional to

$$\mathrm{Im}\, i^{2-\eta} = -\,\mathrm{Im}\, e^{-i\pi\eta/2}$$

$$= \sin\frac{\pi}{2}\eta \approx \frac{\pi}{2}\eta\,, \qquad (5)$$

within the linearized approximation. (We treat η as a small number and neglect terms of order η^2). Thus the spectral function is simply $\pi\eta\,|p|^2/2$. For κ finite this behavior can be expected to set in for $|p| >> 3\kappa$, which provides the asymptotic or saturation condition for the spectral function. The approach to saturation can be characterized by a "healing function" $F(|p|/\kappa)$,

$$\operatorname{Im} g^{-1}(p^2) = \frac{\pi\eta}{2}|p|^2 F(|p|/\kappa), \tag{6}$$

where $F \to 1$ as $|p|/\kappa \to \infty$. The curve labeled as FS in Fig. 5 shows the slow approach to saturation predicted by our phenomenological theory. The abrupt discontinuous jump labeled by FB corresponds to the spectral function for the well-known Fisher-Burford approximant,[7] which is often used for representing the correlation function. The FS spectral function takes into account the connection with the specific heat singularity, whereas the FB spectral function does not. Figure 6 is a sketch of $g^{-1}(p^2)$ along the real axis corresponding to the FS spectral function. This has some curvature, as contrasted with the straight Ornstein-Zernike lines of Fig. 3. The detailed nature of $g^{-1}(p^2)$ is most conveniently revealed by writing it as

$$g^{-1}(p^2) = 1 + \frac{p^2}{\kappa^2}\left(1 + \frac{p^2}{9\kappa^2}\right)^{-\eta/2}\left[1 + \eta s\left(\frac{p}{\kappa}\right)\right]. \tag{7}$$

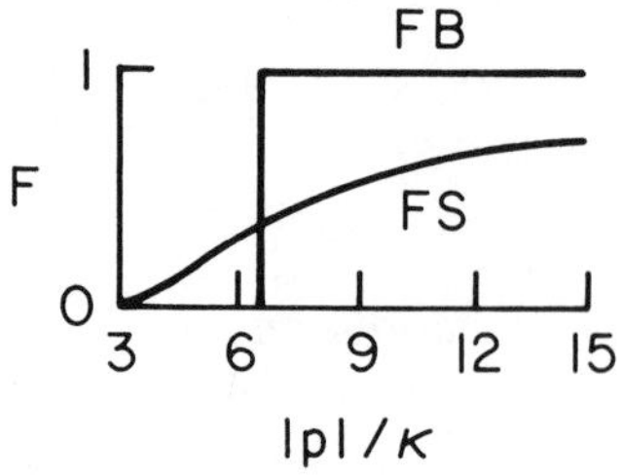

Fig. 5. Healing function $F(|p|/\kappa)$ vs. $|p|/\kappa$. F describes how the spectral function rises from threshold at $|p| = 3\kappa$ to its full asymptotic form for $|p| >> 3\kappa$. The healing function used in this work varies smoothly (FS), while the Fisher-Burford approximant corresponds to an abrupt discontinuity (FB).

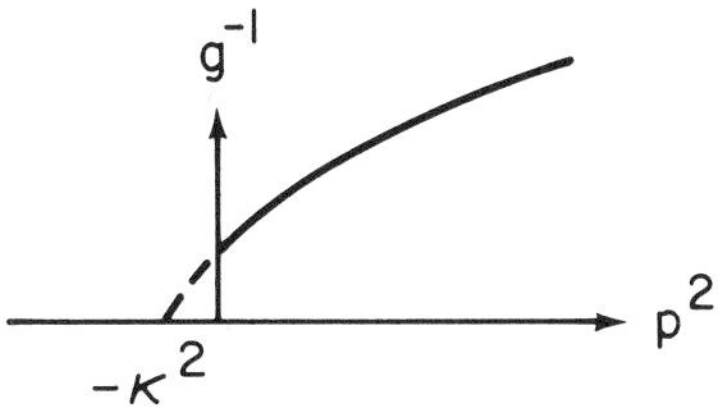

Fig. 6. Ornstein-Zernike plot beyond the Ornstein-Zernike approximation (schematic only). The curvature results from including the contribution to Im $g^{-1}(p^2)$ along the cut shown in Fig. 4. The asymptotic behavior is $g^{-1} \sim p^{2-\eta}$ rather than $g^{-1} \sim p^2$ as for the straight Ornstein-Zernike lines in Fig. 3.

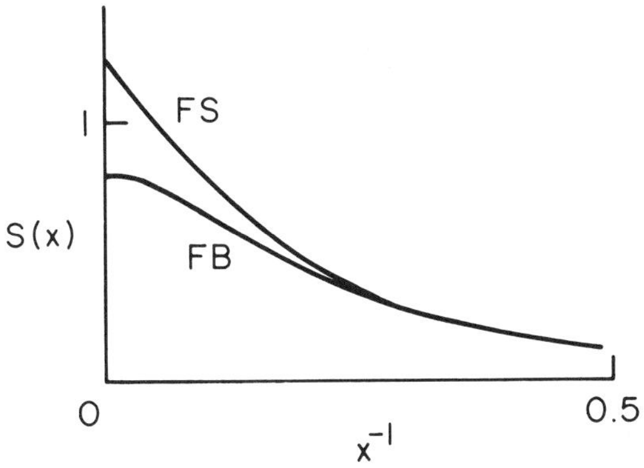

Fig. 7. Scaling function $s(p/\kappa)$ vs. κ/p. $s(p/\kappa)$ specifies the way in which $g^{-1}(p^2)$ approaches its asymptotic form (see Eq. (7)). In the limit $p \to \infty$ (left-hand side of this figure), the value of $s(\infty)$ determines the factor of proportionality in Eq. (4). Thus the magnitude of this factor is significantly greater for the approximant discussed here (FS) than it is for the Fisher-Buford approximant (FB).

The first factor in the last term in Eq. (7) has the correct threshold at $p^2 = -9\kappa^2$ and also satisfies the saturation condition of Eqs. (5) and (6). The remaining factor provides a correction term reflecting the gradual rise of F in Fig. 5 to its saturation value of unity. The function $s(p/\kappa)$ is determined by a dispersion relation where the spectral function is proportional to $1 - F$.

Thus the correlation function is completely specified by the function s(x), which is shown in Fig. 7 for the two different spectral functions of Fig. 5. In order to bring out the difference in the two different curves we have plotted them vs. x^{-1}. The two approximants are indistinguishable in the range $x < 3$, but at $x \approx 3$ the curves begin to separate. The difference becomes largest for $x \to \infty$, where the limiting FS value is significantly larger than the FB value. These large values of x are difficult to attain in light scattering experiments, although with present techniques this may begin to be possible. They are, however, easily attained in neutron scattering experiments, where this difference in the phenomenological theories may become readily apparent.

REFERENCES

1. R. A. Ferrell and D. J. Scalapino, Phys. Rev. Letters 34, 200 (1975).
2. E. Lieb, private communication.
3. A. M. Polyakov, Zh. Eksp. Teor. Fiz. 55, 1026 (1968), and 57, 271 (1969) [Sov. Phys. JETP 28, 533 (1969), and 30, 151 (1970)].
4. A. A. Migdal, Zh. Eksp. Teor. Fiz. 55, 1964 (1968) [Sov. Phys. JETP 28, 1036 (1969)].
5. Although shown in Fig. 1 as a straight vertical line, this line could be tilted, or even curved. The accurate characterization of the true line of symmetry remains an unsolved problem at the present time.
6. L. S. Ornstein and F. Zernike, Proc. Acad. Sci. Amsterdam 17, 793 (1914), and Phys. Z. 19, 134 (1918).
7. M. E. Fisher and R. J. Burford, Phys. Rev. 156, 583 (1967). See also D. S. Ritchie and M. E. Fisher, Phys. Rev. B5, 2668 (1972).

ON THE INFLUENCE OF IMPURITIES ON LIGHT SCATTERING AT PHASE TRANSITIONS

A. P. Levanyuk, V. V. Osipov, A. A. Sobyanin

Institute of Crystallography
USSR Academy of Sciences
Moscow USSR

A pronounced anomaly of light scattering in the region of a phase transition in the solid state, "opalescence" at the $\alpha \rightleftarrows \beta$ transition in quartz, was first observed twenty years ago [1]. However, at present proper interpretation for a number of basic experimental data for quartz as well as for other substances [2, 3] is still absent. Initially, the rise of the intensity of light scattering in the phase transition region was assigned to thermal fluctuations of an order parameter and of parameters connected with it [4-6]. Later some experimental data were obtained which bear witness to the static nature of optic inhomogeneities, causing the "opalescence" in quartz [7]. The recent specification of the theory [8] shows the incorrectness of the former conclusions about the possibility of explaining the observed effect by the increase of thermal fluctuations. Though the latter leads to increasing light scattering, this increase is substantially weaker than the experimentally observed one [1, 7]. In this connection it is natural to discuss the possibility of other mechanisms of light scattering and, in particular, the scattering from lattice defects. The latter subject was broached in papers [9, 10], but, to our knowledge, there are no consistent considerations. An attempt of such a consideration dealing with point defects is discussed below. The results obtained contradict in some respects the known intuitive statements [9, 10].

For definiteness let us consider a second order phase transition with the change of the structure caused by the relative displacement of some sublattices of the crystal. This displacement (noted as η) has the meaning of an order parameter of the phase transition. Let us consider now an impurity atom or native point defect. Such a defect produces in its vicinity pronounced distortions of the crystal. As is known from the theory of defects, the part of these distortions which corresponds to certain elastic ("acoustic") deformations slowly decreases as the distance from the defect rises. The characteristic length of this decrease is much larger than the lattice constant, and, therefore, one can describe this distortion in the continuum-media approximation. Other distortions, in particular "optical" ones, usually decrease at a distance of the same order as interatomic spacings. Therefore, these distortions must be considered in the framework of the theory that takes into account the discrete structure of a crystal. However, this is not the case in the vicinity of second order phase transitions. The optic distortion part, namely that corresponding to η, decreases at r_c (correlation radius of η), which tends to infinity if $T \to T_c$ (T_c is the temperature of transition). In other words, the dimension of the distorted region in the vicinity of the defect increases, if $T \to T_c$. It is natural to expect that under such conditions the impurity contribution to the refractive index of the crystal and intensity of light scattering from the impurity also change.

In accordance with the above remarks we shall describe the temperature dependent distortions near the impurity (spatial distribution of η) in the continuum-media approximation, taking into account the presence of the impurity in formulating the boundary conditions; i.e., considering that the value of η at the impurity position ($\eta_0 + \eta_s$) differs from that corresponding to large distances from the impurity (η_s). The value of η_0 depends on the interaction between the impurity and crystal line lattice and must be found self-consistently. We shall illustrate this procedure by considering a certain model. We shall show that this model is of a rather general nature.

Let us single out an impurity atom with its nearest neighbors - "an impurity quasimolecule". The configuration of such a quasimolecule will be described in terms of the normal coordinates, placing their origin at the positions of the displaced host atoms, if the impurity is at the most symmetrical position in the symmetrical phase ($T > T_c$). The remaining part of the crystal will be considered as a continuous medium, in which there is a field $\eta(r)$, caused by the impurity. Naturally, the quasimolecule configuration depends on the value of η at the impurity position $(\eta_0 + \eta_s)$. We shall take this into account in writing the expression for a change of free energy, caused by a change of configuration of the quasimolecule and the crystal:

$$\Delta\Phi = V(q, \eta_0 + \eta_s) + \int [F(\eta) - F(\eta_s)]dV ,$$

$$V(q_1 \eta_0 + \eta_s) = \frac{a}{2} q^2 + \frac{b}{4} q^4 + g(\eta_0 + \eta_s)^2 - fq(\eta_0 + \eta_s) \qquad (1)$$

In this formula q is the normal coordinate of the quasimolecule, that transforms according to the same irreducible representation of the symmetry group of the quasimolecule as η; $V(q, \eta_0 + \eta_s)$ takes into account the energy connected with a change of the quasimolecule configuration and the energy of its interaction with the crystal line environment; the last term describes the energy of crystal distortion. We shall express the free energy density of a crystal F in the ordinary form:

$$F = F_0 + \tfrac{1}{2} A\eta^2 + \tfrac{1}{4} B\eta^4 + \tfrac{1}{2} D(\nabla \eta)^2 . \qquad (2)$$

Before further analysis let us discuss briefly the proposed model of the impurity center. It should be noted, first of all, that the model, being understood literally, is not applicable to an arbitrary impurity. Indeed, it may occur that among the normal coordinates of the quasimolecule there are no normal coordinates transforming in the same way as η transforms. It is natural to expect such a situation when coordinates of an atom substituted by the impurity are not included in symmetrical coordinates, corresponding to the soft mode or to a mode of the same symmetry. In this case in formula (1), f = 0, and the interaction between the quasimolecule and the η-field is described by a higher order term $f_1\eta^2 q_1$, where q_1 is the normal coordinate of a completely symmetrical vibration of the quasimolecule. However, this classification of impurities really is not of great significance. If one includes in the quasimolecule, for example, atoms located at the second coordination sphere, then the normal coordinates of this quasimolecule may include the coordinates which have the same symmetry as η. One can also point out a macroscopic analog of a quasimolecule. If in the vicinity of an impurity the value of density ρ differs from the density at a large distance from the impurity, then in the vicinity of the impurity the coefficients A and B in Eq. (2) are also altered. For a small region, the value of η has no spatial variation inside this region and may be considered as an analog of q. In this case the region plays the role of a quasimolecule. The sum of free energy of the region and its surface energy is then the analog of $V(q, \eta_o + \eta_s)$. For example, the change of density may be caused by the electrostriction near a charged impurity. Thus, the above proposed model of the impurity center is of a rather general character and may describe, with minor modifications, a wide class of defects.

Let us return to the analysis of the structure of the impurity center. The spatial distribution of η in the vicinity of the impurity is a solution of an equation following from (2),

$$A\eta + B\eta^3 - D\vec{\nabla}^2\eta = 0 \,. \tag{3}$$

The boundary conditions are: $\eta = \eta_o + \eta_s$, if $r = d$; $\eta \rightarrow \eta_s$ if $r \rightarrow \infty$ (r is the distance from the impurity, d is the effective radius of the impurity; d is of the order of the interatomic spacing). The solution which fits these boundary conditions, being substituted in (1), leads to the expression for $\Delta\Phi = \Delta\Phi(\eta_o, q)$. The equilibrium values of η_o and q may be found by minimization of $\Delta\Phi$ with respect to η_o and q. It is easy to show that for $T > T_c$ there are two possibilities: $\eta_o = q = 0$ and $\eta_o \neq 0$, $q \neq 0$. (For $T < T_c$, η_o and q always differ from zero). The first case corresponds to an impurity, which is located at a symmetrical position in a symmetrical phase; the second case is realized when such a position is unstable and, therefore, the impurity is located in an asymmetrical position. The latter case is of the greatest interest from the point of view of light scattering investigations, and we shall consider just this case. It is essential that the equilibrium values of η_o and q are nearly temperature independent in spite of the pronounced temperature dependence of coefficient A in (1): $A \sim (T - T_c)$. For an estimate one can consider that η_o, q are of the order of atomic values.

Equation (3) is not exactly soluble. Approximately, for $\eta^2 > |A|/B$, the solution can be written in the form:

$$\eta(r) = (\frac{D}{2B})^{\frac{1}{2}} \frac{1}{r} (\ell n \frac{2}{d} + \frac{D}{2B} \frac{1}{\eta_o^2 d^2})^{-\frac{1}{2}}. \quad (4)$$

At $r \simeq r_c$ ($r_c = (D/A)^{\frac{1}{2}}$) η is comparable with $\sqrt{|A|/B}$ and at $r > r_c$ it tends to η_s exponentially. For further rough estimates we assume that

$$\eta(r) = \eta_o d/r \text{ at } r < r_c;\ \eta(r) = \eta_s \text{ at } r > r_c . \quad (5)$$

Let us now estimate the contribution of impurities to the optical dielectric constant ϵ of a crystal. Bearing in mind the application of the theory to structural phase transitions such as the $\alpha \rightleftarrows \beta$ transition in quartz, we set [4-6]

$$\epsilon = \epsilon_o + a\eta^2. \quad (6)$$

Then

$$\Delta\epsilon = Na \int (\eta^2 - \eta_s^2)\, d^3r \simeq 4\pi Na\eta_o^2 d^2 r_c \quad (7)$$

where N is the local concentration of impurities. The intensity of light scattering caused by an inhomogeneous distribution of impurities or fluctuations of defect concentration is

$$I = VQ \langle (\Delta\epsilon - \langle \Delta\epsilon \rangle)^2 \rangle \simeq Q(4\pi)^2 \eta_o^4 a^2 d^4 r_c^2 \langle N \rangle. \quad (8)$$

Here V is the scattering volume, $\langle\ \rangle$ denotes satistical average, and Q is a coefficient of proportionality. It was taken into account that $\langle (\Delta N)^2 \rangle = \langle N \rangle / V$. According to (8) the intensity of scattering rises as r_c^2, i.e., $I \sim |T - T_c|^{-1}$ in the region of applicability of the Landau theory. The formula (8) is applicable in the non-Landau "critical" region too. The only modification here is the other temperature dependence of r_c:

$$r_c \sim |T - T_c|^{-\nu}, \quad \nu \simeq 2/3.$$

However, the impurity scattering in the vicinity of the phase transition is not unlimited. Its rise stops when r_c is approximately equal to the mean distance between impurities. This becomes rather evident, if one takes into account that for $r < r_c$ the function $\eta(r)$ does not contain any temperature dependent parameter. Although in deriving formula (8) we actually assume that $\langle N \rangle r_c^3 < 1$, nevertheless, the maximum of the intensity of the impurity scattering can be probably estimated, if one puts $\langle N \rangle r_c^3 = 1$ in (8). In this case we obtain

$$I_{max} \simeq Q(4\pi)^2 (a\eta_o^2)^2 d^3 (\langle N \rangle d^3)^{1/3}. \tag{9}$$

Let us compare this value with the intensity of noncritical thermal scattering from the density fluctuations at $T = T_c$, $I_\rho = Q\rho^2(\partial\epsilon/\partial\rho)^2 k_B T_c/\lambda$. We have:

$$\frac{I_{max}}{I_\rho} \simeq (4\pi)^2 \left[\frac{a\eta_o^2}{\rho(\partial\epsilon/\partial\rho)}\right]^2 \frac{\lambda d^3}{k_B T_c} (\langle N \rangle d^3)^{1/3}. \tag{10}$$

Numerical estimation of the value of I_{max}/I_ρ is difficult because of the absence of information about the microscopic parameters of the impurity center, η_o and d. If one assumes that $a\eta_o^2/\rho(\partial\epsilon/\partial\rho) \sim 10^{-1}$, $d \sim 10^{-8} - 10^{-7}$ cm $\langle N \rangle \sim 10^{18}$ cm^{-3}, $\lambda \sim 10^{12} - 10^{11}$ erg/cm^3, $T \sim 10 - 10^3$ ^oK, then it is easy to obtain:

$$I_{max}/I_\rho (T = T_c) \sim 10^{-2} - 10^5. \tag{11}$$

Therefore, in principle, the great increase of intensity of light scattering near the phase transition may be caused by impurities.

Before discussing the $\alpha \rightleftarrows \beta$ transition in quartz let us note that the impurity contribution to the mean value of $\epsilon(T)$ has a maximum in the phase-transition region. However, this maximum may not be observable due to the "background" of the monotonic rise of ϵ as a function of T, corresponding to the pure crystal. In any case, the absence of the observable maximum in the temperature dependence of the refractive index in quartz near the $\alpha \rightleftarrows \beta$ transition (see the data of Baranskii, cited in [1]), established the upper limit for the concentration of impurities, located at asymmetrical positions. If we assume that $a\eta_0^2 \sim 0.1$, the absence of an observable maximum in $\epsilon(T)$ is possible, if $\langle N \rangle \lesssim 2\text{x}10^{17}$ $(d_0/d)^3$ cm^{-3}, where d_0 is the lattice constant of quartz (d_o=5x10^{-8} cm). Taking into account that for quartz $\rho\, \partial\epsilon/\partial\rho \sim 1$ and $\lambda = 10^{12}$ erg $\cdot$ cm^{-3}, it is easy to show that the ratio of the maximum intensity of impurity scattering to the intensity of thermal scattering at room temperature is approximately equal to 135 $(d/d_0)^3$. Therefore, if the "critical opalescence" in quartz is caused by impurities, the dimension of the impurity-distorted region (d) must approximately be equal to a few interatomic distances.

Let us now discuss briefly the contribution from impurities located at symmetrical positions. In the symmetrical phase ($T > T_c$) such impurities do not cause the long-range distortions of the crystal lattice corresponding to η and therefore do not contribute to the light scattering anomaly. In the asymmetrical phase ($T < T_c$) the distortions corresponding to η arise. However, their magnitude (the value of η_0) decreases proportionally to η_s as $T \rightarrow T_c$. In the region of applicability of the Landau theory the temperature dependences of $\eta_s \sim (T_c - T)^\beta$ and $r_c^{-1} \sim (T_c - T)^\nu$ are identical ($\beta = \nu = \frac{1}{2}$), and the increase of the intensity does not occur. The anomaly of scattering from such impurities is only an intensity jump at the phase-transition point. The values of β and ν do not coincide for the critical region and for the phase transition corresponding to the tricritical point. In this case the intensity of light scattering increases, if $T \rightarrow T_c$, but slower than for impurities located in asymmetrical positions.

It seems that the light scattered from impurities has a smaller degree of depolarization than the one scattered from thermal fluctuations. To elucidate this fact let us consider the case of an isotropic solid. The depolarization of the light, scattered from density fluctuations, arises in this case due to the shear strains which accompany the density fluctuations. The shear strains arise near the defect also, their total contribution to ϵ being equal to zero due to interaction over a rather large region, including the defect. As a result, the light scattered from impurities is completely polarized in this case. Naturally, this conclusion is approximately valid for substances with small optic and elastic anisotropy. At the $\alpha \rightleftarrows \beta$ transition in quartz in the region of great intensity of scattering a pronounced decrease of depolarization has been observed [1]. It seems to be evidence of the increasing contribution of light scattering from impurities to the observed anomaly.

The proposed theory may be verified, in principle, by investigating the scattering anomaly for samples with different impurity concentrations. However, it should be mentioned, that the temperature width of the region of anomalous scattering varies with the change of impurity concentration more significantly than the value of maximum intensity. One should keep in mind also that only those impurities, which at $T > T_c$ cause the local variation of η at the points of their locations, contribute to the light scattering anomaly. In addition, the anomaly of light scattering may be caused not only by impurities, but also by native defects of a crystal. Near the point of a high-temperature phase transition (for example, near the $\alpha \rightleftarrows \beta$ transition in quartz) the concentration of these defects may be sufficiently great to explain the experimental data. Furthermore, the concentration of native defects may increase in the vicinity of the phase-transition point. This effect is caused by the lowering of the energy of formation of defects due to the lowering of the elastic moduli of a crystal in the phase transition region.

In the discussions during the Symposium some other points of view on the nature of critical opalescence at the $\alpha \rightleftarrows \beta$ transition in quartz were suggested. M. A. Krivoglaz believes in particular, that the Frenkel heterophase fluctuations can be considered as the cause of the opalescence in quartz near the phase transition, which is, according to some experimental data, a transition of first order. However, such fluctuations can exist under very peculiar conditions only (very small surface energy of the interphase boundary, but a sufficiently great difference between the structures of different phases), and we see no reasons for realization of these conditions in quartz. According to H. Z. Cummins (see also paper [7]), the cause of the "critical opalescence" in quartz is the light scattering at the boundaries of the Dauphiné twins, i.e., at the boundary of domains, corresponding to various signs of η. According to some experimental data the number of these domains increases near the phase-transition point. However, to our knowledge there are no estimates for this mechanism in literature. V. L. Ginzburg (see also [11]) believes that the formation of the nuclei of the stable phase in the region of overheating or overcooling may be the cause of the "opalescence". It should be stressed, however, that the nucleus formation in solids is impeded due to the arising of elastic stresses. The contribution of the latter effect to the energy of nucleus formation is proportional to the volume of the nucleus. This makes impossible the nucleus formation inside the "old" phase, if the overheating or overcooling is not great enough [12]. For first order phase transitions, which are near the tricritical point, such an overcooling and overheating may be enough for reaching the spinode. In a real experiment the formation of a new phase will occur in this case only at the surface of crystals or at other macroscopic inhomogeneities.

Finally, it should be noted, that for ferroelectric or ferromagnetic transitions and for transitions with the order parameter linearly coupled with acoustic deformations, the solution for $\eta(r)$ must be found separately. This fact is connected with the necessity of taking into account the long range forces (electric, magnetic and elastic fields). According to our preliminary estimates, the main results of the present theory remain valid in this case too.

REFERENCES

1. I. A. Yakovlev, T. S. Velichkina and L. F. Mikheeva, Dokl. Akad. Nauk SSSR, 107, 675 (1956) Sov. Phys.-Dokl. 1, 215 (1956); Kristallografiya, 1, 123 (1956) Sov. Phys.-Crystallogr. 1, 91 (1956); I. A. Yakovlev and T. S. Velichkina, Usp.Fiz.Nauk, 63, 411 (1957).
2. O. A. Shoustin, ZhETF Pis.Red. 3, 491 (1966); P. D. Lazay, J. H. Lunacek, N. A. Clark, G. B. Benedek in "Light Scattering in Solids", ed. G. B. Wright, Springer-Verlag, New York, p.593.
3. E. F. Steigmeier, E. Anderset, G. Harbeke, Proc. of NATO Advanced Study Institute of Anharmonic Lattices, Structural Transitions and Melting, April 1973, Ostaoset, ed. T. Riste, Noordhoff, Leiden, 1974.

4. V. L. Ginzburg, Dokl. Akad. Nauk SSSR, 105, 240 (1955).
5. V. L. Ginzburg and A. P. Levanyuk, Islledovaniya po eksperimental'noi i teoreticheskoi Fizike (Investigations in Experimental and Theoretical Physics (Collection of Papers in Memory of G. S. Landsberg), AN SSSR, 1959 p. 107, J. Phys. Chem. Solids 6, 51 (1958).
6. A. P. Levanyuk and A. A. Sobyanin, Zh. Eksp. Teor. Fiz., 53, 1024 (1967) Sov. Phys.-JETP, 26, 612 (1968).
7. S. M. Shapiro, H. Z. Cummins, Phys. Rev. Lett., 21, 1578 (1968).
8. V. L. Ginzburg, A. P. Levanyuk, Phys. Lett. 47A, 375 (1974), A. P. Levanyuk, Zh. Eksp. Teor. Fiz., 66, 2255 (1974) (Sov. Phys.-JETP, 39, 1111 (1974)).
9. F. J. Bartis, Phys. Lett., 43A, 61 (1973); J. Phys. C6, 295 (1973).
10. J. D. Axe, G. Shirane, Phys. Rev., B8, 1965 (1963).
11. S. M. Shapiro, Thesis, 1969, unpublished.
12. B. Ya. Lyubov and A. L. Roitburd, Dokl. Akad. Nauk SSSR, 131, 552 (1960) (Sov. Phys.-Dokl. 5, 382 (1960)); B. Ya. Lyubov, Kineticheskaya teoriya fazovykh prevrashchenii (The kinetic theory of phase transitions), Metallurgiya, 1969.

COMMENT

(by C. M. Varma)

Dr. B. I. Halperin and I have also recently investigated the effect of defects on phenomena near structural phase transitions, in particular the appearance of a central peak in the dynamic structure factor. We have found it important to distinguish between defects that have symmetry such that they couple linearly to the order parameter and those that couple quadratically to the order parameter. A central peak is always produced in the former case. If further the defects of the former type have a relaxational motion at a slow rate, central peaks with finite width, as have been reported, are found. We also find that in this case the phonon frequency does not go all the way to zero (in mean field theory) as the phase transition is approached; instead, the weight in the central peak diverges as $T \rightarrow T_c$.

CONCLUDING REMARKS

V. L. GINZBURG

(May 30, 1975)

As a co-chairman of the Symposium's final meeting I was asked to make a few concluding remarks.

It seems hardly reasonable to do any essential summary at a Symposium of this kind. Anyhow, I would not like to do this, as recently I practically have almost not concerned myself with the problem of light scattering, neither was I following the literature and, thus, I feel I have no right to do any summing-up. What I am going to touch upon concerns mainly the history of studying light scattering in crystals - the field I had an opportunity to observe for almost forty years. The point is that I was a student at the Physical Faculty of the Moscow University and, as such, I became, as I would say, a disciple of L. I. Mandelshtam's school. The name of L. I. Mandelshtam is not so well-known in the West, though he was an outstanding physicist. His fame was prevented by his personal modesty and, evidently, his shyness and also by such hard conditions as World War I, the Civil War and a long period when close contacts of Soviet scientists with foreign colleagues were very limited. However, L. I. Mandelshtam's role in the development of Soviet physics is universally recognized. He participated and headed intensive scientific activity in the field of radiophysics and vibration theory (together with N. D. Papaleksi, A. A. Andronov, et al.) in theoretical physics (with I. E. Tamm, et. al.) and, finally, in collaboration with G. S. Landsberg and his colleagues, he studied light scattering. Just 50 years ago G. S. Landsberg first separated molecular light scattering in crystals (in quartz) while observing the change of scattering intensity with temperature. Afterwards L. I. Mandelshtam and G. S. Landsberg began searching for that structure in the scattering spectrum of quartz, which nowadays is called the Mandelshtam-Brillouin doublet. They made an attempt to split this doublet accurately with the use of a spectrograph, but failed; however, combinational light scattering was discovered. For many years to follow this scattering has been investigated, as well as a number of other scattering effects in crystals, liquids and gases. My memory still retains small rooms crammed with self-made apparatus. Some of these rooms had walls painted black to reduce background intensity. No light sources of sufficient power were then available. Exposure times were as high as dozens and even hundreds of hours. But in spite of all that, as I have said, a great deal had been done.

Now, I would like to note why we use the term "combinational light scattering" and not the "Raman Effect". I suspect, that is my feeling, that in the West there exists an opinion we do it for some specific reasons. But in fact it is not so. We just know - we have first-hand knowledge of it - that L. I. Mandelshtam and G. S. Landsberg discovered combinational scattering at least absolutely independently and simultaneously with Raman and Krishnan and, besides, on quite different objects (Mandelshtam and Landsberg - for quartz, Raman and Krishnan - in a liquid). Therefore, we consider it unfair that only Raman had been awarded with the Nobel prize, and also the use of the term "Raman Effect" seems to be unjust. I personally can even say more: I am quite sure that Mandelshtam and Landsberg had discovered combinational scattering before Raman and with a much greater certainty, to say nothing of much better understanding of the nature of the phenomenon observed. But there is no need to speak now about this in more detail, and I have no intention to offer our foreign colleagues to make any changes in the terminology accepted. After all, these terms are only markstones, as if road-signs, which allow to show with maximum efficiency what is really meant. It is difficult to make any changes now. I intended only to eliminate any possible misunderstandings by explaining the point of view of those who are using and will use only the term "combinational scattering".

Going back to history, I would like to note that about 20 years ago - in the fifties - it seemed that light scattering stayed in the string of carts of physics and became a not-today problem due to the difficulties of further studying it effectively. But, as it is well-known, in the early sixties the situation changed radically, when lasers appeared, i.e., powerful light sources with very narrow line width. Everybody knows the result - within the last

10-12 years a lot of new interesting results have been obtained. I cannot say what will happen in 50 years, but it is not a very urgent problem for us. Anyway, nowadays we have lots of problems. Not all of them can be solved even with "fine" methods available. It is sufficient to be reminded of light scattering studies in the vicinity of phase transition points in crystals. Here both better spectral resolution and better thermostating are required, etc. In general, we may be sure that there will be a sufficient number of topics for at least 10 or 20 Symposia of this kind.

In conclusion I would like to express my hope that such symposia will be successfully held in the future as this very Symposium proved to be undoubtedly well-timed and useful.

CONCLUDING REMARKS

JOSEPH L. BIRMAN

(May 30, 1975)

It is my pleasant duty to assist in closing our First Joint Seminar. Let me first express on behalf of the entire American group our thanks to the National Science Foundation-USA and the Academy of Sciences-USSR for their general support including providing the material basis for our Seminar: also the National Academy of Sciences-USA for its support. Most particularly, I wish to thank our host in Moscow, Prof. S. L. Mandelstam, Director of the Institute of Spectroscopy, Academgorodok (Podolsky r-n), and Chairman of the Commission of Spectroscopy - USSR, who assumed the overall responsibility of our Moscow Seminar.

To come closer now to our work of the last week, I wish to express our deep appreciation to the USSR Organization Committee: Academician V. L. Ginzburg, Prof. S. A. Akhmanov, Academician A. S. Davydov, Academician K. K. Rebane, and last but certainly not least, to our indefatiguable Chairman, Prof. V. M. Agranovich. We know also that many other Soviet physicists have lent their efforts to our Seminar and we thank them all.

Our day-to-day life in Moscow was made exceptionally smooth through the efforts of Prof. V. M. Agranovich and his devoted associates. Prof. Agranovich and I share the dubious honor of having successfully survived a multitude of detailed crises so that our Seminar would exist--not the least one was trying to communicate clearly over the telephone (at 0600 hours New York time) between Akademgorodok and New Rochelle, New York!! It would be unfair to single out one person from the younger group working with Prof. Agranovich, but let us thank especially Dr. Bobrov and ask him to thank his colleagues for us. The work of the interpreters was magnificent, and cheerful too! To them also our appreciation. In this vein I should also thank our Soviet colleagues for using English so frequently--may I permitted to remark that English has become the Lingua Franca of Science!

Now we should turn to the scientific work of our Seminar. At the outset, let me, following the wise example of Academician Ginzburg, not attempt to summarize the scientific presentations. Fortunately, we are agreed upon the desirability of having a Proceedings of the First Seminar and this should convey the more formal presentations as well as some partial flavor of the informal discussions. But I should like to recall a few of my own personal, and vivid impressions. At the very beginning of our Seminar, and setting the tone of the free and frank exchanges which followed was the discussion of light scattering in the vicinity of phase transitions. Never to be forgotten, as the discussion between Prof. Ginzburg, Prof. Cummins, Dr. Levanyuk, Dr. Fleury, Prof. Kagan, and others heated up, was the sight of Prof. Ginzburg rising up and pointing a finger at Prof. Cummins and saying with a powerful voice and mock anger "YOU HAVE DESTROYED MY THEORY"! But of course immediately the discussion turned to what new experiments (on quartz, on KDP, and other materials) and what new theory (for the mysterious "central peak" inter alia) is needed. Just such a stimulating and immediate personal interchange is in my view the raison d'etre of our Seminar! But this was merely one example of how our Seminar permitted individual scientific contacts to grow. Some other memorable discussions for me took place on the fascinating new topic of Electron-Hole Drops between Dr. Worlock, Dr. Bagaev, Prof. Keldysh and Prof. Falicov (presenting the new Berkeley photographs of the drops taken a few days earlier)--This continued right through Friday late afternoon at the Lebedev; on interacting elementary excitations between Prof. Ruvalds, Prof. Pitaevsky, Prof. Levinson; on Academician Zel'dovich's new (to us) proposal on multiple Compton scattering as a mechanism by which the electromagnetic radiation distribution was established following the "big-bang": this talk certainly illustrated our theme of "light scattering in a generalized sense in condensed (astrophysical!) matter; and finally on problems of spatial dispersion in crystal optics between Prof. Ginzburg, myself, Prof. Davydov, Prof. Agranovich and Prof. Mills--in which we may (or again, may not!) have finally settled the "a.b.c." problem.

Our visits to the laboratories of the Institute of Spectroscopy, of Moscow State University, and of the Lebedev Institute, really permitted members of our group to discuss the substance of the experimental results, i.e., the real world, which underlies theoretical constructs and analysis. In some cases the theorists excused themselves and retreated to the blackboards or pencils and paper--but this of course is an occupational malady of theorists!

To take a broad overview, I believe that our Seminar did enable a group of USA physicsts to meet a comparable group of Soviet physicists and engage in materially beneficial exchange of information as well as frank discussions of all subjects of interest. It permitted scientists to form, or renew, close scientific contacts, which it is hoped will deepen in the future. It was, therefore, a step in the direction of free interchange of ideas and scientists, which manifestly benefits the development of our Science.

We look forward to the continuation of the existence of this frame, and that circumstances will permit a continuation of our Seminars! We look forward to the participation in future Seminars of those colleagues whom we met at this First Seminar! We look forward to the Second Seminar! We hope circumstances will permit the Second Seminar to occur according to our plan in the United States in our Bicentennial Year of 1976!

Let me close also on a personal note by remarking what a thrill it has been for me as a theorist to present these remarks, and others too, here, in this hall at the Institute of Physical Problems, where Prof. L. D. Landau often spoke. Let this continue to be our inspiration and challenge for the future.

Thank you.

AUTHOR INDEX

TOPICAL INDEX*

*The first page of the paper or comment in which the listed topic appears is indicated.